Intellectual Property in Electronics and Software

A Global Guide to Rights and Their Applications
Second Edition

Consulting Editor **Nicholas Fox**

Consulting editor
Nicholas Fox

Managing director
Sian O'Neill

Intellectual Property in Electronics and Software: A Global Guide to Rights and Their Applications, Second Edition
is published by

Globe Law and Business Ltd
3 Mylor Close
Horsell
Woking
Surrey GU21 4DD
United Kingdom
Tel: +44 20 3745 4770
www.globelawandbusiness.com

Printed and bound by CPI Group (UK) Ltd, Croydon, CR0 4YY

Intellectual Property in Electronics and Software: A Global Guide to Rights and Their Applications, Second Edition

ISBN 9781787422049
EPUB ISBN 9781787422056
Adobe PDF ISBN 9781787422063
Mobi ISBN 9781787422070

Table of contents

Introduction

Nicholas Fox

The software, communications and electronics industries are among the most innovative and competitive in the world. This robust competition means that developers and manufacturers must leverage their IP rights to sustain competitive advantage. Unsurprisingly, therefore, the public profile of IP rights in electronics and software is high and keeps getting higher. Rarely has intellectual property been so much in the news, be that as part of the global mobile phone wars between Apple and Android or as the driver of billion-pound mergers and acquisitions as household names such as Facebook and Google seek to consolidate their patent portfolios.

1. Facebook

Facebook is a case in point. Prior to March 2012, Facebook had 160 or so patents. At the same time, Microsoft, Yahoo! and Google were thought to have had at least 1,000 patents each. The simple numbers themselves disclose that Facebook had adopted a different policy with respect to patents from its competition.

Then, in March 2012 Yahoo! brought an action against Facebook, alleging infringement of 10 of its patents covering critical website functions, such as internet privacy, advertising and information sharing. Later that same month, Facebook acquired 750 patents from IBM, covering technologies such as software and networking. In the following month, Facebook paid $550 million to Microsoft for 650 patents and patent applications that Microsoft was itself to acquire from AOL.

Many have speculated that the principal reason behind Facebook's acquisitions was to give it a stronger bargaining position in any negotiated settlement with Yahoo! This may be so, but we should not discount the fact that Facebook had filed for its IPO in February and potential investors could already have begun to ask questions about Facebook's relative lack of patents when compared to its peers.

In any event, when Facebook brought a counterclaim against Yahoo!, it did so in part on the back of some of its newly acquired patents – a move that provoked Yahoo! to allege that the patents in question had been acquired for the purposes of "retaliation", thus failing to meet the US Patent Office's "Duty of Disclosure, Candor, and Good Faith".

Then in July, Facebook and Yahoo! settled their various actions. Interestingly, no money changed hands. Instead, the companies announced what they called a strategic alliance, which included a new advertising partnership, expanded distribution arrangements and a patent cross-licence.

There is no question that the acquisitions increased Facebook's firepower in dealing with Yahoo! and improved its position in the settlement negotiations. At the same time, Yahoo! was almost universally vilified by commentators for its approach in the dispute.

I will leave the last words here to Ted Ullyot, Facebook's general counsel, who said, when Facebook launched its counterclaim against Yahoo!: "We do so in response to Yahoo's short-sighted decision to attack one of its partners and prioritize litigation over innovation."

2. Google

Turning then to Google, in May 2012 Google acquired Motorola Mobility for $12.5 billion. Mobility holds 17,000 patents (or thereabouts). There were a number of compelling reasons for the acquisition, but as Google's director of Android business development John Lagerling put it: "It's mostly about the patents ... you can sort of disarm this huge attack against Android."

Lagerling was referring to Apple's patent-related attack on hardware manufacturers making Android smartphones. In granting clearance for the transaction, the FTC said: "Google will meet its prior commitments to allow competitors access – on fair, reasonable and non-discriminatory terms – to patents on critical standardized technologies needed to make popular devices such as smart phones, laptops and tablet computers, and gaming consoles."

Due to FRAND obligations imposed by competition law, which require that standards-essential patents be available for licence under "fair, reasonable and non-discriminatory" conditions, the acquired rights would be available only for use as a shield rather than a sword in any future licensing battle. Even so, the sums involved served to emphasise the value of a patent portfolio.

3. Future trends

Several trends are likely to maintain or increase the importance of intellectual property in the electronics and software fields, in particular developments in FRAND licensing, the ever-increasing competition in the mobile phone arena, patent trolls and, should it ever come into existence, the long proposed European Unified Patent Court.

3.1 FRAND

FRAND licensing concerns the interface between IP law and competition law.

Interoperability is frequently essential in the electronics and software fields. This requires operators to agree standards. As such standards must be open to all, theoretically standards should be free from all IP rights. However, in practice, it is inevitable that any standard is based on certain innovations made by key developers. In order to prevent such developers enjoying a monopoly position, as a *quid pro quo* for certain innovations being adopted as part of a standard, developers agree that all rights essential for putting a standard into practice are made available freely on FRAND terms.

Although there has been a consensus that FRAND licences must be available, there has been little agreement on what FRAND terms actually are. It is only recently with cases such as *Microsoft v Motorola*[1] in the United States, *Huawei v ZTE*[2] in the European Court of Justice on referral from the German Courts, and *Unwired Planet v Huawei*[3] in the United Kingdom that the courts are finally having to decide on what is meant by "fair, reasonable and non-discriminatory" and the circumstances (if any) in which injunctions might be granted based on a standards-essential patent.

Until the uncertainty is resolved, standards-essential patents have the potential to be one of the key drivers of value in the electronics and software fields.

3.2 The mobile phone wars

A second factor which is driving global IP litigation in this area is the global mobile phone wars. Famously, in relation to the battle between the Apple and Android operating systems, the late Steve Jobs declared that he was "willing to go thermonuclear war on this". The nuclear missiles in such a war are IP rights, and having and maintaining the necessary arsenal is a key strategic element for the leading players.

When he made his famous 'nuclear war' remark, Jobs was speaking from a position of knowledge. He had experienced how the Apple operating system for desktop computers had dwindled against the onslaught of Microsoft Windows. It is only through the tactical use of IP rights that Apple could protect itself against history repeating itself in the mobile era.

Even though the *Apple v Samsung* patent litigation was settled, high stakes mobile phone litigation remains a core component of high-tech patent litigation worldwide.

However, the mobile phone wars merely reflect a wider reality in the electronics and software fields – Moore's Law – the observation of Gordon Moore, co-founder of Intel, in 1965 that the numbers of transistors per square inch on integrated circuits doubled every year. Although that observation was

---

1 US District Court, Western District of Seattle, Case 10-CV-01823-ORD.
2 *Huawei Technologies Co Ltd v ZTE Corp*, ZTE Deutschland GmbH, Case C-170/13.
3 *Unwired Planet v Huawei* [2017] EWHC 711 Pat.

subsequently amended to a doubling every 18 months, the underlying observation that computing power is and has been growing exponentially remains true. This has driven waves of innovation in the electronics and software fields, be it the waves of improvements in mobile phones or the waves of improvements in photolithography as chip manufacturers struggle to keep up with the need to squeeze ever more computing power into ever smaller packages. With each such wave there are inevitably winners and losers and litigation inevitably follows.

3.3 Patent trolls

It also is unsurprising, given the very large sums involved, that the high-tech patents have also attracted the attention of patent trolls. It was estimated[4] that in 2012 more than half of all US patent infringement suits were brought by patent monetisers. However, the 'patent troll' label is a flexible one, which arguably could be extended to any non-practising entity and therefore be applied to both university departments and indeed former innovators which have fallen on hard times.

However, the existence of the patent trolls is merely an example of the increasing competition which exists for quality IP rights. If quality rights are available for sale, it is possible that they can attract additional bidders and therefore prices will increase. Of course, it is just this element of competition and the general paucity of quality patent portfolios which explain the very high figures that companies are prepared to pay, either to acquire rights to enforce or to acquire defensive portfolios to protect themselves against attack.

Steps have been taken to limit the extent of patent litigation by non-practising entities; in particular, there have been procedural changes in the United States (where most such litigation has been undertaken to date) which have made such litigation more expensive and therefore less lucrative. However, even then it remains the case that whenever tech firms are able to generate significant profits from exploiting new technologies, it is inevitable that third-party patent owners will seek to obtain a slice of those profits if it can be argued that their patents are infringed.

3.4 The Unified Patent Court

A potential new European Unified Patent Court would add a further element to this mix. After over 40 years of discussions, an agreement on a Unified Patent Court was finally signed on 19 February 2013. Ratification of the agreement was expected to be completed sometime in 2015, with the new court opening its

---

4 Robin Feldman, Thomas Ewing and Sara Jeruss, "The AIA 500 Expanded: Effects of Patent Monetization Entities", UC Hastings Research Paper 45, 9 April 2013. Available at SSRN: http://ssrn.com/abstract=2247195.

doors four months later. However, the reality was different. As this book is published in 2019, the court still has not opened.

Ratification took longer than expected. Then in the Summer of 2016, the United Kingdom narrowly, but decisively, voted in favour of exiting the European Union, putting continued UK participation in the court into doubt. Although the United Kingdom has since then ratified the Unified Patent Court agreement and the UK government has a stated intention of continued involvement in the Court, it is inevitable that progress has been delayed as the United Kingdom's involvement in the court becomes part of the ongoing negotiations between the United Kingdom and the other EU member states as the United Kingdom negotiates its departure from the European Union. Additionally, ratification has been subject to a constitutional challenge in the German courts. That challenge is also ongoing and is unlikely to be resolved until sometime in 2020.

Either way, the existence of the Unified Patent Court is in doubt. However, should the Unified Patent Court come into existence, it will have a significant impact on any business which is involved in patent litigation or licensing. The existence of the new court will greatly simplify the enforcement of patents in Europe. In contrast to the current system, in which individual national court actions are required to enforce European patents in each country, when the Unified Patent Court comes into existence it will be possible to enforce or revoke such rights across the entirety of the European Union (with the exception of Spain, which has opted out of the system) in a single court action before the Unified Patent Court.

As the Unified Patent Court removes the need for multiple parallel court actions, the cost of patent litigation will be reduced. Since patent enforcement will become easier, it is likely that more companies will be able to present a credible threat of litigation. As a result, the new court may result in companies adopting a more aggressive approach to licensing and therefore the effective value of patent rights will increase with the introduction of the new court.

Electronics and software companies are likely to lead the way in shaping how the new court functions. Provisions in the agreement allow rights holders to opt out patents from the exclusive jurisdiction of the court. It is likely that these opt-out provisions will be widely used in the bio/pharma field to protect companies' key assets from what will be an untested court. However, in contrast to the bio/pharma field, where an innovative drug or treatment is normally the subject of only one or two key patents, in the computing and electronics fields many patents will likely cover the same technology, which means that there is less importance on protecting individual patents. This in turn means that opt-outs in computing and electronics are likely to be filed far less often. Rather, the opportunities for Europe-wide enforcement are likely to appeal and much of the initial work of the court is likely to be in the electronics and software fields.

4. IP strategy – the right rights at the right time for the right price

It is against this background of ever-increasing emphasis on the importance of IP rights that innovative electronics and software companies must reassess and develop their IP strategies.

The essence of any such strategies must be acquiring the right rights at the right time for the right price. Of course, what the right rights are and what the right price is will be directly connected with a company's business strategy and will change over time.

In many cases, unregistered rights such as trade secrets, copyright, and unregistered designs are of greatest importance to a company. It is often erroneously said that such unregistered rights come into existence 'for free', compared with registered rights (eg, patents, registered trademarks, registered designs) which involve legal and official fees. In truth, as anyone that has sought to rely on unregistered rights will be aware, this is a false dichotomy. If unregistered rights are ever to be relied upon, steps must be taken to ensure that the ownership and existence of such rights can be proven. This means that companies must invest money in their business processes to record new rights coming into existence and to ensure that they retain good title to those rights.

This approach is not only good business practice, but also has many practical advantages. If a company needs to enforce its rights, this is possible only if it can establish that it owns rights which actually exist. Given the unregistered nature of the rights, their scope can be unclear. However, if they are properly recorded, it is far easier to establish what, for example, a company considers to be its trade secrets and therefore it is far easier to remind departing employees what it is that they can and cannot share with a new employer. Last, but certainly not least, a well-established records system will undoubtedly add value to a company. IP rights form a significant element of the value of innovative companies. If such a company is looking for investment or its founders are looking for an exit strategy, such value can be identified to outsiders only if it is identifiable and properly recorded.

5. Patent strategy

In contrast to unregistered rights, registered rights require a deliberate decision on the part of a company to apply for and maintain rights. In the electronics and software fields, this normally means patents and designs.

Patents are possibly the most high-profile and potentially most valuable forms of intellectual property, because they can protect a company even where an innovation or improvement is developed independently elsewhere. However, in the electronics and software fields, patents need to be managed very differently from those, for example, in biotech and pharma, because of the way in which innovation impacts on different technologies. In electronics and software, most innovation is incremental – a small improvement here, a small improvement there – which collectively leads to a greatly improved product. This means that – in contrast to the biotech and pharma fields, where it may take many millions of pounds to establish that a particular form of treatment is effective and safe – innovation is far easier in the electronics and software fields. Any change which in some way improves performance is potentially patentable. So any change which makes processing quicker or faster, or which reduces computer resources, can potentially be protected.

Since innovation is so much easier in the electronics and software fields, it is not the case that every innovation which might be patentable should be. Companies must be far more selective about what they protect. It is thus essential that some kind of invention triage be performed before resources are invested in filing a patent application.

'Innovation capture' programmes formalise this approach. Typically, in any innovative company, any problem which requires engineers to spend time and effort in solving it is potentially a suitable candidate for patenting. Assuming that a firm's engineers are any good, if a problem takes time to solve, then that should be an indicator that the solution is a clever one. Of course, this assumes that time is being devoted to trying to solve a problem, rather than – as is frequently the case – mere administration, maintenance or simple bug fixing which is time consuming but necessary in order to get a new product to work.

Having invested time in solving a problem, the next step is normally to try to see whether the solution is already known. This can be a relatively straightforward piece of research on the internet, using online patent search engines. Performing a number of keyword searches will regrettably often identify that time and effort has been expended on merely reinventing what was already known. If this is the case, then further expense on attempting to protect the unprotectable can be avoided.

If, however, what appears to be a good idea also appears to be novel, the information gathered by undertaking an initial search will provide valuable input into the next stage of the assessment. In order to be worth protecting, the investment in patent protection must increase the value of the company. Therefore, it is vital that the overall commercial importance of an invention be assessed at an early stage. A brilliant but commercially unimportant innovation is not worth the time and effort involved in trying to obtain a patent. Unless a patent is in some way going to add to the bottom line, protection should not be pursued.

Here it is important to look at the connection between a company's revenue streams and the nature of the possible innovation. If a development is made to a key product, it may be worth spending money on what otherwise might be an innovation of limited importance. It is not how clever an idea is that counts; it is whether filing a patent is worthwhile.

In many ways, particularly for smaller companies, an IP strategy can act as a litmus test for demonstrating wider business competence. If a company has an IP strategy which is properly aligned with its wider business plan, this suggests that the management of the company is likely to be competent in other areas as well. The value of any intellectual property is directly connected to the value of the business to which it relates. Intellectual property that is unconnected to a product will rarely have much value. However, a successful business with a successful product will underpin that success if it has the appropriate intellectual property in place.

6. About this book

This book has been developed with the needs of electronic and software firms in mind as they and their advisers navigate the maze which is global IP law. Although much of the law has been harmonised, important differences exist between jurisdictions on key elements of IP law which are relevant to electronics and software companies. There are differences as to the boundaries of what is and is not patentable – ranging from the concept of 'technical effect' applied in Europe to the 'anything under the sun made by man' approach in the United States. The boundaries of what can or should be protectable under trade secrets law vary in different jurisdictions, as does the approach of various countries to the protection of computer programs by copyright and the extent to which copyright and unfair competition law can be used to protect against the copying of the design of physical products.

Essentially, this book has three parts. In the first part, "Multi-jurisdictional issues", the global issues facing software and electronics companies are considered. This begins with a chapter discussing the interplay between the various international agreements of importance, followed by a pair of chapters discussing standards setting and FRAND licensing and the challenges presented by open source software.

The next two parts of the book focus on Europe. First of all, an overview of European patent law is given, together with an overview of the issues that arise when enforcing IP rights in Europe and a chapter on the prospective Unified Patent Court. This is followed by four chapters discussing the law in the key jurisdictions of France, Germany, Italy and the United Kingdom in greater detail. This approach has been chosen because the European Union and the European Patent Convention have harmonised many areas of IP law in Europe, so that it is sensible to speak in terms of a European body of law. However, at the level of detail, and particularly when it comes to enforcement, important national differences still exist.

The final part of this book, "National perspectives – non-European", returns to a global theme and considers the approaches adopted in key non-European jurisdictions: Australia, Canada, China, Hong Kong, India, Japan, South Korea and the United States. The countries covered have been chosen because of their contribution to patenting in the electronics and software fields and, in the case of India, its importance as a centre for outsourcing the production and maintenance of software.

An introduction to international IP instruments relevant to electronics and software

Nick Bassil
Kilburn & Strode LLP

Intellectual property relating to inventions in the electronics and software field is governed by an inter-related system. The law has developed in different ways in the various areas of IP protection as technology has evolved. When the initial legal frameworks were being put together the explosion of innovation in relation to computer-implemented inventions could not have been predicted.

The current patchwork of protection is therefore not necessarily how an ideal system would be devised. It is also an international system. Globalisation might be seen as a relatively modern phenomenon, but in the field of intellectual property the roots of international harmonisation go back to the late 1880s.

Innovation in the field of electronics covers inventions relating to large industrial devices as well as smaller pieces of high-tech equipment. Since the devices are often now controlled by sophisticated computer programs that affect the way they work, there is debate over whether they can be protected via patent protection. The historical context for the protection of computer programs did not envisage the uses to which they would be put, which has led to some interesting questions in more recent times. Software therefore falls to being protected under a variety of different systems depending on the nature of the innovation and the national law or treaty concerned.

There are also useful innovations that arise from the way certain methods are performed in terms of the steps carried out, so-called 'business methods'. Inventions relating to business methods are often grouped under the heading of software since many of the processes use computer software, although applications in other technical fields do exist. The use of databases and IT solutions is aiding the continuing discoveries in relation to genomics and synthetic biology. However, business method-type innovations, while certainly providing useful benefits, have also generated questions over the extent to which such inventions are protectable.

Companies operating in this technical area often deal in fast-moving markets where a product life-cycle is only one to two years. Compared to the time taken to prosecute a patent application to grant, such time-scales sometimes mean that traditional routes of patent protection are less relevant than in other technology sectors.

Therefore, other forms of protection such as copyright or trade secrets may be more important in this sector. Copyright in particular has certain attractions in terms of the length of time needed to secure protection, but set against that is the need to prove copying to demonstrate infringement. There are other so-called 'related' rights which are analogous to copyright which also exist in relation to databases and integrated circuit designs. Registered design protection protects the investment of a company in the design process, and other IP rights may also be of relevance to the electronics and software industry – some of these are covered in this chapter.

Despite attempts at international harmonisation over many years now, a large array of international treaties remain, all with different and often overlapping memberships. This is particularly true in Europe, where a complex web of international agreements and treaties now exists.

This chapter attempts to provide an overview of the main IP instruments relevant to the electronics and software industries in the international field. By definition, some simplification has had to be adopted and the precise historical background is not covered in detail.

At the outset, it should also be pointed out that IP rights are exclusive or exclusionary rights. The grant of a registered IP right does not provide any permission to the owner to exercise or to use the right, merely the ability to prevent others from carrying out activities without permission that would otherwise infringe the registered right.

1. General instruments

1.1 Paris Convention

The Paris Convention (or more properly the Paris Convention for the Protection of Industrial Property) was signed on 20 March 1883. It has been the subject of several revisions and is now administered by the World Intellectual Property Organisation (WIPO) which is a specialised agency of the United Nations.

The most recent revision of the Paris Convention was signed in Stockholm on 14 July 1967 and was further amended on 28 September 1979.

The Paris Convention was set up to enable applicants for IP rights in one country to obtain a corresponding filing date in a country other than their home country, simply through a mutual recognition of the filing of the initial application.

When the convention was first established, it was designed to address a very practical consideration about how to put on file corresponding applications in different countries at the same time, in order to establish a 'priority date' with respect to any competing third-party applications. With the advent of modern communications such as email and online (web-based) filing this rationale has become less important, but the convention still provides applicants with some very useful features.

Once an applicant for a patent has established a priority date, the date becomes the point in time against which the patentability of all subsequent applications claiming priority are judged with respect to the 'state of the art'. The state of the art is the term used to describe the sum total of publicly available materials that qualify as prior art (eg, published patent applications, scientific papers, textbooks). In some countries the list of prior art includes oral disclosures, poster displays and internet publications.

Many patent applications are based on an initial scientific observation or prototype device which led to the invention claimed. However, in a fast-moving research field, it is also important to file an early patent application to secure a filing date. The danger in filing an application too early is that the invention will not be well enough developed to provide the necessary support and sufficient written description.

The Paris Convention priority system enables an applicant to file an initial priority application based on a limited disclosure of the invention and then to file a follow-up patent application, claiming priority from the first filing, which can include new experimental results and/or a written description.

The system does require that the later patent application relates to the same invention; otherwise the right to claim the benefit of the priority date will not exist. In Europe there is now a stricter interpretation about the prior art effect of intervening disclosure published between the priority date of the first application and the filing date of the subsequent application if the content of the invention claim has changed between the two patent filings.

The ability to claim priority from one local application in a subsequent foreign application has enabled the growth of the IP system in its modern form and is the basis for the various current international systems. The priority system established by the Paris Convention is now an essential part of the process of filing and obtaining IP rights. For a patent application, it gives an applicant a period of 12 months to carry out further experiments or design additional prototypes before making a full patent filing. Given the costs and complexities of the patent system, such a delay is advantageous as an applicant can also carry out further due diligence by way of searching for any potentially relevant prior art (if not already carried out before making the priority filing).

The Paris Convention is now used to support a fairly standard patent application filing strategy, in combination with the Patent Cooperation Treaty (PCT) which provides applicants with up to 30 months before significant costs start to arise with the filing of separate national patent filings (see below).

For trademarks and registered designs (sometimes referred to as 'industrial designs'), the priority period is six months. For such applications, there is usually no need for further work on the application itself, but the delay provided by the priority period enables the application filing strategy to be decided upon and put into effect.

1.2 World Intellectual Property Organisation Convention

The World Intellectual Property Organisation (WIPO) is an organisation of the United Nations based in Geneva, Switzerland. WIPO is responsible for maintaining certain treaties relating to IP rights. There are 26 treaties in all, including the Paris Convention, Berne Convention, PCT, Hague Agreement and WIPO Copyright Treaty. WIPO in its modern form is governed by the WIPO Convention which is open to members of the United Nations.

1.3 Other agreements giving rise to recognition of priority claims

The modern history of the Paris Convention and its administration by WIPO have shown some limitations to the Convention concerning states or territories that are not recognised by the United Nations.

Some territories, such as Taiwan, have opted to conclude separate bilateral deals with important trading partners, and an alternative system which provides for claims to priority to be recognised has grown up with the World Trade Organisation (WTO).

(a) Agreement on Trade Related Aspects of Intellectual Property Rights

The Agreement on Trade Related Aspects of Intellectual Property Rights (TRIPs) arose from the negotiations under the General Agreement on Tariffs and Trade (GATT) in 1994 (at the end of the so-called 'Uruguay Round'). It is a treaty administered by the WTO.

The provisions of TRIPs apply to all members of the WTO and relate to all major types and classes of IP rights, including confidential information, as well as to procedures for enforcement of rights.

Membership of the WTO is different from the Paris Convention (and also from other international intellectual instruments). Most industrialised countries are members, with the remainder (eg, Iran, Belarus and Serbia) and other developing nations (eg, Sudan, Ethiopia) as observers. A few countries are neither members nor observers and are unlikely to achieve either status in the foreseeable future (eg, North Korea and Somalia).

The importance of TRIPs is that it has helped drive international attempts to achieve harmonisation in both the process of obtaining IP rights and the enforcement of such rights.

For states that have not ratified the Paris Convention (for whatever reason), membership of the WTO provides an alternative procedure for claiming the benefit of a domestic filing date in a subsequent foreign or international patent application (filed via the PCT).

In addition to providing the basis for priority claims, the WTO also provides for minimum standards for IP protection and enforcement.

(b) *Bilateral agreements*

Since the international systems for recognition of claims to priority are administered by WIPO (a United Nations organisation), or the World Trade Organisation, there is a problem for states which exist *de facto*, but which have no *de jure* international recognition.

Taiwan occupies a slightly anomalous international position. The island is self-governing and refers to itself as the Republic of China. The mainland People's Republic of China sees the island as a breakaway province. Taiwan is unable to be a member of the various international conventions which recognise the People's Republic of China as the sole government for the territory of China.

However, Taiwan has negotiated bilateral agreements with many national governments in order for national patent applications to include a right to claim priority in Taiwan. The Taiwanese Patent Office has also concluded a similar agreement with the State Intellectual Patent Office (SIPO) in China on a similar basis.

Similarly, in Kosovo (which is not recognised by all states and consequently is also not a member of the United Nations), it is possible to file an application and to have a right to priority recognised from an earlier filed application in a Paris Convention or WTO state under the relevant IP laws in Kosovo.

2. Patent instruments

2.1 Patent Cooperation Treaty

The PCT is an international treaty administered by the World Intellectual Property Organisation.

The PCT is a system for simplifying the procedures for filing corresponding patent applications in different countries. The PCT provides applicants with a mechanism for filing a single so-called 'international patent application' at a local national patent office (or directly with the International Bureau of WIPO). The application is examined for formalities requirements and is then subject to a search and preliminary examination procedure which results in the issue of an international search report and a written opinion.

The international patent application never proceeds as far as being granted by WIPO, but the results of the search and the comments in the written opinion mean that an applicant who uses the PCT can determine whether it is worth proceeding with the filing of separate local national patent applications subsequently.

Filing an international patent application under the PCT is therefore a very popular filing strategy for applicants in all technical fields. In the life sciences field, too, where specifications can sometimes be quite lengthy (with corresponding translation costs to consider), the ability to receive a preliminary search and opinion from a central source can be very useful in developing a patent filing strategy.

An international patent application filed under the terms of the PCT will pend for a period of at least 30 months from the declared priority date (or filing date if no earlier priority date is claimed). The applicant may claim the benefit of a priority date of an earlier application by virtue of the Paris Convention or the provisions of GATT/TRIPs under the World Trade Organisation.

There is a facility for filing amendments to the claims after receipt of the international search report. In addition, it is possible to file a response to the comments in the written opinion, provided that a request for international preliminary examination is filed.

Since the PCT is administered by WIPO which is a UN organisation, membership is only open to states which are full members of the United Nations. Taiwan therefore falls outside the scope of the treaty. In addition, there remain many states which have not signed up to the PCT. However, the growing importance of trade agreements under the WTO has meant that more states have joined the PCT in recent years.

A PCT patent application may be filed in the languages specified by the national patent office where the application is filed (the receiving office), or by the International Bureau of WIPO. Subsequently, the PCT application must be translated into one of the following languages for publication (if not already filed in one of them): Arabic, Chinese, English, French, German, Japanese, Korean, Portuguese, Russian or Spanish.

At the end of the so-called international phase of the PCT application, the applicant must decide where to file national or regional phase patent applications based on the PCT application. The content of such applications will be the same as the PCT application and in many states the outcome of the PCT search and examination procedure will be taken into account by the local patent office where the national or regional patent application is filed.

The PCT system enables an applicant to select a variety of national patent filings in combination with some regional patent filings. A regional patent application is made to a central patent office which administers a special treaty or arrangement for the central examination and grant of a patent application. At present, the PCT offers applicants the opportunity to designate regional patent filings via the European Patent Convention (EPC) at the European Patent Office (EPO), the Eurasian Patent Convention at the Eurasian Patent Office (EAPO), the African Regional Intellectual Property Office (ARIPO) and the African Intellectual Property Organization (OAPI).

However, there are some restrictions in the states available via a PCT patent application. For example, while it is possible separately to designate the United Kingdom via a national or European patent application, other European states have closed the national route and patent protection is not available there via a European patent application from a PCT application.

In the electronics and software fields, unlike the life sciences, normally patent protection is only sought in a limited number of jurisdictions. This means that frequently the main attraction of the PCT is its ability to defer costs (rather than its wide geographical scope). The PCT system is particularly effective where the commercial significance of an invention can be established before the 30-month deadline for entry into the national/regional phase as, if a decision to abandon an application is reached within that time-scale, only commercially significant inventions will incur the costs of national/regional entry.

2.2 European treaties

(a) European Patent Convention (EPC)

After World War II, various agreements relating to trade were made in Europe in order to assist the recovery of Europe's national economies and to bind the countries together so as to reduce the chances of another conflict emerging.

In relation to intellectual property, the main agreements of interest are the Treaty of Rome (and subsequent treaties of the European Economic Community (EEC) which later become the European Union) and the EPC.

One of the agreements that was finalised relatively early was the Treaty of Rome in 1957, which laid the basis for the European Union. However, until the coming into force of the Single European Market Act in 1987, the European Union was not concerned with IP rights. Instead, a parallel agreement was executed in 1973 between various countries in Western Europe, which became known as the EPC.

The separate development of the European Union and the EPC has meant that it is possible for different states to belong to each organisation. All member states of the European Union belong to the EPC.

It is important to note that the EPC is not a treaty of the European Union. The contracting states to the EPC include countries which belong to both the European Union and EPC, but there are also states which are party to the EPC only which have yet to join the European Union (or which in some cases may never do so). However, for states now wishing to be members of the European Union, it is a requirement that they also accede to the EPC as part of the application process. The announcement by the United Kingdom that it intends to leave the European Union does not, though, affect its membership of the EPC.

The purpose of the EPC is to give to a single authority – the European Patent Office (EPO) – powers to examine and grant patents across Europe. However, although a single patent application is filed at the EPO, the terms of the EPC mean that the patent which is granted only takes effect in the national European states party to the EPC as initially designated in the application, and where appropriate procedures have been taken on grant to validate the patent at the national level.

The EPO has offices in The Hague, Berlin and Munich. European patent applications may be filed at any of the offices of the EPO, or through a local national patent office (for an initial filing only). Divisional European patent applications must be filed directly with the EPO.

The filing of a European patent application at the EPO is therefore equivalent to taking an option on a collection of related national patents. Rather than filing separate individual national patent applications at national patent offices, with all associated issues of local language translation and the inevitable duplication of search and examination, the European patent application provides a convenient mechanism to obtain grant of a patent which can be validated on grant in the required territories.

Contracting states: As of 1 January 2019, the contracting states to the EPC are Albania, Austria, Belgium, Bulgaria, Switzerland, Cyprus, Czech Republic, Germany, Denmark, Estonia, Spain, Finland, France, the United Kingdom, Greece, Croatia, Hungary, Iceland, Ireland, Italy, Latvia, Liechtenstein, Lithuania, Luxembourg, Malta, Monaco, the Netherlands, North Macedonia, Norway, Poland, Portugal, Romania, San Marino, Slovenia, Slovak Republic, Sweden, Serbia and Turkey. Montenegro has now been invited to accede to the EPC.

Extension states: A number of other states have concluded 'extension agreements' with the EPO as a preliminary step to becoming full EPC contracting states. Such states may be added to a European patent application on payment of an extension fee per country required.

As of 1 January 2019, the states with current extension agreements with the EPO are Bosnia & Herzegovina and Montenegro.

Other states have signed agreements to recognise the grant of a European patent. Similar to an 'extension' state, an official fee is payable to the EPO on filing the European patent application to secure the option of completing the requirements to secure validation of the patent on grant in the states required. At present, the EPO has signed such validation agreements with Morocco, Moldova, Tunisia and Cambodia which are now all in force.

The territorial extent of a European patent includes certain dependent territories (outside Europe) of the contracting states of the EPC. A European patent may also be used as the basis for a local patent registration in Hong Kong and/or Macau.

A granted European patent which is validated in the United Kingdom may be used to obtain local patent registration in certain territories dependent on the United Kingdom, as well as some former UK colonies and dependent territories.

(b) *The London Agreement*

One of the key advantages of using the EPO is avoiding the need for local language translation. A European patent application may be prosecuted in English, in French or in German (one of the three). Later, upon grant, it is necessary to file a translation of the granted claim in the two languages not used for prosecution of the application.

The national validation procedure after grant by the local national patent offices requires the completion of various formalities according to local laws. The formalities may include a full translation of the patent specification as granted, the claims as granted, filing a power of attorney and/or recording a local address for service. However, since the coming into force of the London Agreement in 2008 increasing numbers of countries are waiving the need for translations to be filed or are limiting the filing requirements to claims translations.

As of 1 January 2019, those EPC states which have implemented the London Agreement are: Albania, Croatia, Denmark, Finland, France, Germany, Hungary, Iceland, Ireland, Latvia, Liechtenstein, Lithuania, Luxembourg, Monaco, Netherlands, North Macedonia, Norway, Slovenia, Sweden, Switzerland and the United Kingdom.

(c) *Unitary European patent and Unified Patent Court*

The European Union has discussed a unitary European patent for many years. The project was initially referred to as the 'Community Patent Convention', but has more recently become known as the 'unitary European patent'. Following a positive vote in the European Parliament in December 2012, the draft legislation has been approved and came into force as EU Regulation 1257/2012 from 20 January 2013. However, it will apply only from the date on which the related Agreement on a Unified Patent Court enters into force.

The development will provide a single unitary European patent which would include all EU member states (with the current exception of Spain). Upon grant, there would be no need for any national validation procedure in each state, as for the European patent application currently administered by the EPO; instead, a simplified validation procedure would require more limited formalities, including only one translation.

The unitary European patent will be administered by the EPO. Under the system, a European patent application will be filed in the normal way with a designation of all contracting states to the EPC (and consequently all EU member states). The application will be searched and examined just as at present for European patent applications. Upon grant of the application, the applicant will have one month to elect for the patent to have 'unitary effect' in the participating EU member states (ie, currently all EU member states except Spain). Patent protection in the other designated states will be obtained in the usual way as at present, by completing the national validation procedures within three months of grant.

The system for electing for unitary protection will therefore coexist with the current system for obtaining patent protection in the EPC contracting states, as well as the separate national patent systems.

Given the expected high level of maintenance fees for unitary patents, the unitary patent may be of limited interest in practice. For applicants who validate European patents in all EPC states, the unitary patent may be useful. Applicants who require patent protection in only a smaller number of EU member states can continue as at present with the current system of national validations.

The unitary patent and Unified Patent Court project encountered several stumbling blocks in its gestation. First, Italy and Spain opposed the system, since it would not recognise their official national languages. A legal challenge before the European Court of Justice was rejected and Italy later decided to become part of the system. A second problem to emerge related to the unified central court system to be set up to hear complaints of patent infringement and/or validity. However, this was eventually overcome with the approval of the Agreement on a Unified Patent Court, which has been negotiated as an international agreement between the participating member states of the European Union.

As of 1 January 2019, many of the participating member states of the European Union have ratified the UPC Agreement which is necessary for the Unitary patent system to come into effect. However, ratification by Germany and the United Kingdom along with France is required in order for the UPC Agreement to come into force which is a precondition for the introduction of the unitary patent. The announcement that the United Kingdom will leave the European Union has cast a cloud over the implementation of the system. Nevertheless, at the time of writing, the United Kingdom had completed the necessary acts to ratify the UPC Agreement. In the meantime, a case has been lodged with the Constitutional Court in Germany which questions the ability of Germany to ratify the UPC Agreement. The outcome of the constitutional challenge and the United Kingdom's relationship with the European Union after Brexit has the capacity to change or even derail the introduction of the system. However, there remains a significant amount of political momentum to see the new system implemented.

As of 1 January 2019, the EU member states were Austria, Belgium, Bulgaria, Croatia, Cyprus, Czech Republic, Germany, Denmark, Estonia, Spain, Finland, France, the United Kingdom, Greece, Hungary, Ireland, Italy, Latvia, Lithuania, Luxembourg, Malta, Netherlands, Poland, Portugal, Romania, Slovenia, Slovak Republic and Sweden. The United Kingdom has announced its intention to leave the European Union. At the time of writing, the departure of the United Kingdom has been announced to be with effect from 31 October 2019. The date has already been extended and could be again subject to agreement by the remaining EU states. However, the process is untested and the final relationship between the United Kingdom and European Union after this date remains to be determined. A transitional period is envisaged in the event that a withdrawal agreement can be concluded between the European Union and the United Kingdom which will preserve many of the features of EU membership until the terms of trading arrangements between the United Kingdom and the European Union are settled.

2.3 Other regional patent agreements

Other regional patent agreements exist. Some, like the EAPO, ARIPO and OAPI agreements, can be accessed via a PCT patent application. Others, like the Gulf Cooperation Council (GCC), require an application to be filed at the same time as filing a PCT application (with the option of making a claim to priority under the Paris Convention).

(a) Eurasian Patent Convention

The Eurasian Patent Convention is analogous to the EPC and covers some former states of the USSR. It provides for a single patent application to be filed which, once granted, extends to those states selected by the proprietor which are members of the Eurasian Patent Convention. The present member states are the Russian Federation, Belarus, Kazakhstan, Armenia, Azerbaijan, Turkmenistan, Tajikistan and Kyrgyzstan. Moldova exited from the Eurasian Patent Convention with effect from 26 April 2012.

The Eurasian Patent Convention is a special regional patent agreement within the meaning of the Paris Convention and the Patent Cooperation Treaty, so patent applications may be filed at the Eurasian Patent Office (EAPO) from a Patent Cooperation Treaty application.

(b) Gulf Cooperation Council (GCC)

The GCC is a regional patent system enabling a single patent application to be filed at the GCC Patent Office in Saudi Arabia. A GCC patent once granted is effective in all designated states. The GCC is not a party to the Patent Cooperation Treaty, but does recognise claims to priority under the Paris Convention.

As of 1 January 2019, the member states of the GCC are: Saudi Arabia, Bahrain, Kuwait, Oman, Qatar and the United Arab Emirates.

(c) ***African Intellectual Property Organization (OAPI)***

The African Intellectual Property Organization (OAPI) is a regional patent system which covers mostly francophone countries in Africa. It is a party to the Patent Cooperation Treaty and can be designated in a Patent Cooperation Treaty application.

As of 1 January 2019, the member states of OAPI are: Benin, Burkina Faso, Cameroon, Central African Republic, Chad, Comoros, Congo, Cote d'Ivoire, Gabon, Guinea, Equatorial Guinea, Mali, Mauretania, Niger, Guinea Bissau, Senegal and Togo.

(d) ***African Regional Intellectual Property Office (ARIPO)***

The African Regional Intellectual Property Office (ARIPO) is an alternative regional patent system in Africa and includes principally anglophone states. It is complementary to the OAPI system. It is also a party to the Patent Cooperation Treaty and can be designated in a Patent Cooperation Treaty application.

As of 1 January 2019, the member states of ARIPO are: Botswana, The Gambia, Ghana, Kenya, Liberia, Lesotho, Malawi, Mozambique, Namibia, Rwanda, Sao Tome and Principe, Somalia, Sierra Leone, Sudan, Eswatini, Uganda, United Republic of Tanzania, Zambia and Zimbabwe.

4. Copyright instruments

4.1 Berne Convention for the Protection of Literary and Artistic Works

The Berne Convention is concerned with the protection of copyright works. The treaty is administered by WIPO. Copyright is relevant to the field of electronics and software in respect of computer programs which fall to being considered as 'literary works'. The definition of a computer program as a literary work flows from the fact that it is a written work, in a code or computer language. Where the computer code is original it will by definition qualify for protection as a copyright work as a result. Other items in the electronics and software field likely to be covered by copyright include scientific papers, books and manuals of procedure, among other things. Copyright in the electronics and software industries can therefore actually be quite important and should not be neglected.

The Berne Convention was first established in 1886 and for a time was only adopted by certain states, with developing nations adhering to the Universal Copyright Convention (UCC) separately administered by UNESCO. However, more recently, since most states are members of the WTO (or aspire to become members), the provisions of the Berne Convention have assumed a greater importance and the UCC has become less relevant.

The point of treaties such as the Berne Convention and the UCC is to provide for mutual international recognition of copyright protection for original published copyright works in the signatory states by their respective nationals. The recognition of such rights is therefore included in the national law of the signatory states relating to copyright.

Many states have amended their laws to vary the protection afforded to copyright works. Typically, the protection is not applied for, but subsists automatically through the act of creating an original qualifying work. However, the existence of the right is only ever tested in court in actions for infringement.

4.2 WIPO Copyright Treaty

The WIPO Copyright Treaty 1996 (WCT) is a special agreement under the Berne Convention which entered into force in 2002. Any contracting state to the WCT (even if it is not bound by the Berne Convention) must comply with the substantive provisions of the 1971 text (Paris) of the Berne Convention (1886). Furthermore, the WCT mentions two subject matters to be protected by copyright, which are:

- computer programs, whatever may be the mode or form of their expression; and
- compilations of data or other material (databases), in any form, which by reason of the selection or arrangement of their contents constitute intellectual creations.

The WCT also contains provisions with respect to the rights of authors. Specifically, the treaty provides for:

- the right of distribution through sale or transfer of ownership;
- the right of commercial rental; and
- the right of communication to the public.

Each of these rights is defined as an exclusive right, subject to certain limitations and exceptions. Significantly, the right of rental includes computer programs (except where the computer program itself is not the essential object of the rental).

The right of communication to the public is the right to authorise any communication to the public, by wire or wireless means, including "the making available to the public of works in a way that the members of the public may access the work from a place and at a time individually chosen by them". This covers in particular on-demand, interactive communication through the internet.

5. EU legislative initiatives

The European Union has responsibility for the single European market in goods and services and as a result has initiated some legal developments which have created new IP rights.

5.1 Community design

In addition to national laws concerning registered designs, the European Union has enacted Regulation 6/2002 to allow for the application of a Registered Community Design (RCD). An application for a community design is made at the European Union Intellectual Property Office (EUIPO), formerly known as the Office for Harmonisation in the Internal Market (OHIM). The right, once granted, is a unitary right effective across the entire European Union.

Design protection is relevant in the context of the electronics industry for almost all products sold to the public. Many such products are the result of a lengthy design process and the design element can be highly important and very valuable.

In the context of a registered design application, a 'design' is taken to mean the appearance of the whole or a part of a product resulting from the features of, in particular, the lines, contours, colours, shape, texture, and/or materials of the product itself and/or ornamentation. A 'product' is taken to mean any industrial or handicraft item, including packaging, get-up, graphic symbols or typographic typefaces, but excluding computer programs.

Applications may be filed within one year of being published at a national patent office, the EUIPO or at the Benelux Patent Office.

The regulation also provides for a system of unregistered Community design rights with protection of three years from the date of publication of the design.

5.2 EU trademark (formerly Community trademark)

In addition to national laws on registered trademarks, the European Union has also enacted laws to allow for the application of a European Union trademark (formerly known as a Community trademark). The basic legislation was introduced in EU Regulation 40/94. This has been updated over time and most recently replaced by EU Regulation 2017/1430, which codifies and updates the basic law. An application for a European Union trademark is made at the EUIPO. The right, once granted, is a unitary right effective across the entire European Union.

In the context of a trademark application, a 'trademark' is a mark or sign capable of distinguishing the goods or services of an applicant for a defined class of goods or services, without being descriptive or deceptive, provided that the representation of the trademark is clear, precise, self-contained, easily accessible, intelligible, durable and objective. The previous rules prior to 1 October 2017 required that the trademark be capable of being represented graphically. However, this has now been updated. Trademarks therefore include shapes, sounds, colours and smells as well as the more usual words, names, logos and signs. A three-dimensional or shape trademark therefore has some overlap with a registered design.

Trademarks are naturally extremely valuable IP rights. Patents, registered designs and copyright all last only for a limited number of years, whereas in theory a trademark, if used properly, can last forever.

5.3 Database rights

As IT technology became more widespread in the late 1980s and early 1990s it became clear that large amounts of data were being stored in computer files in the form of databases. The skill and effort in compiling a database was not felt to be adequately protected through the standard copyright definitions, so the European Union instigated a 'database right' through Directive 9/96/EC. The directive was an instruction to national states in the European Union to harmonise domestic legal provisions in accordance with the terms of the directive. Database right provides protection for a database when it is created, and through subsequent substantial modifications.

5.4 'Brussels Regime' on jurisdiction in civil and commercial matters

Within the member states of the European Union and the states of the European Free Trade Association, there has been a recognition of the need to regulate the conduct of litigation in civil matters. This has led to the bringing into force of three separate legal instruments.

These instruments are the Brussels Convention (the Brussels Convention of 27 September 1968 on Jurisdiction and the Enforcement of Judgments in Civil and Commercial Matters); the Lugano Convention (Convention of 16 September 1988 on Jurisdiction and the Enforcement of Judgments in Civil and Commercial Matters); and the Brussels Regulation: EC/44/2001 (Council Regulation (EC) No 44/2001 of 22 December 2000 on Jurisdiction and the Recognition and Enforcement of Judgments in Civil and Commercial Matters), the so-called 'Brussels I Regulation'. In order to cope with the introduction of the Unified Patent Court, the Brussels I Regulation has been replaced by Regulation EU/1215/2012.

5.5 Harmonisation of enforcement procedures in Europe

The European Union has also introduced legislation designed to bring an approximation of laws into effect with respect to the enforcement procedures for IP rights across the various member states. The Enforcement Directive (Directive 2004/48/EC of the European Parliament and of the Council of 29 April 2004 on the enforcement of IP rights) was an instruction to the member states to amend national laws as necessary to achieve the effects set out in the directive and has now been brought into force.

5.6 Departure of the United Kingdom from the European Union (BREXIT)

The stated intention of the United Kingdom to leave the European Union (British-exit or Brexit) has implications for unitary IP rights granted for the territory of the European Union. The final agreement by which the United Kingdom leaves the European Union will need to address these issues. Right-holders will need to be given a mechanism to retain protection in the United Kingdom after the final date of departure.

6. Semiconductor topography rights/Washington Treaty on Intellectual Property in Respect of Integrated Circuits

In the electronics field, considerable effort can often be expended in the surface design of integrated circuits. However, such surface ornament often fails to be protected adequately under copyright or design law and there have been moves to protect the layout of integrated circuits via a new IP right sometimes referred to as 'semiconductor topography right' or 'mask right'. The right is analogous to copyright in that it prevents copying but not independent creation of a similar work.

Since different national states had approached this matter independently in order to protect local industries, there were moves to achieve international harmonisation through the Washington Treaty on Intellectual Property in Respect of Integrated Circuits of 1989.

The treaty is administered by WIPO, as are the other international IP treaties. However, at the time of writing the treaty looks to be moribund and is not in force – see www.wipo.int/treaties/en/ShowResults.jsp?lang=en&treaty_id=29 and www.wipo.int/treaties/en/ip/washington/. The Washington Treaty is, though, mentioned within TRIPs as the source of the relevant minimum requirements for the protection of integrated circuit designs.

7. Other IP instruments

The Madrid System for trademark registrations and the Hague Agreement for registered designs provide international systems for obtaining trademark or design registration in an analogous manner to patents under the PCT (but with some key differences).

However, many applicants still opt to use local national filings for strategic reasons, rather than use the Madrid System or the Hague Agreement.

7.1 Madrid System for international trademark applications

The Madrid System has two main elements. First, there is the Madrid Agreement Concerning the International Registration of Marks of 1891; and second there is the Protocol Relating to the Madrid Agreement of 1989 (for states unable to accede to the Madrid Agreement).

The Madrid Agreement and the Madrid Protocol provide a system for international trademark registrations to be applied for and granted.

7.2 Hague Agreement Concerning the International Registration of Industrial Designs

The Hague Agreement dates from 1925 and has been subject to several amendments. The current versions in force are the 1999 Act, the 1960 Act and the 1934 Act. In essence, it provides a system for international design registration, much like the provisions of the PCT system for patents.

Standard setting, competition law and FRAND licensing in Europe

Natalie Coan
Caitlin Heard
Louisa Mottaz
Chris Watson
CMS Cameron McKenna Nabarro Olswang LLP

1. Standard setting

Standardisation is the establishment of a common set of characteristics for goods or services (eg, in relation to their quality, performance, efficiency, labelling, uniformity, professional conduct or interoperability). Standards are based on voluntary cooperation between industry, public authorities and other interested parties and seek to set out "technical or quality requirements with which current or future products, production processes, services or methods may comply".[1] Standardisation plays a particularly important role in the information and communication technologies (ICT) sector, where the need for interoperability and economies of scale and scope are critical to promote innovation and competition, especially as technologies converge. Standardisation is likely to play an increasingly important role with the development of the Internet of Things (IoT), and in a growing number of industries developing an array of 'networked' products that rely on existing and evolving interoperable information and communication technologies.[2] The European Commission, in its communication "Setting out the EU approach to Standard Essential Patents", acknowledges that without interoperability enabled by standards, "40% of the potential benefits of the IoT systems would not be reaped".[3]

Most IP issues arising from technical standards concern patents. It is evident that technical standards are likely to comprise features that are the subject of patents, and there is little point developing a standard if its implementation can

1 European Commission Guidelines on the applicability of Article 101 of the Treaty on the Functioning of the European Union to horizontal cooperation agreements, OJ 2011 C11/1, 14 January 2011, https://eur-lex.europa.eu/LexUriServ/LexUriServ.do?uri=COM:2011:0311:FIN:EN:PDF, paragraph 257.

2 Definition taken from the Internet of Things Global Standard Initiative Recommendation ITU-T Y.2060 (06/2012).

3 European Commission Communication COM(2017) 712 final, 29 November 2017, "Setting out the EU approach to Standard Essential Patents", which cites McKinsey, 2015.

be blocked by a patent holder. In practice, patent holders who participate in standard-setting agree to grant licences to each other and to third parties, and if this commitment is not honoured there can be competition law consequences.

1.1 Technical standards

Examples of technical standards include the MPEG-2 digital video compression standard, the 3, 4 and 5G standards for mobile telecommunication services and the AAC standard for audio coding.

Standards can be facilitated or developed by a standard-development organisation (SDO),[4] or can emerge spontaneously without any form of agreement: so-called '*de facto* standards'. The focus in this chapter will be on standards developed by SDOs.

There are many different types of SDOs. These can be broken down into the following:

- 'Formal' SDOs – these set standards based on institutionalised and consensual procedures in which all interested stakeholders can participate. There are three main sub-types:
  - international SDOs, such as the International Organization for Standardization (ISO);
  - European standards organisations (ESOs) or other regional standards organisations, such as the European Telecommunications Standardisation Institute (ETSI),[5] responsible for developing standards in the ICT sector; and
  - national standardisation bodies, such as the British Standards Institute.

Typically, these formal SDOs have numerous members who may contribute to and influence the standards set by that body. For example, ETSI has over 800 direct members drawn from more than 66 countries worldwide.[6] Its members include technology developers, network operators, handset manufacturers and many government and regulatory bodies who meet regularly with ETSI to agree standards. There are only three formal SDOs that are officially recognised by the European Commission as providers of European Standards, namely: CEN (European Committee for Standardisation); Cenelec (European Committee for Electrotechnical Standardisation); and ETSI.[7]

---

4 SDOs may also be referred to as standard setting organisations (SSOs).

5 European Committee for Standardisation (CEN), European Committee for Electrotechnical Standardisation (CENELEC) and ETSI are recognised as the European Standards Organisations under Council Directive 98/34/EC which sets a procedure for the provision of information in the field of technical standards and regulations (OJ L 204, 21 July 1998).

6 See www.etsi.org/index.php?option=com_content&view=article&id=23:current-members&catid=16:membership&Itemid=124&jjj=1569848109567.

7 Regulation (EU) 1025/2012.

- 'Informal' standardisation consortia – these are set up by groups of companies to draft sectoral standards. It is these types of organisations, as opposed to the ESOs, that have arguably emerged as the world-leading bodies in the development of ICT standards.[8] For example, standards covering internet protocols were established by the Internet Engineering Task Force, a team consisting of network designers, operators, vendors and researchers,[9] and the Blu-ray standard was developed by the Blu-ray Disc Association, a group representing makers of consumer electronics, computer hardware and motion pictures.[10]

2. IP and standards

'Standard essential patents' (SEPs) are those that are infringed, as a matter of technical inevitability, by implementation of the standard. In effect, they cannot be designed around (although SEPs can be essential to an optional part of the standard). Such patents are of enormous value given the high sales volumes of standards-compliant consumer and other devices, on which royalties must be paid.[11] In the ICT sector, each device typically features inventions patented by multiple competitors, who cross-license each other.[12] The relative size of each company's stack of essential patents can make a difference in licence negotiations, although its precise correlation to the amount of any royalty payable will vary according to circumstances. However, over-declaration of SEPs is a known problem – one which has been wrestled with by the UK court in *Unwired Planet v Huawei* (discussed further below), and the European Commission, which has made several proposals of information tools to assist licensing negotiations, including a higher degree of scrutiny of declarations of essentiality.[13]

Because of the inevitability of infringement by a large number of businesses, so-called non-practising entities (sometimes referred to as 'patent assertion entities' or 'trolls') also purchase essential patents purely for the purpose of generating royalties.

2.1 IP policies of standards development organisations

SDOs need to engage with intellectual property for two principal reasons, because:

---

8 Annex 2 to the Commission Staff Working Document entitled "The Challenges for European Standardisation", pp14 and 15.

9 See www.ietf.org/about.

10 See www.blu-raydisc.com/en/Technical/FAQs/HistoryandAssociation.aspx.

11 Harhoff *et al* point out the increasingly strategic use of patents to block rivals and extract concessions, as a consequence of the increased complexity of modern technology, with potential anti-competitive consequences: "The Strategic Use of Patents and its Implications for Enterprise and Competition Policies", Tender for ENTR/05/82 Final Report, 8 July 2007.

12 See Biddle, White and Woods, "How Many Standards in a Laptop? (and other Empirical Questions)", Soc Sci Res Network, 10 September 2010, available at SSRN: http://ssrn.com/abstract=1619440 or http://dx.doi.org/10.2139/ssrn.1619440, identifying 251 standards embodied or utilised in a laptop computer.

13 European Commission Communication COM(2017) 712 final.

- a standard will be inoperable if implementation is blocked by patent rights; and
- standards and essential patents have competition law implications.

SDOs therefore encourage:

- disclosure of patents that are believed to be essential, or potentially essential, to relevant standards or proposed standards; and
- commitment to licence such patents (usually on fair, measurable and non-discriminatory 'FRAND' or reasonable and non-discriminatory 'RAND' terms).

Some SDOs such as the Institute of Electrical and Electronics Engineers (IEEE), for example, have also provided guidance on the interpretation of the FRAND commitment in respect of IEEE standards.[14]

The licensing commitment is not conditional upon the patent being proven essential. However, a patent holder that has declared a belief that specific patents are essential, or potentially essential, to a standard should subsequently be able to make a correcting statement if that interpretation changes – for example, if a patent is declared by a court (or conceded by the patentee) to be non-essential.

The common patent policy for ISO/International Electrotechnical Commission (IEC)/International Telecommunication Union (ITU), which is also endorsed by CEN and CENELEC (although neither CEN nor CENELEC has any involvement in licence negotiations), encourages the early disclosure and identification of relevant patents, and states that patented technology may be incorporated into standards subject to provisos. In summary, the patent holder must provide a statement, in a prescribed form, of willingness to negotiate irrevocable licences on a non-discriminatory basis on FRAND terms and conditions (although not necessarily free of charge). This licensing declaration remains in force as long as a patent remains essential unless superseded by a more favourable declaration (eg, offering licences free of charge). Identification of specific patents is encouraged but is only mandatory for patentees who refuse to grant licences. CEN/CENELEC has also published a Workshop Agreement entitled "Core Principles and Approaches for Licensing of Standard Essential Patents"[15] which sets out six core principles to help drive FRAND licensing practices, and addresses the 'best practice' that parties might choose to adopt in order to resolve SEP licensing issues in compliance with the FRAND obligation.

ETSI's IP rights policy has been the subject of much scrutiny in telecoms patent disputes in Europe. Where essential intellectual property is identified, the owner is given three months to offer an irrevocable undertaking that it is

14 IEEE-SA Standards Body Bylaws, Clause 6, https://standards.ieee.org/about/policies/bylaws/sect6-7.html.
15 CWA95000 dated June 2019.

prepared to grant irrevocable licences on fair, reasonable and non-discriminatory terms and conditions (FRAND), failing which the work on the relevant parts of the standard may be suspended. ETSI members refusing to give licence commitments must explain themselves to the Director General unless a viable alternative technology exists. The ETSI policy is expressed to be governed by French law, as is the licensing declaration in the prescribed form. The declaration form appended to the ETSI IP rights policy states that the irrevocable undertaking to grant a licence on FRAND terms only takes effect to the extent that the IP remains essential to the technical standard or specification. Therefore, while the declaration is technically irrevocable, it is only effective in so far as the patent is in fact, and so long as it remains, essential. The enforceability of the ETSI FRAND undertaking, and the interpretation of French law, have been considered in *Unwired Planet v Huawei*[16] by both the UK High Court, and the Court of Appeal (discussed further below).[17] At trial, Birss J (not challenged on appeal in respect of this issue) found that an undertaking given to ETSI has international effect, and that the FRAND undertaking is contractually enforceable by third parties pursuant to the French doctrine of "stipulation pour autrui".[18] Birss J recognised that the enforceability of the FRAND undertaking under French law "is not a clear cut question", but found that:

> *The reason [the doctrine of stipulation pour autrui] should be applied is because the FRAND undertaking is an important aspect of technology standardisation. Holders of essential IPR are not compelled to give a FRAND undertaking but it serves the public interest that they make it clear whether or not they are doing so, and it serves the public interest that if they do, the undertaking is public, irrevocable and enforceable. To avoid hold up, implementers need to know that they can hold SEP owners to a FRAND obligation.*[19]

2.2 Patent pools

A patent pool involves multiple undertakings agreeing to 'pool' their technologies and license them to each other or to third parties. This may be associated with the development of a standard or may arise separately.

Typically, a patent pool is created through an agreement between licensors owning essential patents. The licensor/parties to a pool cannot insist on a licence that bundles SEPs with non-SEPs in the portfolios.[20] The licensors appoint a patent pool administrator[21] to oversee the operation of the patent

---

16 [2017] EWHC 711 C Pat.
17 The decision in *Unwired Planet v Huawei* is subject to a pending appeal to the Supreme Court.
18 Paragraph 139 of the High Court decision.
19 Paragraph 146 of the High Court decision.
20 Birss J in *Unwired Planet v Huawei*, at paragraph 787, "I am in no doubt that a patentee subject to a FRAND undertaking cannot insist on a licence which bundles SEPs and non-SEPs together".
21 Examples include MPEG LA, One-Blue, Sisvel and Via Licensing.

pool. It is the role of the administrator to evaluate essentiality, identify and market to licensees, execute and enforce uniform licence agreements, collect royalties and distribute revenue fairly to participating licensors. As this involves effort and expense, it will not be appropriate unless the advantages over purely bilateral licensing are clear – for example, to overcome patent 'thickets' whereby individual licence negotiation is uneconomical.

3. Competition law and the establishment of standards and patent pools in Europe

3.1 EU competition law principles

Article 101(1) of the Treaty on the Functioning of the European Union (TFEU) prohibits, as incompatible with the internal market, agreements between undertakings that may affect trade between member states and which have the object or effect of preventing, restricting or distorting competition within the common market.

Agreements that infringe Article 101(1) are void and unenforceable in respect of the provisions that restrict competition (Article 101(2)). Exemption under Article 101(3) may be possible, where the beneficial aspects of the agreement outweigh its anti-competitive effect, ie, the agreement contributes to technical or economic progress, allows consumers to get a fair share of the benefits, does not impose restrictions unless indispensable and does not substantially eliminate competition.

Article 102 of the TFEU prohibits the abuse by one or more undertakings of a dominant market position. The following examples of abuse are given: "directly or indirectly imposing unfair purchase or selling prices or unfair trading conditions", "limiting production, markets or technical development to the prejudice of consumers", or "applying dissimilar conditions to equivalent transactions with other trading parties, thereby placing them at a competitive disadvantage". This is not an exhaustive list of all forms of abuse that have been developed in case law.

The European Commission has the ability as the enforcement authority to impose fines on parties that infringe Articles 101 and 102 of the TFEU of up to 10% of their annual worldwide group turnover.

In September 2018, the Commission announced that it was investigating five automotive manufacturers (Audi, Daimler, BMW, VW and Porsche) for allegedly engaging in five-way unlawful discussions aimed at coordinating their behaviour with regard to the development of 'clean car technology' which led to delay in launching various technology. The Commission sent out its Statement of Objections (formal notice of action) on 5 April 2019 setting out its case against the alleged infringers. This case is seen as a departure from traditional competition law investigations, which typically focused on price fixing and market sharing abuses, to extend the application of competition law enforcement to purely technological discussions. In this case, the Commission has distinguished technical collaboration aimed at innovation and development which is not necessarily anti-competitive and from unlawful technical collaboration which aims to restrict the launch or roll-out of technological advancements.

3.2 Application to standards

Standardisation may give rise to competition concerns by, for example, facilitating collusion and/or the exchange of competitively sensitive information, or leading to discrimination and foreclosure of competitors or other deceptive conduct such as patent hold-up. However, the Commission takes the view that, under certain conditions, a standardisation agreement is not likely to produce restrictive effects on competition. Such conditions are set out in the Commission's guidelines on the applicability of Article 101 of the TFEU to horizontal cooperation agreements (the guidelines)[22] and effectively create an informal safe harbour to avoid the scope of Article 101(1) of the TFEU. According to the safe harbour, standard-setting will normally not restrict competition if the following four conditions are met.

(a) Participation in the standard-setting is unrestricted

The rules of the SDO must be unrestricted to ensure that all competitors in the market or the markets affected by the standard are able to participate in the process of selecting the standard. The relevant SDO would also need to have objective and non-discriminatory procedures for allocating voting rights and, if relevant, objective criteria for selecting the technology to be included in the standard.[23]

(b) The procedure for adopting the standard is transparent

The standard-setting process must be transparent, so that stakeholders can effectively inform themselves of upcoming, ongoing and finalised standardisation work in good time at each stage of the development of the standard.[24]

(c) The standard is non-binding

The members of an SDO must remain free to develop alternative standards or products that do not comply with the agreed standard.[25] In the case of *Philips v VCR*, the standards in question related to the manufacture of VCR equipment,

---

22 See footnote 1 above.
23 Paragraph 281 of the guidelines.
24 Paragraph 282 of the guidelines.
25 Paragraph 293 of the guidelines.

and obliged the parties to manufacture and distribute only cassettes and recorders conforming to the VCR system licensed by Philips. The parties were prohibited from changing to manufacturing and distributing other video cassette systems. The Commission held that such prohibitions constituted a restriction of competition.[26]

(d) ***Access to the standard is available on fair, reasonable and non-discriminatory (FRAND) terms***

The rules of the SDO must ensure effective access to the standard on FRAND terms.[27]

3.3 Application to patent pools

Historically, patent pools have been viewed in principle as anti-competitive as such arrangements have the potential to raise the barriers to market entry or to create market power. However, in the last decade, competition authorities have become more lenient towards the formation of patent pools as it is recognised that they have the potential to provide clear pro-competitive benefits.[28] For licensees, patent pools provide a one-stop shop to access patents under a single licence, as opposed to negotiating licences separately with each patent owner. For licensors, such pools provide a convenient way to generate income.

Nonetheless, the structure and constitution of patent pools may have the potential to restrict competition by excluding potential members from the pool, and the exchange of sensitive information could facilitate collusion.

The safe harbour conditions outlined above apply generally in the context of patent pool agreements. In addition, the Commission recognises that a patent pool presents fewer problems when the technologies in question are complementary, ie, when all the technologies are required to produce the product or carry out the process to which the technologies relate, rather than when the technologies are competitive (substitutes): where patents pooled are substitutes, there is no price competition. As a general rule, the Commission considers the inclusion of substitute technologies as a violation of Article 101(1) of the TFEU and, in the case of pools comprising to a significant extent substitute technologies, it is unlikely that there would be an exemption under Article 101(3).[29] The assessment of whether a patent pool is anti-competitive is

---

26 Commission Decision of 20 December 1977 in Case IV/29.151 – video cassette recorders, OJ L 47/42, 18 February 1978, paragraph 23, https://eur-lex.europa.eu/legal-content/EN/TXT/PDF/?uri=CELEX:31978D0156&from=EN.

27 Paragraph 283 of the guidelines.

28 In 2000, the Commission approved the technology agreement between the major electronics companies Hitachi, Matsushita Electrical, Mitsubishi Electrical, Time Warner and Toshiba to pool their respective patents covering applications of Digital Versatile Disc (DVD) standard technology.

29 Paragraph 219 of the European Commission Guidelines on the application of Article 81 of the EC Treaty to technology transfer agreements (the Technology Transfer Guidelines, OJ C 101/2, 27 April 2004, https://eur-lex.europa.eu/legal-content/EN/TXT/PDF/?uri=CELEX:52004XC0427(01)&from=EN).

ongoing, since a particular technology included in the pool may become non-essential (meaning there are substitutes) during the lifetime of the pool and may turn a non-infringing pool into an infringing one.[30]

On the other hand, a pool composed only of essential technologies, ie, where there are no substitutes for the technology either inside or outside the pool and where the technology is needed to develop the product or process for which the pool was set up, will generally not restrict competition, irrespective of the market position of the parties.[31]

3.4 Abuse of dominance

Infringements under Article 102 TFEU are likely to arise in the ICT sector, where the proprietary technology used in a standard will tend to increase the market power of the holder of the essential IP rights. However, there is no presumption that holding or exercising IP rights constitute the possession or exercise of market power.[32] Whether a standard confers market power must be assessed on a case-by-case basis (see Section 3.10, Investigations by EU authorities for further details of abuse of dominance cases in this area).

3.5 Evolution of SDOs' IP rights policies on European competition compliance

SDOs have developed IP policies to enable them to identify essential intellectual property at an early stage of the standard-setting process, thereby reducing the risk of rights holders restricting access to an adopted standard.

In 2000, the Commission launched an investigation into ETSI's IP rights policy following concerns that the policy did not protect against the risk of patent ambush (see further paragraph 3.8 below).[33] The policy in question required members to use reasonable endeavours to inform ETSI in a timely manner of any essential intellectual property of which they became aware. The Commission was concerned that the policy did not encourage early disclosure but enabled rights holders to conceal essential intellectual property until the standard was developed. In response to the Commission's concerns, ETSI strengthened its policy, among other things by removing a reference to 'becoming aware' of an essential patent as a condition for disclosure.[34]

The lessons learned in the ETSI case are reflected in the revised guidelines which provide more guidance on how best to develop IP policies. In particular, the guidelines note that an IP rights policy must require participating rights holders to provide an irrevocable commitment in writing to offer to license their essential intellectual property to all third parties on FRAND terms, prior to

---

30 Paragraph 222 of the Technology Transfer Guidelines.
31 Paragraph 220 of the Technology Transfer Guidelines.
32 Paragraph 269 of the guidelines.
33 Case 37926 (Sun/ETSI), http://ec.europa.eu/competition/publications/cpn/2011_1_1_en.pdf.
34 Press release 12 December 2005, IP/05/1565.

the adoption of the standard. Participating rights holders must also commit to ensuring that any third party to which the participant transfers its intellectual property is bound by the FRAND commitment.[35]

On 9 October 2012, ETSI reported that the ETSI IP rights special committee had embarked on discussions on whether ETSI should make further policy changes, including:

- changes to provide guidelines or principles for compensation elements under the FRAND commitment;
- whether to permit declarations to be conditional upon reciprocity;
- the availability of injunctions; and
- changes to ETSI's IP rights policy to bring its approach on the transfer of a FRAND commitment in line with the Commission's guidelines.[36]

3.6 FRAND licence requirements

The Commission's objective is to ensure that all essential intellectual property incorporated in a standard will be available to all users on FRAND terms and to prevent IP rights holders from impeding the implementation of a standard by refusing to license or by doing so on unfair, unreasonable or discriminatory terms.[37] This was reconfirmed by Joaquín Almunia, the European Commissioner of competition, in a speech on 10 February 2012 when he said:

> *Standardisation processes must be fair and transparent, so that they are not in the hands of established firms willing to impose their technologies. But it is not enough. We must also ensure that, once they hold standard essential patents, companies give effective access on fair, reasonable and non-discriminatory terms.*[38]

It is unclear whether this is intended to include companies that did not participate in, or even know about, the standard. There is also no clear definition of the terms 'fair', 'reasonable', and 'non-discriminatory' either individually or taken together, and there has been no European case law to establish such definitions (although the courts have to assess factually what these terms mean in the context of the individual cases considered). Furthermore, few SDOs provide an interpretation of FRAND terms when setting industry standards as they are under no obligation to do so. On 11 November 2011, Apple wrote to ETSI arguing that FRAND principles were not being consistently adhered to and suggested the creation of a FRAND licensing framework.[39]

---

35 On 30 November 2011, the ETSI IPR policy was updated to grant power on the director general to request the owner of an essential IP right to give, within three months, an irrevocable undertaking in writing that it is prepared to grant irrevocable licences on FRAND terms under such IP right; and in case of transfer or assignment, the member is under a duty to use reasonable efforts to notify the assignee or transferee of the undertaking made to ETSI with regard to the essential IPR.

36 See www.isacc.ca/isacc/_doc/GSC/GSC16bis-IPR-03.ppt.

37 Paragraph 287 of the guidelines.

38 Speech entitled "*Quo vadis* Europa? New Frontiers of Antitrust 2012", 10 February 2012, http://europa.eu/rapid/press-release_SPEECH-12-83_en.htm.

39 See www.scribd.com/doc/80899178/11-11-11-Apple-Letter-to-ETSI-on-FRAND.

The lack of certainty in this area has led to antitrust investigations initiated by the Commission. On 1 October 2007, the Commission opened antitrust proceedings against Qualcomm, a US chipset manufacturer, following allegations that Qualcomm breached Article 102 by licensing its patents essential to the WCMDA standard on non-FRAND terms.[40] The complaints further alleged that charging non-FRAND royalties would lead to higher handset prices, a slower development of the 3G standard, and a slower adoption of the 4G standard. On 24 November 2009, the Commission closed its investigation without reaching any formal conclusions, following the withdrawal of complaints.[41]

The Commission has provided guidance on the FRAND criteria in the guidelines, which state that, in case of dispute, the assessment of whether fees charged for access to intellectual property are unfair or unreasonable should be based on whether the fees bear a reasonable relationship to the economic value of the intellectual property. Assessing the economic value and cost of essential intellectual property is extremely difficult, especially when one takes into account the past research costs of rights holders that did not result in commercially exploitable products. The Commission therefore puts forward the following four methods for determining a 'reasonable relationship':

- Compare the licensing fees charged for the relevant patents in a competitive environment before the industry has been locked into the standard (*ex ante*) with those charged after the industry has been locked into it (*ex post*);[42]
- Obtain an independent expert assessment;[43]
- Rely on previous unilateral *ex ante* disclosures of most restrictive licensing terms in the context of the specific standard-setting process.[44] The Commission recognises that it is important for parties involved in the selection of a standard to take an informed decision based on the advantages and disadvantages of alternative technologies, from a technical and pricing perspective. Unilateral disclosure prior to the adoption of a standard will not normally lead to a restriction of competition within the meaning of Article 101(1), provided that it does not serve as a cover to jointly fix prices.[45] A distinction is made between *ex ante* disclosures of licensing terms and *ex ante* negotiations between the licensor and potential licensees, with the latter scenario likely to raise concerns of price-fixing and therefore a breach of competition law; or

40 Commission MEMO/07/389 http://europa.eu/rapid/press-release_MEMO-07-389_en.htm.
41 Commission MEMO/09/516 http://europa.eu/rapid/press-release_MEMO-09-516_en.htm.
42 Paragraph 289 of the guidelines.
43 Paragraph 290 of the guidelines.
44 *Ibid.*
45 Paragraph 299 of the guidelines.

- Draw on royalties charged for the same intellectual property in other comparable standards.[46]

The guidelines leave open the question of what other elements, apart from royalty levels, may be relevant to determining FRAND licence terms. It must be noted that SDOs are not required to verify whether the licensing terms of participants fulfil the FRAND commitment. Participants must assess for themselves whether their licensing terms and, in particular, the fees they charge, fulfil the FRAND commitment. In contrast, as will be discussed further in this chapter, national courts tend to take the lead in setting out practical guidelines in this area of law. By way of recent example, the High Court in the United Kingdom went a step further in its ruling in *Unwired Planet v Huawei*,[47] setting out in detail factors to be taken into account when determining whether a licence term is FRAND.

3.7 Transfer of essential patents

To ensure the effectiveness of the FRAND commitment, the Commission recommends that rights holders ensure that any company to which they transfer intellectual property is bound by that commitment, for example through a contractual clause between the buyer and seller.[48] ETSI's Guide on Intellectual Property Rights states that it is desirable for licensing undertakings to transfer automatically to the new owner, but recognises that this may not be the effect of the law in every jurisdiction.[49] In one case highlighting the Commission's expectations, IPCom agreed to adopt Bosch's earlier commitment to grant irrevocable licences on FRAND terms, following discussions with the Commission.[50]

3.8 Patent ambush

A 'patent ambush' can occur where a company intentionally conceals that it has essential intellectual property for a standard and asserts its intellectual property only when the standard has been agreed. In such cases, the rights holder may be in a position to exploit its dominance by charging excessive royalties to those now locked into the technology.

A prominent example of alleged patent ambush arose in the Commission's

46 *Ibid.*

47 [2017] EWHC 711 Pat.

48 Paragraph 285 of the guidelines and see, eg, Guidelines for Implementation of the Common Patent Policy for ITU-T/ITU-R/ISO/IEC, Part I, Section 7, http://isotc.iso.org/livelink/livelink/fetch/2000/2122/3770791/ Common_Guidelines_01_March_07.pdf.

49 ETSI Guide on Intellectual Property Rights, Version adopted by Board#94 on 19 September 2013, www.etsi.org/images/files/IPR/etsi-guide-on-ipr.pdf.

50 MEMO 09/549, http://europa.eu/rapid/press-release_MEMO-09-549_en.htm?locale=en. It is not, however, clear that such expectations apply to a divisional application filed by the transferee. See *IPCom v Nokia*, Mannheim District Court, Germany, February 2011 (7 O 100/10) in which the patentee was held not bound by a predecessor's declaration in respect of a new divisional patent.

investigation into Rambus. Rambus's patents related to Dynamic Random Access Memory (DRAM) chips. In December 2002 Hynix Semiconductor, a competitor of Rambus, complained to the Commission that Rambus engaged in intentional deceptive conduct by not disclosing the existence of patents and patent applications in the development by the Joint Electron Device Engineering Council (JEDEC), a US-based standard-setting organisation, of an industry-wide DRAM standard subsequently applied in around 95% of personal computers.

On 30 July 2007, the Commission notified Rambus of its Statement of Objections. The Commission came to the preliminary conclusion that Rambus may have infringed Article 102 of the TFEU by abusing a dominant position in the market for DRAMs. In particular, the Commission was concerned that Rambus had engaged in a patent ambush by intentionally concealing that it had patents and patent applications which were relevant to technology used in the JEDEC standard, and subsequently claiming unreasonable royalties for the use of those patents.

On 9 December 2009, the Commission accepted commitments by Rambus to address the Commission's concerns. In particular, Rambus agreed to a worldwide cap on its royalty rates for certain DRAM patents for a period of five years.[51]

Further, in a Dutch case concerning alleged infringement of a patent that was essential to the JPEG standard, The Hague District Court suspended a previously granted injunction on the grounds that the patentee had abused its patent rights and was estopped from enforcement as a result of having failed to disclose its patent when participating in the standard-setting process.[52]

3.9 Royalty stacking

Royalty stacking arises when numerous licences are required, under essential patents, to make a product, resulting in an unacceptably high royalty burden. This is a particular concern in the ICT sector, where technical standards incorporate numerous patented technologies.

One way to overcome royalty stacking is through *ex ante* disclosure. This requires those involved in the standard-setting process to commit to licensing terms before the protected technology is incorporated in the standard. The Commission recommends *ex ante* disclosures of licensing terms, to ensure that technology royalties are fair.[53] ETSI contemplates voluntary *ex ante* disclosures of licensing terms by licensors of essential intellectual property, but does not itself

51 IP/09/1897, http://europa.eu/rapid/press-release_IP-09-1897_en.htm.

52 *Koninklijke Philips Electronics NVV v LG Electronics Benelux Sales BVV*, 11 June 2007, 287026/KG ZA 07-528. The court also held that the position was the same whether the patentee had actively proposed parts of the standard or was present merely as an observer.

53 Paragraph 290 of the guidelines.

become involved in these, other than as a depository of information as to how to access such terms.[54]

From a competition law perspective, *ex ante* disclosure can have pro-competitive benefits. For example, it promotes price competition between technologies before a standard becomes locked in and allows informed decision-making. However, without appropriate safeguards, it also has the potential to be anti-competitive, especially where royalty rates are negotiated.

3.10 Investigations by EU authorities

Initially, the Commission was reluctant to become involved in standardisation disputes, considering that it might not be able to grant the remedies sought by complainants.[55] However, the Commission has launched a number of Article 102 TFEU investigations in the ICT sector, albeit most of them were either dropped or resolved with the parties by way of commitments. These include:

- In October 2007, the Commission announced its investigation of Qualcomm,[56] a chipset manufacturer. Qualcomm was accused of imposing terms for the licensing of its intellectual property that were unfair, unreasonable and discriminatory. The Commission considered this case to raise important issues with regards to the pricing and terms for the license of technology once standardised but ultimately decided not to investigate further in view of its priorities. Therefore, no decision in this case was taken as to whether Qualcomm's behaviour may have breached Article 102 of the TFEU (see Section 3.6 above for further details).
- In July 2007, the Commission sent a Statement of Objections to Rambus[57] for claiming unreasonable royalties and the case was settled in 2009 when Rambus offered commitments to address the Commission's concerns (see further details of the case discussed above in Section 3.8).
- In December 2012, the Commission sent a Statement of Objections to Samsung[58] for allegedly abusing its dominant position, thus breaching Article 102 of the TFEU, by seeking injunctive relief against Apple for infringing its SEPs for 3G mobile devices. The issue was that Samsung had given a commitment to ETSI to license its patents on FRAND terms and that Apple was willing to license the patents on these terms. In 2014,

54 See www.etsi.org/website/aboutetsi/iprsinetsi/Ex-ante.aspx.

55 Speech by Magdalena Brenning, DG COMP, European Commission, "Competition & Intellectual Property Policy, Implications of Late or No IPR Disclosure in Collective Standard-Setting", 17 June 2002, http://ec.europa.eu/competition/speeches/text/sp2002_037_en.pdf, p2, section 3.

56 Qualcomm, Commission MEMO/09/516.

57 Commission press release IP/09/1897. Case COMP/38.636 – Commission decision of 9 December 2009.

58 Case AT.39939 – Samsung – Enforcement of UMTS standard essential patents. Commission press release IP/12/89, http://europa.eu/rapid/press-release_IP-12-89_en.htm. The Commission sent Samsung a Statement of Objections on 21 December 2012.

the Commission ultimately accepted binding commitments from Samsung not to seek any such injunctive relief from any court within the EEA for a period of five years.

- In April 2014, the Commission found that Motorola had abused its dominant position. As with the *Samsung* case, Motorola had agreed with ETSI to license its GPRS standard essential patents but had sought injunctive relief from the German courts against Apple for allegedly infringing the patent. However, the Commission used its discretion and decided not to impose a fine in this instance given the lack of precedent decisions at the EU level in this area.[59]
- In October 2014, the Commission sent a Statement of Objections to Honeywell and DuPont alleging that they had breached Article 102 of the TFEU by failing to disclose the existence of patents for a new refrigerant used for car air conditioning and refusing to grant licences on FRAND terms. The Commission in this case noted the importance of ensuring that arrangements relating to IP rights contribute to faster or better development of new technologies rather than to their delay, an approach which has been at the heart of the recent autotech case (discussed below). Ultimately, the Commission closed its investigation against Honeywell and DuPont.[60]

Almunia, the Competition Commissioner until 2014, reinforced the Commission's relatively light touch involvement by stating that the Commission remains "determined to use antitrust enforcement to prevent the misuse of patent rights to the detriment of a vigorous and accessible market".[61] Nonetheless, the Commission prefers industry participants to resolve their issues without its intervention wherever possible.[62]

Generally, the Commission continues to support the view that national courts are best placed to assess in each case whether FRAND negotiations have been concluded in good faith and on a reasonable basis and to decide whether an injunction may be warranted. Therefore, EU involvement in this area is mostly in the form of references by national courts to the European Union. For example, the European courts have allowed injunctions to proceed in the following key jurisdictions where SEP and FRAND disputes tend to arise more frequently: *St Lawrence v Deutsche Telekom and HTC* (Germany, District Court,

---

59 COMP/C-3/39.986 – Motorola – Enforcement of ITU, ISO/IEC and IEEE standard essential patents; see Motorola – ITU, ISO/IEC and IEEE standard essential patents. Commission press release IP/12/345, http://europa.eu/rapid/press-release_IP-12-345_en.htm.

60 Commission press release IP/11/1560.

61 "*Quo vadis* Europa? *New Frontiers of Antitrust* 2012".

62 Since this chapter was written, the Commission has issued a Statement of Objections setting out its preliminary view that Motorola Mobility's seeking and enforcing of an injunction against Apple in Germany on the basis of its mobile phone SEPs amounts to an abuse of a dominant position (European Commission Press Release, 6 May 2013, http://europa.eu/rapid/press-release_IP-13-406_en.htm).

2015), *NTT DoCoMo v HTC* (Germany, District Court, 2016), *St Lawrence v Vodafone and HTC* (Germany, District Court, 2016), and *Unwired Planet v Huawei* (UK, English High Court, 2017).

In continuation of the questions asked in the *Samsung* and *Motorola* cases, in *Huawei Technologies Co Ltd v ZTE Corp*,[63] on reference from the Regional Court of Dusseldorf, the European Court of Justice of the European Union (CJEU) confirmed that an action for injunctive relief brought by a dominant SEP holder who has undertaken to grant a licence on FRAND terms could amount to an abuse of dominant position unless certain specific conditions were met. In order to seek injunctive relief, the patent holder must be able to show that the proprietor has alerted the alleged infringer of the violation, specifying how it was in breach and offering a written licence, and the infringer has not responded to the offer and continued to infringe the patent.

3.11 Future EU policy development

The Regulation on European Standards,[64] aimed at making standards more accessible and ensuring interoperability, was introduced in 2012. In the same year, the European Commission and various stakeholders, including companies such as Apple, Google, Microsoft and Samsung, as well as standard bodies, regulators, patent offices, government representatives and academics, held a roundtable. The purpose of the roundtable was to discuss a roadmap for setting out the conditions where it is appropriate for companies that made FRAND commitments to seek injunctions and to clarify the meaning of a 'reasonable' FRAND term; a discussion that continues to garner traction from the Commission and stakeholders alike.

In March 2014, the Commission published a study requested by DG Enterprise and Industry looking at standardisation and licensing practices in four industries of particular relevance: telecommunications, consumer electronics, automotive and electricity – the latter three being new focus areas given the development of the Internet of Things (IoT) discussed in more detail below. Echoing the developments in 2012, the Commission remained concerned about barriers to efficient licensing of SEPs, the lack of clarity around FRAND terms, the increasing number of SEPs, and the lack of transparency and high costs of dealing with numerous SEP owners. Following the study, the Commission also issued a public consultation which ultimately led to the 2016 "Communication on Standardisation Priorities for the Digital Single Market".[65]

---

63 *Huawei Technologies Co Ltd v ZTE Corp*, ZTE Deutschland GmbH, Case C-170/13.

64 See Council of the European Union press release (PRES/12/409) at http://europa.eu/rapid/press-release_PRES-12-409_en.htm, http://register.consilium.europa.eu/pdf/en/12/pe00/pe00032.en12.pdf and Regulation (EU) 1025/2012 of 25 October 2012 on European standardisation, OJ L 316/12, 14 November 2012, https://eur-lex.europa.eu/LexUriServ/LexUriServ.do?uri=OJ:L:2012:316:0012:0033:EN:PDF.

65 See https://ec.europa.eu/digital-single-market/en/news/communication-ict-standardisation-priorities-digital-single-market.

This communication stated three areas for improvement:

- opaque information on SEP exposure;
- the lack of clarity around FRAND; and
- the uncertainty in enforcement of SEPs.

Following the *Huawei v ZTE* case in early December 2017, the Commission published a further communication called "Setting Out the EU Approach to Standard Essential Patent".[66] In particular, this communication focused on the importance of a transparent SEP environment to allow for future developments, such as the IoT, where appliances, vehicles and everyday objects will be interconnected. The development of the IoT highlights more than ever the tension between SEPs which allow owners of R&D to benefit from monopolistic rights over their creations and the need for standards to ensure collective use of technology. This also means that a new group of stakeholders beyond the typical ICT sector such as manufacturers of household appliances and cars must now grapple with the challenges of SEP licensing. For this purpose, the European Committee for Standardization and the European Committee for Electrotechnical Standardization have published a paper[67] on behalf of SEP holders, including Ericsson and Nokia, discussing six key principles and guideline Q&As for licensing SEPs intended for the expansion of 5G and the IoT.

The most recent EU development in this area involves Nokia, and relates to interconnected technology for smart cars. In this case, Daimler and car parts manufacturers Valeo and Continental brought a complaint in March 2019 to the Commission against Nokia relating to its licensing terms for telecommunication SEPs used in car parts. The investigation is still in early stages and the Commission has recently sent questionnaires to stakeholders in the market to gather facts on licensing conditions and to decide whether to proceed with a formal case against Nokia. Although the legal issues in this case may not be entirely novel, the case promises to raise interesting questions with regards to the access to telecommunications patents in other industries and a key precedent for the applicable rules for IoT.

To further support efforts to bring clarity in this area, in 2018 the Commission announced[68] the creation of an Expert Group on SEPs made up of independent experts and members representing stakeholder interests. The group is aimed at increasing the Commission's expertise and know-how around the determination of fair and reasonable FRAND licensing. It is hoped that it will also contribute to developing a more transparent SEP environment.

---

66 COM(2017) 712 final (SEP Communication).

67 See www.cencenelec.eu/news/workshops/Pages/WS-2019-013.aspx.

68 See http://ec.europa.eu/transparency/regexpert/index.cfm?do=groupDetail.groupDetail&groupID=3600&news=1.

4. Declarations of essentiality

4.1 Number of declared essential patents

The number of patents declared as essential to a typical standard is growing rapidly. By way of example, the Commission commented that by 2017 over 23,000 patents were declared essential to the Global System for Mobile Communication standard and the 3G or Universal Telecommunications System standards.[69] However, studies have shown that only between 10% and 50% of declared patents are actually essential.[70] The impact of this on licence negotiations between the patentee and an implementer could be huge as declarations create a *de facto* presumption of essentiality. Although the value of an essential patent portfolio does not lie solely in the number of essential patents held by a particular patentee, it must be recognised that licence rates are negotiated by counting patents and this can result in a tendency to over-declare. Indeed, the English High Court has referred to patent counting as "inevitable" in FRAND licensing negotiations.[71]

SDOs such as ETSI publish their databases of patents declared as essential or potentially essential but they do not check or certify essentiality, and there is normally no mechanism for anyone other than the patentee to change or withdraw such declarations. The EU Commission in its 2017 Communication on the EU approach on Standard Essential Patents noted that these databases may lack transparency and "are not subject to scrutiny regarding the essentiality of the declared patent, which can evolve in the course of the standard adoption procedure". Although the ETSI's IP rights information statement form does require the rights holder to specify the relevant parts of the standard when declaring a patent with a view to reducing false declarations, this is clearly not enough. The Commission believes that quality and accessibility of the information recorded in the SDO databases should be improved. Suggestions include allowing all declared information to be searchable based on the relevant standardisation projects and providing links to patent office databases in addition to stricter scrutiny on compliance with declaration obligations. This should serve to reduce over-declaration and could go some way to resolving issues relating to the valuation of patented technologies and SEP licensing, which will be key to fostering a smooth and predictable framework for SEPs. The implementation of the measures will need to be proportionate and so the Commission has suggested a gradual implementation, for example by starting with new and key standards only (eg, 5G). This issue has been noted in litigation in the English High Court

---

69 Communication from the Commission to the European Parliament, The Council and the European Economic and Social Committee setting out the EU approach to Standard Essential Patents, Brussels 29 November 2017, COM (2017).

70 *Ibid.*

71 *Unwired Planet v Huawei* [2017] EWHC 711 (Pat), at paragraph 806(11).

between Nokia and InterDigital, in which both parties asserted that "too many patents are notified to ETSI as essential and so the strength of the respective patent portfolios is exaggerated in subsequent licensing or cross-licensing negotiations".[72] In response to Nokia's application for declarations of non-essentiality in respect of InterDigital patents alleged to have been wrongly declared essential, InterDigital asserted that Nokia also over-declared, and argued that this debarred Nokia from relief. The Court agreed that if the claimants had themselves deliberately or recklessly over-notified, it might be "submitted with force" that they should not benefit from declaratory relief in respect of a system that they had abused.[73] However, such conduct was not alleged against Nokia and the Court found that over-declaration was not relevant.

More recently, the English High Court has considered the reasons behind the issue of over-declaration[74] and recognised that it may be the result of a number of factors. Under the ETSI IP rights policy, members are required to declare essential patents in a timely fashion which arguably results in members erring on the side of caution and declaring as essential patents which ultimately fall on the other side of the line. In other words, the ETSI IP rights policy encourages an attitude of "if in doubt, declare it". Members must also draw ETSI's attention to any of that member's patents "which might be essential",[75] and the IPR licensing declaration includes a statement that "the IPR(s) disclosed ... may be or may become essential". Thus, it may not be clear whether a declaration of essentiality is speculative or intended to be definitive. Further, the definition of 'essential' under ETSI's IP rights policy includes cases where different ways of implementing a necessary technical solution would all infringe patents, in which case each such patent is considered 'essential'.[76]

A further difficulty arises from the fact that in any given patent family the national designations can differ in scope and it would be disproportionate for a patentee to be required to consider each national patent against standards which will inevitably change over time.

4.2 Determination of essentiality by the national courts in European proceedings

The English courts will entertain and determine applications for binding declarations as to whether the claims of a patent are non-essential, or essential,[77] to a particular standard. In *Nokia Corporation v InterDigital Technology*

72 *per* Pumfey J, [2007] EWHC 445 (Pat) at paragraph 4.

73 *Ibid*, at paragraph 23.

74 *Unwired Planet v Huawei* [2017] EWHC 711 (Pat).

75 Section 4.1.

76 The IEEE policy states that patents are essential if technical implementation of the standard cannot be done in any other commercially feasible way (6.1). The licensing obligation only applies to the specifically essential claims of the patent.

77 Either the patentee or an alleged infringer may have an interest in a declaration of essentiality. For the alleged infringer, such a finding as between themselves and the patentee may be the basis for a claim that the patentee is obliged to grant a FRAND licence, or has already done so. It is also possible to seek a declaration that it is impossible to implement the standard in a way that infringes the patent in suit.

Corporation,[78] Nokia initially sought revocation of three InterDigital patents. In light of assertions of essentiality by InterDigital (in the form of its declarations to ETSI, as well as oral assertions by its counsel in the proceedings), Nokia argued by way of a 'squeeze' that the terms of the standard should force a particular breadth of construction of the claims on InterDigital.

To support its position on construction, Nokia also sought a declaration that the patents were not essential to the standard. In the Court of Appeal, it was held that such an approach to construction was not (as alleged by InterDigital) an impermissible attempt to construe the claims by reference to subsequent conduct. The Court of Appeal also held that such declaratory relief did not amount to a declaration of non-infringement pursuant to Section 71 of the Patents Act 1977,[79] but that the issue of essentiality was sufficiently clear to invoke the court's inherent jurisdiction: "You take the standard, you take the patents, you take the definition of 'essential' and you see whether there is any way around or not."[80] As to the question of whether such a declaration was appropriate in the absence of an assertion of infringement, the court found that essentiality was a serious commercial issue to be resolved between the parties.

In subsequent proceedings between the same parties, Nokia sought only declarations of non-essentiality, in respect of 31 patents declared essential to the 3G standard by InterDigital.[81] In applying to strike out Nokia's claims, InterDigital argued that they had made no assertion of a claim of right, so that the court lacked declaratory jurisdiction, and that the court should not exercise its discretion to hear what was in effect an application of no commercial consequence. Although the earlier case had differed because InterDigital's counsel had positively asserted essentiality in the proceedings, Nokia relied on InterDigital's declarations to ETSI as to the essentiality or potential essentiality of the patents. Refusing the strike-out, the court held that "to approach an international standards body and suggest that the use of a particular invention is essential ... necessarily involves a 'formulated claim' against potential users of the standard".[82] In the Court of Appeal,[83] Lord Justice Jacob elaborated on this point and emphasised that anyone seeking such a declaration must have a real commercial reason: "An interest in making 3G telephones which must therefore comply with the standard is clearly sufficient."[84]

78 [2004] EWHC 2920 (Pat), confirmed on appeal [2005] EWCA Civ 614.

79 An incidental effect of this, in light of Sections 74 and 75 of the Patents Act 1977, is that in standalone proceedings for such a declaration, the proprietor may not amend the patent specification. See *InterDigital Corporation v Nokia* [2008] EWHC 749, at paragraph 29.

80 *per* Jacob LJ, at paragraph 21. See also *InterDigital Technology Corporation v Nokia Corporation* [2007] EWHC 1913, at paragraph 6, for an example of the wording of non-essentiality pleadings. The burden is on the party asserting non-essentiality to identify the claim features not mandated by the standard (*Ibid*, at paragraph 14).

81 Alleging invalidity of 31 highly technical patents would not be practical. Four patents remained in issue at trial and the court found three of them to be non-essential. Nokia reserved its position as to the validity of any found to be essential.

82 [2006] EWHC 802 (Pat), at paragraph 24.

83 [2006] EWCA Civ 1618.

84 *Ibid*, at paragraph 20.

While these interlocutory decisions determined that there is jurisdiction to entertain such an action, the court must also determine the utility of granting declaratory relief on the specific facts of the case, and any other relevant discretionary considerations. Relevant factors concerning utility may include, among other things, the size of the UK market and whether a declaratory judgment as to essentiality might be influential on licence negotiations.[85] Such discretionary considerations will often be suited to determination as a preliminary issue in the proceedings.[86]

Parallel opposition proceedings in the European Patent Office (EPO) may result in claim narrowing by amendment, with the effect that an essential patent becomes non-essential. On the question of a stay of the declaratory proceedings pending the EPO opposition outcome, the English court will apply the guidance of the Court of Appeal in *Glaxo Group Limited v Genentech.*[87] A stay was refused in *InterDigital Corporation v Nokia.*[88] The position may be different in proceedings for a declaration of essentiality, as opposed to non-essentiality, since amendments in the EPO in such a case could call into doubt a finding of essentiality by the court.[89]

No such declaratory practice concerning essentiality can be observed in the courts of other significant patent jurisdictions in Europe such as the Netherlands, France and Germany, although they may consider essentiality if relevant to determination of the issues. There is likewise no provision for declarations of essentiality or non-essentiality under the proposed Agreement on a Unified Patent Court, which court would have exclusive jurisdiction to determine validity and infringement of European patents (and of the new EU unitary patent), nor is such provision contained in the draft procedural rules, although provision is made for declarations of non-infringement.

The English High Court[90] has stated its willingness to make a declaration of non-infringement in respect of UK and non-UK designations of a European patent, provided that the court's jurisdiction is well founded under the Brussels Regulation or national law (as the case may be, depending on the domicile of the defendant).

---

85 *Nokia Corporation v InterDigital Technology Corporation* [2007] EWHC 3077 (Pat); and comments thereon by Floyd J in *InterDigital Corporation v Nokia* [2008] EWHC 504. In *Nokia v IPCom GmbH & Co KG* [2012] EWHC 225, the court made a declaration of non-essentiality in respect of the LTE standard at the request of Nokia, despite the standard still being under development and Nokia not yet manufacturing standard-compliant equipment.

86 See, eg, *InterDigital Corporation v Nokia* [2008] EWHC 969, at paragraphs 1 and 2.

87 [2008] EWCA Civ 23.

88 [2008] EWHC 749.

89 Compare also *IPCom GmbH & Co KG v HTC Europe* [2012] EWHC 1567 (stay of English validity proceedings pending decision of EPO Technical Board of Appeal, in light of scheduled trial concerning FRAND licence terms and in light of undertakings offered by the defendant to take a FRAND licence and not to infringe in the meantime).

90 *Actavis Group HF v Eli Lilly & Co* [2012] EWHC 3316.

5. The availability of injunctive relief and determination of FRAND licence terms in European patent infringement proceedings

The availability of injunctions and damages for infringement of essential patents will form the backdrop to licence negotiations concerning both essential and non-essential patents. Where agreement cannot be reached with a prospective licensee, an important question is whether a patentee, having made a declaration of willingness to grant FRAND licences, should be entitled to an injunction. Defendants are in a position to argue that no injunction should be granted, or indeed that they are already licensed, as a result of a FRAND declaration. Claimants may feel that the threat of injunction is necessary to avoid undue prevarication by the counterparty in agreeing reasonable terms.

The owner of essential patents has an advantage when it comes to enforcement: in theory at least, infringement can confidently be asserted against any unlicensed product that is said to comply with the relevant standard. Thus, EU customs authorities, for example, can be provided with a list of licensees and simply asked to detain[91] any products of a certain type (eg, mobile handset or DVD player) not made or sold by a licensee. Further, in some countries, seizure orders can be obtained from an administrative court. Patent management company Sisvel SpA, as licensing agent for elements of the MP3 and MPEG standards, has publicised its use of a wide variety of enforcement techniques, including for example seizures of unlicensed products and advertising material at trade shows and retail outlets in the Netherlands, Germany and Italy, with the assistance of court bailiffs and police.[92] These advantages of essential patents have made their acquisition particularly attractive to non-practising entities, sometimes known as 'patent trolls'. However, the use of customs seizures, and similar *de facto* injunction mechanisms has been criticised in the CEN/CENELEC Workshop Agreement entitled "Core Principles and Approaches for Licensing of Standard Essential Patents".[93]

The English courts and the CJEU have both expressed their agreement that in certain circumstances injunctive relief will be appropriate.[94] When considering whether to grant injunctive relief, the court will be bound by Article 3(2) of the IP rights Enforcement Directive which requires that injunctive relief be effective, proportionate and dissuasive.

---

91 Pursuant to Council Regulation (EC) 1383/2003.

92 See *Sandisk Corporation v Koninklijke Philips Electronics & Others* [2007] EWHC 332, in which Sisvel's enforcement conduct was alleged to amount to abuse of a dominant position through harassment. The claims of abuse were rejected on the grounds that any such abuses could have had no effect in England and Wales, and further that the reasoning in *ITT Promedia NV v Commission* (see 6.1 below) applied.

93 CWA95000 dated June 2019.

94 *Unwired Planet v Huawei* [2018] EWCA Civ 2344.

5.1 The CJEU's decision in *Huawei Technologies Co Ltd v ZTE Corp, ZTE Deutschland GmbH*

In the CJEU's judgment in *Huawei v ZTE*[95] it was held that seeking an injunction in patent enforcement proceedings against a licensee who requires a FRAND licence may in certain circumstances be abusive. While the ruling did not provide a comprehensive checklist as to when seeking an injunction would not be an abuse, it did rule that the SEP owner would be considered to be in abuse of its dominant position if, prior to seeking the injunction, it did not write to the potential infringer setting out the proposed licensing terms.

The judgment concerned the availability of injunctive relief for owners of SEPs (in this case, telecoms patents) bringing infringement proceedings. The decision proceeds on the footing that the SEP holder is in a dominant position, as per the instructions of the referring court, as the existence of a dominant position was uncontested.

In the judgment the CJEU distinguished between two different scenarios:

- infringement proceedings where an injunction or the recall of products is sought; and
- infringement proceedings where the rendering of accounts and an award of damages is sought.

(a) *Infringement proceedings where an injunction or the recall of products is sought*

The CJEU has confirmed the Advocate General's guidance that for a SEP holder to bring infringement proceedings and seek an injunction without abusing its dominant position, the following steps need to be taken before proceedings are issued:

- The SEP holder should notify the alleged infringer of the infringement complained of by designating the infringing SEP in question, and specifying the way in which it has been infringed; and
- After the alleged infringer has expressed its willingness to conclude a licence on FRAND terms, it is for the proprietor of the SEP to put forward a "specific, written offer for a licence on FRAND terms", specifying (i) the amount of the royalty; and (ii) the way in which that royalty is calculated.

If the SEP holder carries out the steps identified above, the alleged infringer has to respond diligently to the FRAND offer, in accordance with recognised commercial practices in the field and in good faith. The alleged infringer must not be seen to be applying 'delaying tactics'.

If the alleged infringer does not accept the offer, the alleged infringer can only avoid an injunction if it carries out the following steps:

95 Case C-170/13.

- It promptly puts forward a written counter-offer on FRAND terms; and
- It provides appropriate security (eg, a bank guarantee). The calculation of that security must include the number of past acts of use of the SEP, and the alleged infringer must be able to render an account in respect of those acts of use.

The CJEU has also confirmed that the alleged infringer is entitled to challenge the validity and/or essentiality of the patent; and the parties may, by agreement, request that the amount of the royalty be determined by an independent third party.

(b) ***Infringement proceedings where the rendering of accounts and an award of damages are sought***

In respect of infringement proceedings where an account of profits, or damages, are sought in lieu of an injunction, the CJEU found that such proceedings do not constitute an abuse of a dominant position.

5.2 Approach to injunctive relief and application of the *Huawei v ZTE* framework in national courts

Article 3(2) of the EU Enforcement Directive[96] provides that enforcement measures and remedies must be "effective, proportionate and dissuasive and shall be applied in such a manner as to avoid the creation of barriers to legitimate trade and to provide for safeguards against their abuse". The requirement for proportionality may mean that an injunction is inappropriate where the patented invention plays a relatively minor role in the product, perhaps especially where pan-European relief is contemplated (eg, under the proposals for a Unified Patent Court and unitary patent).

(a) ***United Kingdom***

Generally, an injunction will be granted by the English court where the infringement of a property right is demonstrated, and where repetition is threatened, provided that the effect would not be oppressive (in the sense of 'grossly disproportionate'); the jurisdiction to award damages in lieu of an injunction is not exercised merely because the defendant is willing to pay.[97] It was observed (obiter) by Justice Floyd in *InterDigital Corporation v Nokia*[98] that:

> *The obligation accepted by Nokia to grant licences on FRAND terms only results in a licence if the parties agree on what a FRAND royalty is or a court determines it.*

---

96 Directive 2004/48/EC of the European Parliament and of the Council of 29 April 2004 on the enforcement of IP rights.

97 *per* Pumfrey J, *Navitaire Inc v easyJet Airline Co Ltd (No 2)* [2005] EWHC 0282 (Ch). *Cf Shelfer v City of London Electric Lighting Co* [1895] 1 Ch 287.

98 [2008] EWHC 749, at paragraph 19.

Until the licence is granted, InterDigital may find it is infringing if as Nokia asserts the opposed patents are essential. Put another way, it will be important to Nokia in any licensing negotiations to be able to hold a threat of an injunction over InterDigital on its essential patents ... it is far from clear that InterDigital are insulated from an injunction on the opposed patents.[99]

As the meaning and effect of FRAND declarations, and related competition law defences, are substantial issues related to the relief to be granted if liability is found, the practice has arisen in the English court of deferring their consideration until after the trial of issues of validity, infringement and (if pleaded) essentiality. This was the case, for example, in *Qualcomm v Nokia*[100] where the court also ordered at an early stage that no interim injunctive relief should be sought or ordered. In the event that the patent is found to be both valid and infringed, the court may decline to order an injunction pending determination of FRAND licence issues. In *Nokia OYJ v IPCom GmbH & Co KG,*[101] Justice Roth noted[102] that while IPCom had initially sought an injunction, the court had felt that this would be inappropriate because the parties acknowledged Nokia's readiness and entitlement to take a licence on FRAND terms. Instead of a trial on complex FRAND and competition law defences to injunctive relief, IPCom agreed to the court scheduling a hearing to determine the FRAND licence terms, however this was settled before trial.[103] The practice should mean that cases in England involving willing licensors and licensees should reach a conclusion, or settle, before the court is required to render a judgment concerning the legal effect of the FRAND declaration. However, there has been one determination of FRAND terms in the United Kingdom to date, and the UK High Court has granted injunctions on SEPs.

In *Unwired Planet v Huawei*[104] the English High Court determined on the facts that Huawei had sufficient prior notice of Unwired Planet's SEPs and it therefore did not fall within the safe harbour of *Huawei v ZTE*. In the High Court trial Birss J had held that the patentee, Unwired Planet, was entitled to insist on the worldwide licence that was FRAND and therefore that an "injunction ought to be granted because Huawei stand before the court without a licence but have

99 *Ibid*, at paragraph 36.

100 [2008] EWHC 329. See also *Telefonaktiebolaget LM Ericsson v Samsung Electronics UK Limited* [2007] EWHC 1047, *Samsung Electronics Co Limited v Apple* [2011] EWHC 2711 and *Nokia Oyj v IPCom GmbH & Co KG* [2011] EWHC 1868. Compare *Koninkiljke Philips Electronics NV & Ors v Harvard International* [2009] EWHC 1600 in which a defence of third-party licence was ordered to be tried first, FRAND licence defences were ordered to be pleaded in detail and validity issues were stayed. The judge declined to require the defendants to make a payment into court as a condition of being allowed to continue their FRAND defence. See also *Philips Electronics NVV v Ingman Limited* (above) ordering patent infringement issues to be tried separately from competition defences (which were stayed).

101 [2012] EWHC 1446 (Ch).

102 At paragraphs 5 and 6.

103 See also *IPCom GmbH & Co KG v HTC Europe* [2012] EWHC 1567 in which, in light of the approach taken in the parallel Nokia proceedings, HTC and IPCom agreed that a similar course would be taken, and the judge stayed the infringement and validity proceedings pending an appeal on validity in the EPO.

104 [2018] EWCA Civ 2344.

the means to become licensed open to them".[105] The Court of Appeal did not disagree and indicated that it may actually have been disproportionate to refuse an injunction,[106] which gives patentees a solution for resolving global disputes against an infringer unwilling to exit the market due to substantial sales in the United Kingdom, rather than enter into a licence on terms set by the UK courts. The type of injunction granted in this case, coined the 'FRAND injunction', was a normal injunction to restrain infringement of the relevant patents, with two added features. First, the injunction will cease to have effect if the defendant enters into that FRAND licence. Secondly, if the licence term is shorter than the life of the relevant patent(s) the injunction was also subject to an express liberty to either party to return to court in the future to address the position at the end of the term. The injunction was also subject to an express liberty to apply in the event the FRAND licence ceases to have effect.

The question of proportionality and injunctive relief was also considered in *TQ Delta v ZyXEL*.[107] ZyXEL argued that as the valid patent was due to expire in a few months the grant of an injunction would have been disproportionate. The Court disagreed. ZyXEL had not paid any royalties to TQ Delta in respect of any SEPs despite an earlier declaration suggesting that ZyXEL would accept a licence on terms held by the court to be RAND (they later refused to accept the outcome of the RAND determination). The court refused to allow ZyXEL to benefit from this 'hold-out' behaviour and granted the injunctive relief, which it considered to be proportionate.

As to damages, the English High Court in *Unwired Planet* confirmed that damages for patent infringement in the UK are compensatory and so, for a SEP owner, this is the FRAND royalty rate for the patent(s) in suit. In *Unwired Planet* the damages were determined to be the sum the patentee would have earned under the FRAND licence from the infringing acts in the UK, ie, UK sales.

(b) *Germany*

In their rulings the German courts have implemented the procedural framework provided for by the CJEU decision in *Huawei v ZTE*. Decisions of the Karlsruhe Higher Regional Court (OLG Karlsruhe) and Düsseldorf Higher Regional Court (OLG Düsseldorf) are particularly pertinent.

In accordance with those decisions, injunctive relief claims (Section 139(1) of the German Patent Act (Patentgesetz – PatG)) for SEPs can only be enforced if the holder of the SEP strictly adheres to the 'Huawei procedural framework', while the infringer does not fulfil its relevant obligations. It should be noted that these individual steps must in principle follow one another consecutively, meaning that

105 [2017] EWHC 711 (Pat), at paragraph 793.
106 [2018] EWCA 2344, at paragraph 276.
107 [2018] EWHC 1515 (Ch).

an obligation to act does not arise until the other party has fulfilled its prior obligations. The German courts have substantiated some of these obligations:

Notifying the alleged infringer of the patent infringement: The notification of an infringement only needs to identify the SEP publication number and state in what manner the SEP is supposed to have been infringed. It is not generally necessary to present claim charts. According to the OLG Karlsruhe, it is not in principle necessary for the SEP owner to present claim-charts, whereas the district court Mannheim (LG Mannheim) so far demands the presentation of claim charts.

Expression of willingness by the alleged infringer to conclude a licensing agreement: The requirements for this expression are relatively low. A general declaration of willingness to conclude a licensing agreement on FRAND terms is sufficient. Establishing that the SEP user does not demonstrate a willingness to conclude a licensing agreement can only be made in cases where it is certain that the alleged infringer is not in fact willing to take a licence.

Licence offer: The courts will fully verify whether the written licence offer of one party meets the FRAND criteria, as only those offers meeting the criteria will trigger an obligation on the opposing party to take the next steps. It is possible to implement the individual procedural steps prior to the court's examination. Both the OLG Karlsruhe and the OLG Düsseldorf are of the opinion – contrary to LG Mannheim – that the Huawei procedural framework can also be applied retrospectively in ongoing proceedings as there are internal procedural options to eliminate the dangers such action may pose for the licensing market. The court does not set forth the terms of a FRAND-compliant offer in the case to be decided. It only checks the offer submitted. In any case, the offer must specifically determine the licence fee and the relevant calculation parameters (relevant reference value, applicable royalty rate, or a staggered royalty if required) as well as the method of calculation. Meaningful provisions must be made on all points which are usually the subject of licence agreements in the relevant sectors. The SEP holder must also explain the main reasons why it considers the proposed remuneration parameters to be FRAND. If the SEP holder has previously granted licences to third parties, it is obliged (depending on the circumstances of the individual case) to justify in particular why the licence fee it intends to pay is FRAND considering (different) older licences.

Since the SEP holder challenged the decision of the OLG Düsseldorf, the German Federal Court of Justice now has the opportunity to respond to the findings. The proceedings are as at the time of writing pending under the Federal Court of Justice Number KZR 36/17.

(c) *Netherlands*

The Huawei procedural framework as established by the CJEU in *Huawei v ZTE* has also been applied by the Dutch Courts in a number of proceedings.

In accordance with those decisions, injunctive relief claims for SEPs can only be enforced if the holder of the SEP adheres to the Huawei procedural framework, while the infringer does not fulfil its obligations. The Dutch courts, however, have noted that the CJEU did not set strict rules through the Huawei procedural framework, but rather stated that these steps should be regarded as guidelines for good faith negotiations between the parties regarding a FRAND licence. Nonetheless, the Dutch Courts will basically follow the Huawei procedural framework's step-by-step plan, whereby the individual steps must in principle follow one another consecutively, meaning that an obligation to act does not arise until the other party has fulfilled their prior obligation.

District Court The Hague: *Archos v Philips* case[108]: In the *Archos v Philips* case, Archos started proceedings requesting a declaratory decision that the Philips (opening) offer should not to be considered FRAND while its own offer should be considered FRAND. These proceedings were started in a response to infringement proceedings brought by Philips concerning certain patents, which Philips had declared as being essential to the UMTS (3G) and LTE (4G) standards.

The District Court of The Hague found that the burden of proof was on Archos, the plaintiff initiating the (declaratory decision) proceedings. The District Court (ultimately) decided that Archos had failed in proving the breach of the FRAND obligations on the part of the SEP holder Philips. In this case, therefore, the District Court never addressed the question of whether the initial licence offer from Philips should be considered FRAND.

The District Court had analysed the history of negotiations between the parties. Philips had initially brought its UMTS and LTE patent portfolio and licence programme to the attention of Archos by letter, explaining that it was possible to take a FRAND licence with regard to these patents and inviting Archos to enter into negotiations. Various meetings took place between the parties. Archos was of the opinion that Philips' offer was not FRAND and responded with an offer of its own, which it had deemed to be FRAND.

Contrary to Archos' position that the Philips offer was not FRAND and not negotiable, the District Court concluded that Archos' failure to negotiate those points and the fact that Archos had held off on further negotiations were unacceptable:

[4.3] … In this, it must be taken into account that until the decision of the

108 *Archos SA v Koninklijke Philips NV, NL*, District Court of The Hague, 10 February 2017, Case No ECLI:NL:RBDHA:2017:1025.

European Court of Justice in the matter of Huawei/ZTE, the initiative for taking a licence was in the case law laid with the user of the standard and was not so much with the SEP holder, at least, section 6.22 of the decision of this court in the matter of Philips/SK Kassetten, as well as the Orange Book decision of the German Bundesgerichtshof were generally understood in that way. The, on this crucial point, contrary decision of the Court of Justice came on 15 July 2015 and in the opinion of the court formed a new moment for negotiation between parties. Philips in accordance with that decision took the initiative with its offer of 28 July 2015. In the meeting subsequently planned on 3 September 2015, Archos however still appeared to slam the door shut by, as has been established without contest, indicating that if Philips wanted more than a few thousands of euros it would have to take legal action against Archos. In that light, it seems unfitting that Archos reproaches Philips to have not been open to negotiation, or at least that position is insufficiently substantiated.

At the same time, the District Court attached importance to Philips' willingness to negotiate.

Both parties agreed that the *Huawei v ZTE* case does not lead to only one proposal being FRAND, but allows for a range of conditions that can be considered FRAND.

In determining the reasonableness and fairness of Philips' royalty proposal, the starting point is that the requested royalty must be proportionate to the value of the patent. The District Court followed Archos' argument, that the value of a SEP must be determined by looking at the nearest alternative or low-cost alternative solution available and the technical and economic advantage that a SEP confers on top of that.

Archos argued that the royalty should not be calculated on the basis of the price of the entire telephone, but only on the basis of the price of the chip in which the relevant technology is contained (the Smallest Saleable Patent-Practising Unit or SSPPU theory). The District Court indicated that there is something to be said for and against both positions, and to be the parties to decide on this issue themselves during their negotiations.

The District Court rejected Archos' 'Patent Hold up' argument (requesting excessive royalties blocks the application of the entire standard) because Archos had not demonstrated that Philips' proposal was non-FRAND. In addition, the District Court considered that requesting a higher royalty for products already sold appeared to be in violation of the *Huawei v ZTE* decision, because the SEP holder must first take action, when enforcing its patents, by informing the user of the infringement of the SEPs.

In the end, none of the defences raised by Archos were successful. The District Court principally followed Philips' position that Archos should simply have negotiated further on those points it did not consider to be FRAND. The Court therefore concluded that Archos had failed to prove that Philips' offer was not FRAND. In contrast, Archos' counter-offer was considered not to be FRAND because it was based on inaccurate facts and was more than 10 times lower than Philips' offer.

Court of Appeal The Hague: *Asus v Philips* case[109]: In this appeal case, Philips claimed that Asus had infringed its UMTS and LTE essential patents. Asus, in its non-infringement defence, claimed that Philips' licence offer was not FRAND and therefore Philips had been abusing its dominant position, as well as claiming abuse of right, absence of pre-contractual good faith and violation of Philips' ETSI FRAND statement.

Asus based its abuse of dominant position claim on the fact that Philips had violated the 'disclosure-obligation' of Article 4.1 of the ETSI IP rights policy. According to Asus, Article 4.1 provides that in the creation of a standard, such as the UMTS standard, the aim is to include as much free technology as possible. By notifying its SEP after adoption of the standard and thereby depriving ETSI of the possibility of seeking a royalty-free alternative, Philips abused its dominant position when enforcing this patent. The Court of Appeal dismissed this argument for many reasons, but primarily because the objective of the ETSI terms is to arrive at a technically optimal standard and not to provide free technology.

Asus also based its abuse of dominant position claim on alleged violation by Philips of the Huawei procedural framework, as provided in *Huawei v ZTE*. The Court of Appeal started its ruling by stating that the CJEU did not set strict rules in the Huawei procedural framework step-by-step plan:

> *The Court of Appeal is of the opinion that the ECJ did not intend to lay down strict rules with the step-by-step plan given in the Huawei/ZTE decision, in the sense that if the SEP holder had not met one of the steps precisely and completely, that it would immediately and necessarily entail that maintaining his SEP would constitute an abuse of a dominant position, regardless of the further circumstances of the case. As expressed by the ECJ in nrs. 55 and 56, when assessing whether enforcement of its patent rights by a SEP holder can be classified as an abuse, the factual circumstances of the case must be taken into account. The steps mentioned by the ECJ in the Huawei/ZTE judgment can therefore rather be regarded as guidelines for negotiations in good faith between the parties regarding a FRAND licence. On the one hand, the SEP holder may be expected to inform the SEP user of his SEPs, to be the first to make a licence offer and to negotiate in good faith about this before initiating a prohibition action, so that the negotiations can be conducted without pressure from a threatening ban. On the other hand, the SEP user must be willing to act and in particular not to apply delaying tactics, which means, among other things, that he must make a written counter-proposal within*

109 *Koninklijke Philips NV v Asustek Computers Inc, Asus Europe BV and Asus Holland BV*, Court of Appeal, The Hague, The Netherlands, 7 May 2019, Case No 200.221.250/01.

a short term period if he does not want to accept the licence offer from the SEP holder. The High Court in England (Birss, J) stated that in the Unwired Planet v Huawei *ruling (of 7 June 2017, [2017] EHWC 711 (Pat), as ratified by the Court of Appeal by judgment of 23 October 2018, [2018] EWCA Civ 2344) as follows:*

> *... the judgment does not hold that if the circumstances diverge from the scheme set out in any way then a patentee will necessarily abuse their dominant position by starting such a claim. In those circumstances the patentee's conduct may or may not be abusive. The scheme sets out the standard of behaviour against which both parties' behaviour can be measured to decide in all the circumstances if an abuse has taken place.*

The Court of Appeal rejected this FRAND defence from Asus primarily because Asus clearly had not acted as a 'willing licensee' during the negotiations. According to the Court of Appeal, to be a willing licensee, it is not sufficient, according to the Court of Appeal, only to join in on meetings and negotiations. Asus had failed to take any initiative for arranging such meetings, and while Asus had frequently requested that Philips deliver technical data, it had failed to have specialists with sufficient technical expertise participate in the meetings. Moreover, Asus terminated such meetings after a few hours as a result of which the technical discussions were incomplete and discussions of the licensing terms were not even started. Asus did not respond to Philips' statements after such meetings and, in particular, failed to provide any input regarding the proposed licensing terms.

According to the Court of Appeal, Asus did not therefore have a constructive attitude during these meetings and even tried to avoid discussing the licensing terms, thereby causing as much delay as possible. This behaviour qualifies as a 'hold-out' as referred to in the *Huawei v ZTE* judgment, which required that the SEP-user should refrain from using such tactics.

As Asus had failed to show willingness to enter into a licence agreement on FRAND terms, Philips was not obliged to take the next step to make a licence offer to Asus. Consequently, Philips was entitled to enforce its patent by requesting an injunction. Nonetheless, Philips offered Asus its standard licence agreement and conveyed its willingness to negotiate if Asus made a counter-offer. However, Asus failed to provide a counter proposal, either before, during, or within a reasonable time frame after the meeting.

Asus further argued that the offer of Philips did not meet the Huawei procedural framework, as Philips did not indicate why that offer was FRAND. The Court of Appeal held that, besides the fact that it had established that Philips was not required to make an offer as Asus had not shown willingness, it does not follow from *Huawei v ZTE* that the SEP holder should state why its offer is FRAND:

*From par 63 [*Huawei v ZTE*], it cannot be inferred that it contains the requirement that the SEP holder explains in his license offer why that offer would be FRAND, just as that condition is not explicitly or implicitly included with regard to the* *counter offer by the SEP user. The paragraphs 63 and 64 should be understood in the light of the questions referred by the German court to the ECJ against the background of the Orange Book decision of the German Bundesgerichtshof, which stipulated that the SEP user should be the first to submit a license proposal. For the reasons given in paragraph 64 (the SEP holder has made a FRAND statement and is in the best position to make a first proposal) the ECJ has opted for a different approach than in Orange Book, considering that it is the SEP holder who must submit the license proposal (if and after the SEP user has shown himself to be a 'willing licensee'). Furthermore, as part of the third question referred for a preliminary ruling, the ECJ was asked whether the license proposal should contain all the provisions that are usually included in licensing agreements in the technical field concerned. Against that background, the phrase "the amount of the royalty and the way in which that royalty is to be calculated" from paragraph 63 is therefore understood that the ECJ considers it particularly important that the offer sets out not only what the royalty fee is, but how it should be calculated. In that context, it is also possible to take into account that there are many modalities for the calculation of the royalty (for example, based on the entire product or only a part of it, percentages, purchase and selling prices, costs, levels, etc.).*

The fact that Asus (finally) made an offer, but only after Philips had initiated infringement proceedings, will not result in this issue being reconsidered by the courts. At most, any counter-offer made after the commencement of infringement proceedings can lead to the obligation on the SEP holder to negotiate in good faith in parallel to the infringement proceedings. It does not lead to an obligation on the SEP holder to stay the proceedings, nor does it make the proceedings by the SEP holder an abuse of dominance. After all, this would give free rein to delaying tactics on the side of SEP users which is contrary to the decision in *Huawei v ZTE*.

The Court of Appeal therefore dismissed the abuse of dominance claims, as well as Asus' other defences and awarded the injunction.

(d) *Italy*

The Italian courts have declined to order interim injunctive relief in cases where the parties were negotiating FRAND licence terms on a variety of grounds. In *Samsung Electronics Co Ltd v Apple Inc*,[110] for example, the court held that *periculum in mora* (ie, necessary urgency based on a risk of irreparable harm) was lacking, especially since the defendant had lodged a payment with the court on account of potential royalties.[111] The decision does not preclude the possibility of a final injunction on the merits.

110 *Samsung Electronics Co Ltd e Samsung Electronics Italia SPA v Apple Inc, Apple Italia Srl, Apple Retail Italia Srl, & Apple Sales International*, 5 January 2012, Tribunale di Milano, n RG 45629-1/2011.

111 See also *Telefonaktiebolaget LM Ericsson v ZTE Italy Srl*, 8 July 2011, Court of Rome (confirmed on appeal, 2 September 2011) and *Ericsson v Onda*, 29 July 2011, Court of Trieste (confirmed on appeal, 11 October 2011). Compare *Koninkilijke Philips Electronics NVV v Computer Support Italcard Srl*, 7 May 2004, GR 2521/2004, where the defendant had made no licence request.

The only case to consider the application of *Huawei v ZTE* is the decision issued by the Court of Turin on 18 January 2016 in the case *Sisvel International SA v ZTE Italy Srl*. By that decision, the Court of Turin rejected the request for preliminary injunctions formulated by Sisvel International SA (owner of a patent covering the UMTS standard) against ZTE Italy Srl (producer of devices implementing the aforementioned standard). One of the reasons for the decision is that Sisvel did not comply with the patent owner's obligations provided for by the CJEU in *Huawei v ZTE*, as a precondition to asking for preliminary injunctions. Indeed, after the interruption of long negotiations with the ZTE parent company aimed at entering into a FRAND licence, Sisvel had started a legal action against the ZTE Italian subsidiary without previously communicating to the latter neither its intention to take the matter to the Court or the essential character of its patent.

(e) ***France***

French case-law regarding SEPs is relatively scarce. In particular, the French courts have not ruled on the compliance of negotiations with the Huawei procedural framework deriving from the *Huawei v ZTE* decision. It is uncertain whether French courts consider such principles only to be guidelines or to be a strict framework. In this respect, Professors Galloux and Azéma consider that the Huawei procedural framework should be regarded as guidelines to avoid the request of a preliminary injunction being regarded as abusive.

In this context, French case law has specific inputs as to the Huawei procedural framework. However, the compatibility of the *Huawei v ZTE* principles with the unfair competition theory has been challenged: a SEP user brought a claim according to which sending a notification letter to distributors would amount to commercial disparagement since this would discredit the SEP user's products. The Court dismissed this claim, considering that notification letters are lawful so long as they are in line with the *Huawei v ZTE* requirements.[112]

French courts will assess whether an asserted patent can be considered as essential to the concerned standard:

- In the *Vringo v ZTE* case,[113] the Paris First Instance Court held that patent EP 1 186 119 was not essential to the UMTS standard. Hence, the infringement claims were dismissed, Vringo being unable to evidence infringement without referring to the standard;
- Similarly, in the *Conversant (formerly Core) v LG Electronics* case, the essentiality of the asserted patents was thoroughly reviewed by the

112 Marseille Commercial Court, 20 September 2016, *Wiko v SISVEL*, No 2016F01637.

113 Paris First Instance Court, 30 October 2015, *Vringo Infrastructure Inc v ZTE Corporation, ZTE France*, No 13/06691.

judges. The First Instance Court dismissed the infringement claims brought by Conversant merely on the basis that the claimant had not sufficiently evidenced the essentiality of the asserted patents for the relevant standards (UMTS, LTE, E-GPRS);
- The Paris Court of Appeals confirmed the first instance ruling: the Paris Court analysed the alleged essentiality of two patents (EP 0 978 210 and EP 0 950 330). In both cases, the Court concluded that an electronic device could implement the concerned standards without infringing Conversant Wireless' patents. The alleged patents are then declared non-essential to the standard in the meaning of ETSI rules.

The availability of injunctive relief for SEPs: When a patent is found to be infringed, the French courts traditionally grant a permanent injunction against the infringer. However, in *Ericsson v TCT Mobile*, the case-management judge refused to grant an injunction in a context where the licence between the parties had expired and negotiations were ongoing: the judge highlighted that an injunction would excessively disadvantage one party in the context of the negotiation of the renewal of the FRAND licence.[114]

Whether the Court will settle the terms of a FRAND licence: The French courts have not yet settled the terms of a FRAND licence. Recently, in *Conversant (Core) Wireless v LG Electronics*, the Paris Court of appeal was asked to set a FRAND royalty rate. The Court heard the arguments of each party on this question and in particular the patentee presented comparable licences to the Court. These debates were conducted in a confidential manner in order to preserve confidentiality: access to the courtroom was restricted to the lawyers of the parties as well as a small number of representatives from said parties. This closed session revolved around the most sensitive exhibit (the Nokia-Qualcomm agreement) and the exhaustion defence. However, the Paris Court did not rule on these claims since the asserted patents had been found non-essential.[115]

6. Determination of FRAND terms by the national courts in European proceedings

6.1 The court's approach to the meaning of FRAND

It is generally thought that it is for the SEP holder and implementer to engage

---

114 Paris First Instance Court, JME, 29 November 2013, *Ericsson v TCT*, No 12/14922.
115 Paris Court of appeals, 16 April 2019, *Conversant Wireless Licensing v LG Electronics*, No 15/17037.
116 Communication from the Commission to the European Parliament, the Council and the European Economic and Social Committee setting out the EU approach to Standard Essential Patents, Brussels 29 November 2017, COM(2017) 712 final.

in good faith negotiations to arrive at an agreement as to what terms are FRAND.[116] However, an agreement cannot always be agreed between the parties and in such circumstances the English courts have recently been asked what constitutes fair, reasonable and non-discriminatory licensing terms. To date, of all the European jurisdictions, only the courts of England have considered the terms of a FRAND licence between a willing licenser and willing licensee.

In *Unwired Planet v Huawei*,[117] the court considered a number of issues relevant to what the FRAND undertaking requires, namely:

- whether there be more than one set of FRAND terms;
- whether the court can set the FRAND rate;
- how to assess FRAND;
- whether there is a hard-edged non-discrimination aspect of FRAND;
- whether the licence scope of FRAND is UK or worldwide;
- whether a patentee can bundle SEPs and non-SEPs into one licence; and
- the relationship between competition law and FRAND.

6.2 Can there be more than one set of FRAND terms?

At first instance, Birss J found that "[t]here is only one set of licence terms which are FRAND in a given set of circumstances".[118] Birss J considered that this conclusion eliminates an issue that he identified in *Vringo v ZTE*,[119] namely what is to be done where each side in a dispute such as this has made a FRAND offer, but those offers are different. The Court of Appeal in *Unwired Planet v Huawei* disagreed with the High Court on this one particular point, and found that more than one set of terms agreed between a SEP holder and implementer could be FRAND. The Court of Appeal considered the Vringo issue to be more of a theoretical problem than a real one and stated that:

> *If the SEP owner and prospective licensee cannot agree upon the terms and royalty rates of a FRAND licence and the question of what is FRAND falls to be decided by a tribunal, whether a court or an arbitrator, then the tribunal will normally declare one set of terms as FRAND and that will be the set of terms the SEP owner must offer to the prospective licensee. If, however, the outcome of the proceedings is that two different sets of terms are each found to be FRAND then in our judgment the SEP owner will satisfy its obligation to ETSI if it offers either one of them. It will in that way be offering an irrevocable licence of its SEPs on FRAND terms.*[120]

6.3 Can the court set the FRAND rate?

Birss J, upheld on appeal, found that FRAND is justiciable and enforceable in the

---

117 [2017] EWHC 711 Pat.
118 Paragraph 806(4).
119 [2013] EWHC 1591 (Pat) and [2015] EWHC 214 (Pat).
120 Paragraph 125.

English court.[121] Accordingly, the English courts, in cases where they have jurisdiction, can (and will) settle the terms of a global FRAND licence having extra-territorial effect.

Post-*Unwired Planet*, in *Conversant v Huawei and ZTE*[122] the Court of Appeal confirmed that the English court has jurisdiction to settle the FRAND terms of cross-border SEP portfolio licences and could grant an injunction in the United Kingdom if the implementer does not take the settled licence. In this action, Conversant sued both Huawei and ZTE for infringing four UK SEPs. Conversant asked for a licence fee for a global portfolio to be determined in the United Kingdom. ZTE and Huawei objected on the basis that the United Kingdom represented only 0.07% of ZTE's mobile device sales, and only 1% of Huawei's global sales on which royalties are claimed. However, the court found this was nonetheless "a real and substantial business".[123] The defendants argued that the English court was not the correct forum, and the Chinese courts would be more appropriate, as most of their profits were generated in China. The court disagreed. The court followed *Unwired Planet* and once again said that a global FRAND licence could be determined by the English court and that it was appropriate to do so here. In reaching its conclusion the court acknowledged that "It is ... not legitimate to characterise the claim as one for enforcement of a global portfolio right. No such right exists".[124] However, "the fact that the dispute concerns UK patents is a matter of substance and not of form. Resolution of the dispute will involve determining infringement, essentiality and validity of UK patents. A UK forum is clearly the most appropriate forum, indeed the only possible forum, for this dispute to be tried".[125] The net effect of this decision is that if a SEP is found to be valid and infringed in the United Kingdom, and the implementer wants to rely on the FRAND undertaking to avoid an injunction, the implementer will find itself subject to a FRAND determination in the United Kingdom, which will most likely result in a global portfolio licence.

This Court of Appeal decision followed an earlier decision of the High Court (Carr J) in the same claims, where the High Court indicated it would be willing to grant anti-suit injunctions to prevent litigation commenced outside of the United Kingdom, to settle the terms of a FRAND licence in China, that would interfere with the English court's ability to determine the SEP infringement dispute in the United Kingdom.[126]

In another judgment handed down in May 2018, the English courts gave further guidance as to the application of jurisdictional gateways in FRAND

121 Paragraph 806(1).
122 [2019] EWCA Civ 38.
123 Paragraph 115.
124 Paragraph 103.
125 Paragraph 40.
126 [2018] EWHC 2549 (Ch).

cases. In *Apple v Qualcomm*,[127] Morgan J found that the jurisdictional gateways had not been met. The key point where establishing jurisdiction failed in *Apple v Qualcomm* was based on the reference in Clause 6.1 of the ETSI IP rights policy to "the owner" of SEPs. It was held that this does not mean that affiliates of the SEP owner should be required to comply with the undertaking given by the SEP owner to offer FRAND licences. The Qualcomm UK subsidiary did not own the relevant patents and therefore Apple's attempt to use that UK subsidiary as an anchor defendant in the proceedings was unsuccessful.

More recently, in *ZyXEL v TQ Delta*[128] the Court of Appeal cancelled a FRAND trial after a finding infringement and essentiality on the basis that the decision in *Unwired Planet v Huawei* does not suggest that a patentee has an independent right to ask the court for a declaration regarding the extent or scope of a licence it is required to offer an implementer, when the implementer has no interest in taking the licence. In this case, ZyXEL had submitted to pay the full damages claimed by TQ Delta for past infringement and to a three-month injunction. ZyXEL also waived its right to a FRAND licence in the United Kingdom under the TQ Delta patents, and so there was no utility in having a FRAND trial that would have taken up 10 days of court time. The Court of Appeal also found that a declaration to the effect that the implementer ZyXEL was not a 'willing licensee' would have no utility because:

- there were no other existing proceedings between the parties anywhere in the world;
- there was no pan-European or global approach to the interaction between the FRAND undertakings and the grant of relief for patent infringement;
- there was no evidence of how a UK court's findings would be treated in foreign jurisdictions; and
- the concept of a 'willing licensee' was not an internationally recognised term of art.

In the circumstances, the Court of Appeal found that it would be an exercise in "jurisdictional imperialism" to impose the UK court's view as to whether ZyXEL were unwilling licensees or holding out in a yet to be determined jurisdiction.

The decisions in *Unwired Planet* and *Conversant v Huawei and ZTE* are both subject to a pending appeal to the Supreme Court, which is due to be heard on 21 October 2019. One of the issues the Supreme Court will consider is whether England is the proper forum for a determination of FRAND terms by reference to a global portfolio of SEPs.

127 *Apple Retail UK Limited and Ors v Qualcomm (UK) Limited and Ors* [2018] EWHC 2711 (Ch).

6.4 How to assess what is FRAND

In *Unwired Planet v Huawei*, Birss J gave the following practical guidance:

The legal effect of the FRAND undertaking relating to a SEP is not that the implementer is already licensed. Its effect is that an implementer who makes an unqualified commitment to take a licence on FRAND terms (settled in an appropriate way) cannot be the subject of a final injunction to restrain patent infringement. Whereas an implementer who refuses to take a licence on terms found by the court to be FRAND has chosen to have no licence, and so if they have been found to infringe a valid patent an injunction can be granted against them.[129]

FRAND characterises the terms of a licence but also refers to the process by which a licence is negotiated. Although an implementer does not owe a FRAND obligation to ETSI, an implementer who wishes to take advantage of the patentee's FRAND obligation, must themselves negotiate in a FRAND manner.[130]

Offers in negotiation which involves rates higher or lower than the FRAND rate but do not disrupt or prejudice the negotiation are legitimate.[131]

An appropriate way to determine a FRAND royalty is to determine a benchmark rate which is governed by the value of the patentee's portfolio. That will be fair, reasonable and generally non-discriminatory. The rate does not vary depending on the size of the licensee. It will eliminate hold-up and hold-out. Small new entrants are entitled to pay a royalty based on the same benchmark as established large entities.[132]

A FRAND rate can be determined by using comparable licences if they are available. Freely negotiated licences are relevant evidence of what may be FRAND. A top down approach can also be used in which the rate is set by determining the patentee's share of Relevant SEPs and applying that to the total aggregate royalty for a standard but this may be more useful as a crosscheck.[133]

In assessing a FRAND rate counting patents is inevitable.[134]

In view of the above, parties or potential parties to SEP litigation which could result in a FRAND determination might wish to consider whether early or pre-action disclosure of comparable licences might assist in early resolution.

6.5 Is there a hard-edged discrimination aspect of FRAND?

Birss J (upheld on appeal) found that "[t]he non-discrimination limb of FRAND does not consist of a further 'hard edged' component which would justify a licensee demanding a lower rate than the benchmark rate because that lower

128 [2019] EWCA Civ 1277.
129 Paragraph 806(5).
130 Paragraph 806(6).
131 Paragraph 806(7).
132 Paragraph 806(8).
133 Paragraph 806(10).
134 Paragraph 806(11).

rate had in fact been given to a different but similarly situated licensee".[135] It was held that while the rightholder cannot discriminate between implementers which are similarly situated, this does not mean that a SEP holder cannot offer one licensee more advantageous terms than another, provided that the terms are fair and reasonable. To adopt a hard-edged approach, which would enable a licensee to demand a rate lower than the benchmark 'fair and reasonable' rate purely because such a rate had been offered to a similarly situated licensee, would not be appropriate. The undertaking given by the SEP holder when it declared its patent as essential to the standard should not prevent it from being able to charge less than the licence is worth if it chooses to do so.

This issue has been appealed to the Supreme Court, which has been asked to determine: "Is the non-discrimination limb of the FRAND undertaking to be interpreted in such a way that it is breached if the licensee's competitor receives better terms?"

6.6 Geographical scope of a FRAND licence: UK or worldwide?

In *Unwired Planet v Huawei*, both the High Court and the Court of Appeal considered the question of whether a FRAND licence can or should be global or jurisdictional. The High Court in *Unwired Planet v Huawei* determined FRAND terms for both UK and non-UK SEPs and the Court of Appeal[136] held that, provided it has been established that the SEP is valid, essential and infringed, the English High Court may oblige prospective implementers to take a global SEP portfolio licence or be injuncted in the United Kingdom.

Birss J was not persuaded that global portfolio licensing necessarily restricted competition. Rather, he found that portfolio licensing was common in the industry. The Court of Appeal summarised his findings:

> *It saved transaction costs for both licensors and licensees and obviated the need to determine a royalty on a patent by patent basis. What was more, the vast majority of patent licences before the court were worldwide licences or at least covered a number of different territories. The judge drew the inference from all of the evidence before him that multi-jurisdictional portfolio licences were in and of themselves unlikely to have anti-competitive effects and that a demand for a worldwide licence was not inherently likely to distort competition.*[137]

On the facts, while Birss J did accept that the portfolio in question was much smaller than that of the large telecoms companies such as Huawei, Samsung and Ericsson, his view was that the portfolio was not so small that it merited different treatment. Birss J noted that Unwired Planet's portfolio was sufficiently large, with a sufficiently wide geographical scope that a willing

135 Paragraph 806(9).
136 [2018] EWCA Civ 2344.
137 Paragraph 36.

licensor and licensee would "regard country by country licensing as madness". He commented that if a multinational had a significant manufacturing base in a country where the portfolio was weak, that could be taken into account in determining the rates.

The decision that a global portfolio licence is FRAND was received with some criticism as some considered that the English courts do not have jurisdiction to interfere with issues relating to foreign property, especially the validity of a non-UK SEP. However, Birss J expressly notes that there is "nothing to prevent a FRAND licensee from challenging the validity of a licensed SEP in any jurisdiction and the licence could and should include an appropriate mechanism to deal with the outcome of such proceedings". Further, in circumstances where the parties are global players it may not be efficient to have a country-by-country licensing approach so that a global licence would be appropriate and would be FRAND. A question does arise as to whether it will always be the case that a global licence is FRAND. For example, where the SEP holder's portfolio is small and the implementer's sales and manufacturing footprint is limited, it may be more appropriate to assess this on a national basis. In those circumstances, and given the Court of Appeal has found that more than one set of terms could be FRAND, the 'Vringo issue' (see "Can there be more than one set of FRAND terms" above) could well become a tangible, rather than theoretical, problem in the context of portfolio versus per patent licensing. This was certainly contemplated in *Vringo v ZTE*[138] by Birss J (prior to *Unwired Planet*) where he stated in the context of availability of injunctive relief (*albeit obiter*):

> *Even if the portfolio global licence is FRAND and it is not an abuse of a dominant position to make the offer, it is not obvious to me why that means that the only options to ZTE in this case are being injuncted under EP 1,212,919 (UK), a territorial patent covering only the UK and only covering one aspect of the standard, or signing up to a global portfolio licence, in effect on all patents (within some relevant class) and for all countries, which ZTE does not want.*
>
> *I will say no more than this. I suspect the fallacy in the reasoning of Vringo at this stage may be that just because it may be so that the global portfolio offer is a FRAND offer, it does not follow that the global portfolio licence on offer is the only set of terms which could be FRAND. It seems to me that there is likely to be a FRAND rate for EP 1,212,919 (UK). I can see that the aggregate of individual FRAND rates for patents taken alone and on a territorial basis may well be far more than global portfolio rates and so a rational defendant may well prefer to take a global portfolio licence rather than a series of individual ones. Moreover I accept, as Vringo urges on me, that global portfolio licences are the kinds of licences industry normally enters into.*

138 [2015] RPC 23.

> *However this is very different from saying that somehow the fact that a global licence on a portfolio of patents is FRAND necessarily means that a defendant in one jurisdiction faced with one patent is forced to take a global portfolio licence in order to stave off a national injunction on that one patent.*[139]

This issue has also been appealed to the Supreme Court, which has been asked to determine: "Is it appropriate for an English court to enforce FRAND terms by reference to a global portfolio of SEPs?"

6.7 Can a patentee bundle SEPs and non-SEPs into the same licence?

In *Unwired Planet v Huawei*, Huawei had argued that the bundling of SEPs and non-SEPs into the same licence was unlawful and posed two main threats: that one cannot tell in such a licence whether the SEP owner would be in compliance with a FRAND commitment; and that the practice of bundling SEPs and non-SEPs into the same licence would potentially eliminate competition on the merits between non-SEP technologies.[140]

Birss J ruled in the High Court decision that a "patentee subject to a FRAND undertaking cannot insist on a licence which bundles SEPs and non-SEPs together".[141] Birss J had accepted Huawei's second point on the competition implications but confirmed that it does not follow from his judgment that it is contrary to competition law to make a first offer which puts SEPs and non-SEPs together, as there is "clear evidence that in some cases the parties agree to a licence which includes both SEPs and non-SEPs together".[142] Accordingly, everything is case and fact-dependent, whereby a licence that bundles SEPs and non-SEPs together does not, by mere fact, take it out of FRAND nor indicate that a patentee has used the market power of the SEPs to secure a licence for the non-SEPs as well.

7. Euro defences

It is widely recognised that EU competition law arguments, the so-called 'Euro-defences', can be used as defences against the unlawful exercise of IPRs. In *ITT Promedia NV v European Commission*[143] the CJEU recognised that:

> *... in specific circumstances, undertakings in a dominant position may be deprived of the right to adopt a course of conduct or to take measures which are not in themselves abuses and which would even be unobjectionable if adopted or taken by non-dominant undertakings. Thus, the conclusion of a contract or the acquisition of a right may amount to abuse ... if that contract is concluded or that right is acquired by an undertaking in a dominant position.*

Nevertheless, in the CJEU's opinion: "As access to the Court is a

139 Paragraphs 106 to 108.
140 Paragraph 785.
141 Paragraph 787.
142 Paragraph 787.
143 [1998] ECR II 2941, 2987, paragraph 139.

fundamental right and a general principle ensuring the rule of law, it is only in wholly exceptional circumstances that the fact that legal proceedings are brought is capable of constituting an abuse of a dominant position."

The Article 101 defence may be used in circumstances where a licence agreement contains restrictive terms. For example, in *IPCom v Nokia UK Limited*, it was argued by Nokia that, among other things, the terms of IPCom's licence were "discriminatory, excessive and unfair" in breach of Article 101.[144] The issue is one of adequate nexus between the exercise of the right and the abuse alleged.[145]

An Article 102 defence is confined to exceptional circumstances "where the exercise of an exclusive right by the proprietor involves abusive conduct".[146] The 'exceptional circumstances' doctrine was first laid down in *Magill*,[147] and imposes a significant limitation on the application of Article 102. According to *Magill*, exceptional circumstances will only exist when the following three conditions are met:

- Access is indispensable to enable an undertaking to carry on business in a market;
- Refusal prevents the emergence of a product for which there is a potential consumer demand; and
- Refusal excludes any competition on a secondary market.[148]

What is covered by 'exceptional circumstances', however, is not a closed category. A refusal to supply may be abusive where it unjustifiably denies access to a product or service that is either "essential for the exercise of the activity" or "a new product whose introduction might be prevented despite specific, constant and regular potential demand on the part of consumers".[149] Article 102 has also been pleaded as a defence in UK proceedings on the basis of a patent holder's alleged refusal to license essential patents on FRAND terms in breach of a FRAND commitment.[150]

In February 2012, the Commission considered whether the threat of injunctions constituted an abuse in the context of Google's proposed acquisition of Motorola Mobility.[151] The Commission concluded that:

... the seeking of enforcement of injunctions on the basis of SEP[152] *is also not, of*

144 *IPCom GmbH & Co KG v Nokia UK Limited* [2009] EWHC 2742 (Pat), paragraph 33.

145 *ICI v Berk Pharmaceuticals* [1981] 2 CMLR 91; *British Leyland v Armstrong Patents* [1984] 3 CMLR 102; and *Sandvik Aktiebolag v KR Pfiffner (UK) Ltd & Others* [1999] All ER (D) 372.

146 *Magill (RTE) & Ors v European Commission* [1995] ECR I 743, 823, paragraph 50; see also Case 238/87 *AB Volvo v Erik Veng (UK) Ltd* [1988] ECR 6211; Case C-418/01 *IMS Health GmbH & Co OHG v NDC Health GmbH & Co KG*, ruling of 20 October 2003; Case COMP/C-3/37.792 *Microsoft v Commission*.

147 *Magill (RTE) & Ors v European Commission* [1995] ECR I 743, 823.

148 *Ibid*, paras 52–8.

149 Case T-504/93 *Tierce Ladbroke v Commission* [1997] ECR II 923, paragraph 131.

150 See *IPCom GmbH & Co KG v Nokia UK Limited* (above); *Telefonaktiebolaget LM Ericsson v Samsung Electronics UK Limited & Another* [2007] EWHC 1047 (Pat); and *Nokia GmbH v IPCOM GmbH & Co KG* [2009] EWHC 1017 (Pat), all interlocutory procedural decisions.

151 Commission ruling Case No COMP/M.6381, GOOGLE/MOTOROLA MOBILITY, 13 February 2012.

152 Standard-Essential Patent.

itself, anti-competitive. In particular, and depending on the circumstances, it may be legitimate for the holder of SEP to seek an injunction against a potential licensee which is not willing to negotiate in good faith on FRAND terms.[153]

The Commission is of the view that where licensing discussions fail, a patent holder may ultimately take its counterparty to court and seek an injunction.[154] However, it was acknowledged by the Commission that the threat of injunction or indeed the actual enforcement of injunctions may significantly impede effective competition by imposing terms *ex post* that one would not have been able to achieve *ex ante*, for example demanding a higher royalty than would otherwise have been agreed. In issuing its statement of objections to Samsung on 21 December 2012, the Commission stated its preliminary view that "Samsung's seeking of injunctions against Apple ... on the basis of its mobile phone standard-essential patents amounts to an abuse of a dominant position".

One aspect mitigating the risk of unreasonable patentee behaviour is the opportunity for the counterparty "to have the terms of the cash-only option licence assessed by an independent third party (whether a court or arbitrator) without the threat of immediately being excluded from the market".[155]

EU Regulation 1/2003[156] is also of note, given the Commission's activity in this area, as regards the possible requirement for a stay of national litigation pending a related Commission decision. Under Article 16:

When national courts rule on agreements, decisions or practices under Article 81 or Article 82 of the Treaty which are already the subject of a Commission decision, they cannot take decisions running counter to the decision adopted by the Commission. They must also avoid giving decisions which would conflict with a decision contemplated by the Commission in proceedings it has initiated. To that effect, the national court may assess whether it is necessary to stay its proceedings.[157]

In Unwired Planet v Huawei, *Birss J had derived the following principles from* Huawei v ZTE*:*

- *In the judgment the CJEU has set out a scheme which both the patentee and implementer can be expected to follow in the context of a dispute about a patent declared essential to a standard and subject to a FRAND undertaking.*
- *In stating that the implementer and patentee must express a willingness to conclude a licence on FRAND terms, the CJEU is referring to a willingness in general terms. The fact that concrete proposals are also required does not mean it is relevant to ask if those proposals are actually FRAND or not.*

153 Commission ruling Case No COMP/M.6381, GOOGLE/MOTOROLA MOBILITY, 13 February 2012, paragraph 126.

154 *Ibid*, paragraph 106.

155 *Ibid*, paragraph 140.

156 Council Regulation (EC) 1/2003 of 16 December 2002 on the implementation of the rules on competition laid down in Articles 81 and 82 of the Treaty.

157 See also *Philips Electronics NVV v Ingman Limited* [1999] FSR 112 in which the court ordered a stay of competition law issues pending the outcome of the defendants' complaint to the European Commission.

- *If the patentee complies with the scheme prior to starting a claim for infringement of that patent which includes a claim for an injunction, then bringing such a claim will not be abusive under Article 102. That is the ratio of the CJEU's decision.*
- *In the circumstances contemplated by the CJEU, bringing a claim for infringement of a SEP which includes a claim for an injunction without prior notice of any kind will necessarily be an abuse of dominant position. Insofar as the decision identifies what is abusive rather than what is not, the decision does not go further than that.*
- *Bringing a claim for infringement which includes a claim for an injunction even with sufficient notice is capable of being an abuse of dominant position. However, the judgment does not hold that if the circumstances diverge from the scheme set out in any way then a patentee will necessarily abuse their dominant position by starting such a claim. In those circumstances the patentee's conduct may or may not be abusive. The scheme sets out the standard of behaviour against which both parties' behaviour can be measured to decide in all the circumstances if an abuse has taken place.*
- *Nor does it follow that if the patentee complies with the scheme such that bringing the action is not* per se *abusive, the patentee can behave with impunity after issue. Again, the scheme sets out standards of behaviour against which both parties' behaviour can be measured to decide if an abuse has taken place.*
- *If the patentee does abuse its dominant position in bringing the claim or in its conduct after issue, that affords a defence to the claim for an injunction. In other words, the proper remedy is likely to be refusal of an injunction even though a patent has been found to be valid and infringed and the implementer has no licence.*
- *The legal circumstances of this case differ from the circumstances assumed by the CJEU in a crucial respect. FRAND is justiciable and the undertaking can be effectively enforced at the suit of the defendant irrespective of Article 102. The defendant does not need Article 102 to have a defence to the injunction claim.*[158]

The Supreme Court in the joined *Unwired Planet* and *Conversant* appeals, due to be heard on 21 October 2019, will also consider the following question: "Does the CJEU's *Huawei v ZTE* decision identify no more than a 'safe harbour' within which a SEP owner would not abuse a dominant position?"

The authors would like to acknowledge the kind assistance of CMS lawyers Jean-Baptiste Thienot (France), Paola Nunziata (Italy), Matthias Eck (Germany), Jacqueline Vallat and Elijha Wong (United Kingdom) and Rogier de Vrey (Netherlands) in the preparation of this chapter.

158 [2017] EWHC 2988 (Pat), at paragraph 744.

Licensing principles of open source software

Heather Meeker
O'Melveny & Myers

1. The birth of open source software

In the 1970s, a top computer programmer named Richard Stallman was working at MIT's Artificial Intelligence Laboratory. The lab's printer kept jamming, and print jobs were piling up. Stallman wanted to write a software routine to notify users when the printer was jammed, but he did not have the source code for the printer's software. He tried to get a copy of the source code, but the manufacturer refused to provide it, because it was reserved as a trade secret; the manufacturer only allowed the software to be distributed in object code form. Stallman decided that every software user should have access to the source code, as a political and moral right. So, in a spark of invention engendered by that necessity, the free software movement was born.

UNIX was then the most popular operating system in academia and large computing environments. But UNIX was licensed under terms that did not allow licensees to share source code with others. As a result, UNIX had become fragmented into many incompatible 'flavours'. To fix this, Stallman began the GNU project, seeking to create an operating system free for all to use, improve and share. To facilitate that goal, Stallman created a copyright licensing model he called 'copyleft'. That model required any redistribution of the software to be under terms that allowed sharing and modification. The licence he wrote to embody that model was the GNU General Public Licence (GPL). The GNU project eventually promulgated a UNIX-compatible operating system under the GPL, and that operating system was called Linux.[1]

In the 1990s, Linux and the GPL were still not well known – viewed as the purview of hobbyists and free software advocates. However, with the recession of the early 2000s, industry latched onto this new model as a way to cut costs and find an alternative to the ubiquitous Windows operating system, which was the only serious contender as an operating system for the smaller, cheaper computers of the day. Over the next decade, free software took a deep hold in industry, as more and more companies enjoyed the benefits of Linux and other

1 For more on the story of the GNU Project, and its relationship to the Linux operating system kernel, see www.gnu.org/gnu/gnu.html.

free software. By the mid-2000s, the so-called 'LAMP stack' – Linux operating system, Apache web server, MySQL database and Python scripting – became the backbone for much computing, particularly web computing. All of these stack elements are open source software.

Today, technologists are fond of saying that open source software has "eaten the world". Indeed, open source[2] software is everywhere, almost all companies mix open source and proprietary software in their computing environments and products, and open source is the backbone of cloud services, which in turn has begun to eclipse on-premises deployment of software. But because open source licensing is so different – in a way, opposite – to traditional software licensing, IP lawyers, business people and engineers are often confused about this legal paradigm. This chapter outlines the basic legal principles for open source licensing and development, and provides basic guidance on managing the risks and rewards of this alternative paradigm.

2. What is open source software?

As illustrated by Stallman's predicament, one must have access to the source code for software in order to make changes to it. Historically, much software has been licensed in object code format only – a format understood by computers, but not by human beings. By reserving the source code as a trade secret, vendors made their customers dependent upon them for support (fixing bugs) and maintenance (making improvements).

Open source licensing removes the walled garden of the vendor and allows the user to undertake its own maintenance and support – or to hire others to do so. Open source advocates claim this freedom results in better software, because "With enough eyeballs, all bugs are shallow".[3]

Open source is both a development model and a licensing model. It is important to understand the distinction, because different risks and benefits of open source software spring from each.

The open source development model is explained through an analogy advanced by Eric Raymond, a seminal author in the open source world, in his article *The Cathedral and the Bazaar*.[4] The closed source or 'proprietary' model is like a cathedral: a large, well-funded organisation conceives of a plan, hires a master builder, funds the development, and controls every aspect of the project from the top-down. The open source model is like the bazaar: anyone can

---

2 The term 'open source' was developed in the 1990s to include the GNU project's free software licensing model, as well as more permissive models, and to promote both neutral terminology and cooperation between the free software movement and industry. For more information, see www.opensource.org.

3 This quotation is attributed to Linus Torvalds, the author of the original Linux operating system kernel, but may be apocryphal.

4 Eric S Raymond, *The Cathedral & the Bazaar: Musings on Linux and Open Source by an Accidental Revolutionary*, O'Reilly Media, 2001. This book is now an evolving project available free of charge at www.catb.org/esr/writings/cathedral-bazaar/.

contribute or change the product, and the 'official' version is the one selected by the market. Open source advocates postulate that this market-driven model will actually result in more standardisation and fewer 'forks' of the software into different incompatible versions.[5]

The licensing model enables, but is not strictly necessary for, the development model. 'Open source' is officially defined by the Open Source Initiative as "software licenced under terms that allow free redistribution and modification, not limited by specific licencees, technologies, fields of use, or products".[6] Accordingly, open source licences cannot contain restrictions limiting the licence in certain ways common in proprietary licensing, such as:

- end use only and not redistribution;
- non-commercial use; or
- use in products consistent with technology standards.

3. Open source as a business model

While at first blush the idea of creating business models based on open source seems counter-intuitive, open source business models have been both common and successful in the 21st century – 2019 was a watershed year in this regard, with business exits such as IPOs and acquisitions of open source companies totalling in the billions, driven materially by the sales of GITHUB and Red Hat.[7] Indeed, open source business today is an undeniably huge industry sector. Open source business models, however, need to be carefully considered and planned. Just how does one make money giving things away for free?

In fact, this is an old question. In business parlance, open source business models all involve – metaphorically – selling 'razor blades'.[8] They are premised on the notion that providing software free of charge will drive sales of complementary goods or services. Thus, successful models include:

- professional service businesses, that customise open source software for their clients' devices or systems;
- support businesses, such as Red Hat, which offers maintenance, support and packaging for its Linux distribution;
- web service businesses, such as Google, Uber and Facebook, which builds substantial software infrastructure to offer web-based services;

---

5 As happened with UNIX and its 'flavours'. Open source projects do fork, but forking of major projects is actually quite unusual. Linux itself has actually consolidated into fewer distributions over time. But this standardisation is driven by market preference, rather than vendor policy.

6 For the full definition, see http://opensource.org/docs/osd.

7 See https://docs.google.com/spreadsheets/d/17nKMpi_Dh5slCqzLSFBoWMxNvWiwt2R-t4e_l7LPLhU/edit#gid=0.

8 This comes from the motto, "Give 'em the razor; sell 'em the blades" credited to the founder of the Gillette Safety Razor Company and the inventor of the disposable safety razor – a well-known meme in the world of business.

- 'widget frosting', such as Amazon.com, which sells Android-based devices that run on the Android open source platform; and
- open core licensing models, such as Elasticsearch, in which the vendor sells commercial licences to software that complements its open source core.[9]

In the early 2000s, venture capitalists began to see the potential of open source as a business model, or at least as an element of attractive business models. By leveraging third-party open source software, businesses can focus their development resources on the highest-value layer of technology. By developing and releasing open source software, businesses can do any of the following:

- Disrupt the market for competing goods by offering a free-of-charge substitute;
- Drive an industry towards standardisation upon standards beneficial to the company; or
- Drive adoption of free products that are complementary or interoperable with the company's commercial products.
- Make extremely efficient use of capital.[10]

The business success of open source has been discussed at length in the press and many publications, and a detailed discussion is beyond the scope of this chapter. However, anyone wishing to understand open source licensing should be aware of a few fundamental ideas.

First, 'free software' is not free of cost. It is free like a puppy is free; the main cost is care and feeding. While it bears no licensing fees, open source software incurs costs to the organisation using it. Open source software sometimes does not enjoy the maintenance and support commitments of a single source. Under open source licences, the software is provided free of any warranty, so any commitments as to maintenance and support or other forms of support (eg, IP infringement indemnities and security patches) are likely to be provided only for a fee, or at self-funded internal costs to the company using the software. People discussing the relative merits of open source often speak of "total cost of ownership", and it is not always clear which alternative – open source or proprietary – will provide the best cost profile. Sophisticated consumers of open source software in industry bear this in mind. This is one reason why Linux has enjoyed limited penetration in consumer desktop computers; it tends to be a do-it-yourself environment that is confusing to lay users. In contrast, Linux is

9 See www.elastic.co/products/x-pack/open.

10 See Joseph Jacks, "Commercial Open Source Software Company Value Creation and Capture Fundamentals", Medium, 26 February 2019, https://medium.com/open-consensus/9-coss-company-value-creation-and-capture-fundamentals-9f0689c32ab0.

ubiquitous in embedded systems like DVRs, game platforms, IoT and mobile devices – closed systems that are walled gardens for users.

Second, open source is an extremely useful and successful model for core software elements, such as the LAMP stack. It is also ubiquitous for applications that provide a high-level platform – from content management tools like WordPress to data crunching packages like Hadoop, to scripting language engines like PERL, PHP and Python. Companies – and the investors that fund them – do not want to put time and money into reinventing the wheel. Open source allows them to stand on the shoulders of giants. But investors want to use their development dollars at the top level of value, and those top-level developments are almost always proprietary. So open source has been less successful penetrating the high-value application market.

4. Open source and copyright

Open source licences are drafted, almost without exception, as conditional copyright licences. The conditions always include some form of notice (copyright notice and licensing notices), and sometimes include copyleft requirements such as making available source code and describing modifications.

This approach has implications for the available remedies for breach of the licences. Because violating the conditions terminates the licence (automatically, in the case of GPL version 2, and with certain notice and cure options, in the case of some other licences), violation of the conditions supports a copyright infringement claim.[11]

Open source licences come in two broad categories. The first is sometimes referred to as 'viral', but is more accurately called 'copyleft' – a pun that refers to the way such licences use the power of copyright law to compel sharing of copyrightable material. The most common copyleft licences are the GNU General Public Licence (GPL); the GNU Lesser General Public Licence (LGPL); the Mozilla Public Licence (MPL); and the Eclipse Public Licence. The requirements of copyleft are as follows. Though the requirements vary slightly from licence to licence, any redistribution of the code must be:

- Available in source code format – this sometimes includes the requirement to provide build instructions, compilation tools, or installation instructions. (See GPLv2 Section 3: "For an executable work, complete source code means all the source code for all modules it contains, plus any associated interface definition files, plus the scripts used to control compilation and installation of the executable".) It is not necessary to distribute source code with every binary distribution; binary distributions must include an offer to provide source code; and

11 See below, *Jacobsen v Katzer.*

- re-licensed under the same copyleft terms – some copyleft licences, such as the MPL, allow binary distribution under proprietary terms, as long as the source code is made available under the terms of the MPL. However, GPL requires the identical GPL terms to be used for binaries. No additional restrictions (such as confidentiality requirements) may be added to a copyleft licence (see GPLv2 Section 6).

All open source licences require a licence notice to be included in redistributions, though the exact requirements vary quite a bit from licence to licence. However, the notice usually requires reproduction of the entire licence text. This is one reason why some distributors of open source code prefer to distribute in source code form when it is not strictly necessary under copyleft requirements – the source code serves as a licence notice file.[12] For binary distributions, licence notices can be time-consuming and tedious to create.[13]

Permissive licences (sometimes called 'non-viral', 'notice' or 'academic' licences) have no material conditions other than notice. The most common include the Apache Software Licence version 2, the BSD Licence and the MIT Licence.

Although not strictly speaking open source licences, many kinds of licences are treated similarly from a business and licensing perspective. The distinction is sometimes more a matter of vocabulary than substance. They include the following:

- Freeware – a term generally used to describe software that is subject to a limited end user licence, but provided free of charge (eg, Adobe Acrobat).
- Public domain – software whose copyright interest has been waived. This term is not properly applied to open source software, which must have a copyright interest to support its licensing conditions.
- Creative Commons (CC) – a group of licences that applies a copyright licensing paradigm similar to open source. However, the licences are intended primarily for non-software content such as books, music, artwork, and audio-visual works. The CC 'sharealike' licence is roughly equivalent to a copyleft licence. The CC 'by' or 'attribution only' licence is roughly equivalent to a permissive licence.

It is useful to pay special attention to software that is released under CC licences. Because copyleft software and content is often released by authors who are not sophisticated about licensing, it is common for the wrong licences to be applied to those releases. Examples of these kinds of wrong licences are:

---

12 This is because source distribution packages usually contain licence files embedded in the source code files. The easiest way to pass on these notices, therefore, is to deliver the source code without removing them.

13 Because code provided under permissive licences can be re-licensed under any terms, the notices required usually serve only an informational purpose and are not licensing notices.

- Non-software copyrightable works released under GPL – this leaves recipients wondering what the language of GPL means for a video or photo – an essentially unanswerable question.[14] Documentation for software is also sometimes released under GPL. There are options such as the GNU Free Documentation Licence, but it is a complex and confusing document. The preferable licensing choice for documentation is probably a CC licence.
- Software released under CC licences – while the CC licences are written generally enough to apply to software, there is no compelling reason to use them when so many well-known open source licences are available.
- Data released under CC or GPL – there are open source data licences, but these are not in common use. Because data enjoys only thin protection under copyright law,[15] assessing the scope of copyleft data licences is very challenging.
- Other 'licence bugs' – a typical example would be a code library released under GPL. This would arguably require all applications that use the library to be subject to GPL. Accordingly, sophisticated licensors usually choose LGPL or one of the 'GPL exception' licences that insulate the applications from licence issues.[16]

A decade ago, many private enterprises balked at using open source software under permissive licences. At least their management and lawyers did, given there was often a deep divide between what 'management' and 'legal' knew the company was using, and what the engineers were actually using. Today, most private companies approve use of open source software under permissive licences, but many baulk at the use of copyleft software, particularly in distributed products. Accordingly, representations in M&A and licensing agreements have shifted somewhat to focus on copyleft software, rather than open source software generally.

Most companies today nonetheless consider ubiquitous stand-alone packages like Linux to be relatively low risk – at least from a copyright perspective. This acceptance has coincided with the growing lifespan of GPL, and along with it, growing comfort with the interpretation of the licence as defined by industry practice.

One of the most confusing elements of open source licensing, to those

---

14 The language of GPL is difficult enough to apply to certain kinds of software; it is written for languages like C and C++. References to phrases like "a complete machine-readable copy of the corresponding source code" (GPLv2 Section 3b) lose their meaning even in the context of high-level scripting languages like PERL or Python. They are impossible to parse when applied to non-software works such as data, documentation, drawings or photographs.

15 The extent varies greatly across jurisdictions – the United States has relatively little protection, and there is more in the European Union.

16 For more on exception licences, see http://en.wikipedia.org/wiki/GPL_linking_exception.

accustomed to proprietary licensing, is that it is a direct licensing model. An open source licence grants rights to the entire world at once, and runs directly from the author (or owner of the copyright in the code) to each recipient. Once granted, the licence is not withdrawn or terminated (except for breach by the licensee) and need not be re-granted to downstream recipients. Accordingly, if a recipient of open source code makes changes, the recipient as author grants a parallel licence to its own changes, and the original author's grant flows directly to the recipient (see Figure 1).

Several corollaries flow from this model. First, if a licence is breached, it is the author/licensor that has a remedy against the recipient, rather than the distributor, regardless of where the recipient appears in the channel of distribution. This remedy is generally thought to be an action based on copyright infringement, and not breach of contract.[17] Second, there is no need to transfer licences from one party to another, obviating issues such as transfer restrictions. Third, if the licence to the intermediate licensee (the contributor) fails, the end recipient retains its licence.

Figure 1. The open source direct licensing model

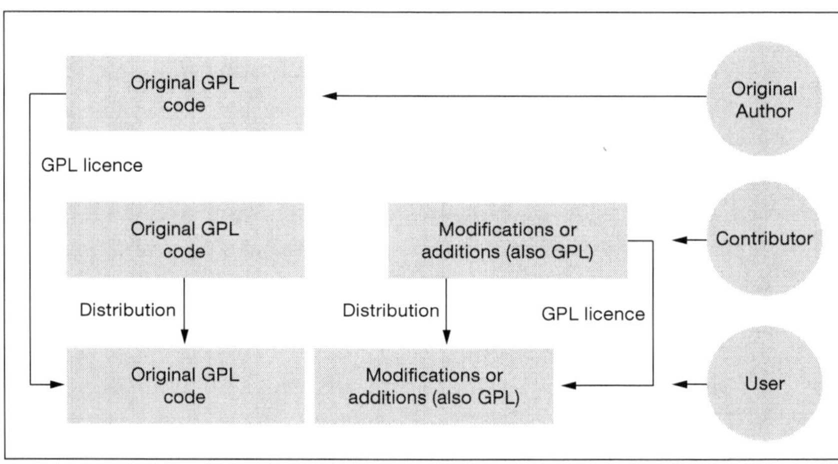

5. Compliance

Most attorneys and business people who embark on the quest to learn about open source licensing have been tasked with compliance for their company. Learning about open source can be daunting, because it turns conventional

17 A detailed discussion of 'licence versus contract' in open source is beyond the scope of this chapter. For more information, see Heather Meeker, *Open Source for Business*, 2017. In the United States, *Jacobsen v Katzer* held that an action for breach of an open source licence sounded in copyright, but did not foreclose the possibility of a contract claim. Generally, remedies for copyright claims are more robust than for contract claims – including the possibility of injunction and statutory damages – so the issue is often moot, as plaintiffs will almost always choose a copyright remedy.

notions of intellectual property backwards, as the term 'copyleft' suggests. This section is a very brief guide to developing compliance within an organisation.

Today, it is customary for sophisticated businesses to have written open source policies. While they vary greatly in length (from hundreds of pages to only a few) and style (from the very formal to the very practical), written policies are more and more common. But the purpose of a written policy is to communicate business strategy and facilitate adoption of the policy – not window dressing. In fact, the worst choice is to have a written policy no one follows.

Companies go through stages of open source compliance. At first, the company may have no notion of what open source software it is using, or what it should be using. This phase is common for small companies, start-ups and non-technology companies – but in most large companies it is not consistent with reasonable business governance. At some point, a company decides it needs open source controls, and begins the process of developing them.

Writing an open source policy is straightforward, but making the strategic decisions behind the policy can be difficult. Some topics typically addressed in the policy are:

- identifying use cases (such as use in products, as development tools, as back-end processing);
- identifying what open source may be used in each use case;
- identifying internal decision-makers for each use case;
- setting business processes for initiating and conveying the decisions within the business;
- describing how notices will be conveyed in company products;
- describing how source code deliveries will be effectuated for copyleft code;
- educating employees about corporate open source policy;
- setting company preferences for open source terms in transactions; and
- setting the process for approving the contribution to third-party open source projects of company engineers.

Open source policies are a combination of corporate strategy (eg, we will never use Affero GPL[18] code) and business process (eg, project leads can approve use of code under permissive licences without separate legal approval). While some companies prefer a formal approach, most companies want their policies to be succinct and written in clear and simple language.[19]

Of course, even the best policy is not useful if it is not properly disseminated to company personnel. This can be particularly challenging in today's

18 The Affero GPL is an ultra-strong copyleft licence. Most companies find its copyleft terms too difficult to administer to be able to use it with comfort.

19 A model policy is available at https://blueoakcouncil.org/company-policy.

technology environment, where global resources are applied to IT development. A policy must be understandable to all relevant staff, taking into account their first languages and levels of training. The best approach is to incorporate open source policy training into general IP training, or general training about the use of third-party technology.

The last point in the list above is actually not about compliance at all. Top engineers today often insist on being able to work with their favourite open source projects, either on their own or on company time. This is because most engineers today come to corporate work from an academic environment that is steeped in open source development and use. Technology companies understand that these days, approving open source participation by company engineers is a recruiting necessity, as well as a way to leverage open source in the business.

Finally, the most sophisticated companies adhere to a standard called Open Chain, which describes processes necessary for open source compliance.[20]

6. Audits

Open source compliance audits usually take place upon acquisition of a company, or as a new but ongoing business initiative for a going concern. In either case, the process of an audit involves collecting information about what open source is being used, then assessing whether the uses are compliant with the applicable licences. The first step is a pure informational challenge; because no payment is required to use open source, its use tends to bypass ordinary corporate business processes. The second step is a legal and technical inquiry, assessing such questions as whether distribution of open source software has been done with proper notices, or whether, for copyleft software, source code is being made available.

In either case, the initial task is a snapshot of code being used on a specific date in a specific product version. However, for going concerns, the audit process is often wrapped into an ongoing compliance program which seeks to gather and assess information about open source use on a continuous basis. Snapshot approaches usually involve a third-party consultant that uses automated means of scanning software to find third-party open source code elements. Ongoing compliance involves installing software development tools that gather licensing information as part of the development cycle. The best-known consultant providing these services is Synopsys (which acquired Black Duck Software). However, these services or tools provide information, not answers. The ultimate questions of whether particular code bases or uses of code are consistent with company policy, and with the company's risk tolerance, are answered by human judgement and not automation.

---

20 See www.openchainproject.org/.

Other tools to help automate compliance include FOSSology, an open source compliance tool that is free of charge and developed by the community, and a new generation of tools that adds process flows to the basic code identification task, such as FOSSA and White Source. Compliance automation tools will undoubtedly continue to evolve as the information problem of tracking and analysing open source use and licence grows.

7. Mergers and acquisitions and other transactions

As use of open source software has become ubiquitous, transactional documents have developed to reflect the need to allocate risk between parties to transactions. Most M&A agreements today contain representations by the target regarding open source software. In the 1990s, these representations stated that the target used no open source software in its business. Today, such a representation would so rarely be true that the clauses have been adjusted to reflect actual practice. They usually require that the target:

- list all open source used in its business (or products); and
- represent that its use is in compliance with the conditions of the relevant licences.

As time moves on, these representations continue to change to reflect a more sophisticated understanding of open source issues. Today, targets sometimes are not required to list open source software covered by permissive licences, or are not required to list open source software except where it is included in a target's product offering.

Most M&A agreements, of course, contain IP non-infringement warranties of some type. Thus, open source representations are often more about disclosure than liability: a warranty that a product or business is not infringing third-party copyrights is essentially a warranty that open source licences have not been violated.

In transactions such as M&A deals, open source licence disclosures are often segregated from disclosures of other third-party inbound licences. But the distinction is artificial, and often results in the unnecessary ministerial task of relocating disclosures from one schedule to another as lawyers separate the truly open source licences from freeware and public domain materials. In fact, much open source software is subject to dual licensing, and therefore could fall under either open source or non-open source licences.

Today, open source representations are also common in licensing and commercial agreements; customers and resellers want to know what open source is in the products they buy, so they can assess their own open source compliance. As a corollary, original equipment manufacturers and resellers often want their suppliers to deliver notices for open source licences suitable for binary distribution – whether or not the supplier is providing the software in binary form. The administrative question of how to deliver a suitable bill of materials for software has been the subject of the 'SPDX' project, sponsored by the Linux Foundation.[21] That project seeks to create a standard and automated means of providing a bill of materials for software that includes many open source elements.

8. Open source enforcement

This section summarises some of the lawsuits to date that have sought to enforce open source licences.

8.1 gpl-violations.org

The initial set of cases brought by gplviolations.org is usually considered the first significant formal enforcement effort for GPL. gpl-violations.org is an organisation spearheaded by Harald Welte, a technologist and open source advocate in Germany. That these initial cases preceded actions elsewhere is not surprising. German courts allow *ex parte* actions for injunctions – a suit for an injunction brought by the plaintiff where the defendant does not participate, and the plaintiff pursues the injunction directly with a court.[22] Some, but not all, of the initial string of cases brought by this organisation related to Welte's own software. The most widely publicised of the cases was against Fortinet UK. A Munich district court granted a preliminary injunction against Fortinet prohibiting distribution of their products absent compliance with the GPL. Fortinet eventually agreed to make certain source code in its products available under GPL.

8.2 The BusyBox cases

Starting in 2007, the Software Freedom Law Center (SFLC), an open source software advocacy organisation, filed a series of lawsuits on behalf of Erik Andersen and Rob Landley, two of the authors of the 'BusyBox' software. BusyBox emulates many standard Unix tools in a small, efficient executable useful for embedded devices. The first suit was filed against Monsoon Multimedia, Inc. That case was settled with release of source code and an undisclosed settlement payment. SFLC next brought suits against Xterasys and High-Gain Antenna. Both soon settled. Further suits were brought between 2007 and 2009 against Verizon Communications, Bell Microproducts, and Super Micro Computer. All settled quickly.

On 14 December 2009, SFLC filed another lawsuit naming 14 defendants including Best Buy, JVC and Samsung. Some of these are still pending.

---

21 For more information, see http://spdx.org.

22 Germany is a jurisdiction particularly friendly to the granting of IP injunctions. See www.managingip.com/?Page=17&ISS=21425&SID=614860.

One case, against Westinghouse Digital Electronics, resulted in a damages award, but this was because the defendant was in liquidation and defaulted on the litigation by failing to answer discovery requests. The judge awarded statutory damages, attorneys' fees, and injunctive relief.[23] Although the order characterises the damages as 'treble', that was only true in the sense that the court awarded three times the upper limit of statutory damages in the absence of wilfulness. The better characterisation is enhanced damages – under 17 USC §504, a court can in its discretion award up to $150,000 per work in statutory damages if the infringement is wilful. The court did not make a finding of actual damages because the defendant failed to respond to discovery and thus never presented evidence to support calculation of actual damages. Also, as described in footnote 39 of the court's order, absent the default, statutory damages might have been unavailable because the plaintiff failed to register the copyright on time. Given that most software authors do not register their copyrights, this could often be a significant limitation in claims enforcing open source licences. The result in this case is interesting mostly because it confirms that injunction and statutory damages are available for open source claims, but the facts were unusual and thus probably difficult to extrapolate to non-default judgments.

In a further development on the same case, a successor entity of Westinghouse Digital Electronics met the same fate. On 8 August 2011 the District Court for the Southern District of New York held Westinghouse Digital, LLC (WD) in contempt for failing to defend its BusyBox case. Westinghouse Digital Electronics, LLC (WDE), with the default judgment standing against it, liquidated its assets after severe business distress and told the court it would not defend the litigation. WD purchased the assets of WDE. The court assessed whether there was "a substantial continuity of identity" between WDE and WD, and found that there was. The court invited evidence on damages and attorneys' fees, and ordered forfeiture of all infringing articles.

8.3 *Jacobsen v Katzer* (2006–2010)

In August 2008, the US Court of Appeals for the Federal Circuit issued the most significant decision on open source licensing in the United States to date – and followed the guidance of the open source community.

The case arose from a complicated set of facts. Both parties developed and distributed software for controlling model railroads – Jacobsen making his 'Java Model Railroad Interface' (JMRI) software available under an open source licence free of charge, and Katzer (via his company Kamind Associates) selling commercial products under proprietary licences. In particular, Jacobsen made available through the JMRI project software called 'DecoderPro', which is used

23 Attorneys' fees were awarded because of the default.

by model railroad enthusiasts to program decoder chips in model trains to control lights, sounds and speed.

Jacobsen received a letter inviting him to license patents owned by Kamind, suggesting the patents were infringed by the JMRI software. Jacobsen filed a declaratory judgment action asking the court to rule that the patent was invalid due to prior art (or failure to disclose prior art including that of Jacobsen himself) or not infringed. Katter's original claim received a great deal of press in the open source world, feeding on fears that the existence of software patents spell death for open source projects. As the patent case progressed, Jacobsen discovered that Katzer had copied some of Jacobsen's open source software and used it in Katzer's proprietary product. Jacobsen's software was licensed under the Artistic Licence, which is a relatively rarely used licence that was originally written to provide rights to the PERL programming language interpreter. The requirements of the Artistic Licence are modest; it is generally considered a permissive open source licence, similar to the Apache, BSD or MIT licences. The Artistic Licence requires, as a condition to exercise of the licence, certain copyright and licence notices, and identification of changes made to the original author's source code. Because these notices were not preserved by Katzer, Jacobsen made a counterclaim, alleging violation of the licence.

In 2007, the District Court for the Northern District of California issued a preliminary ruling in the case, stating that the licence violation constituted a breach of contract. However, the court's ruling did not support a claim for copyright infringement. The court reasoned that Katzer was exercising rights in the copyrightable work, but was licensed to so do and thus was not liable for infringement. Copyright infringement claims can result from breach of the scope of a licence, but the court distinguished violations of licence scope and licence conditions. This distinction is core to open source law because open source licences grant all the rights of copyright – and so it is generally not possible to violate the scope of an open source licence, only its conditions.

Proponents of free software such as the Free Software Foundation have long taken the position that their licences are not contracts. Originally, this position may have been a strategic attempt to avoid contract formation issues that plagued the law of online distribution of the software in the early 1990s, but the position persists notwithstanding the intervening publication of a line of cases supporting enforceability of contracts in the download context.

The issue in this case is sometimes referred to in the open source legal community as the 'licence or contract' issue. In other words, are open source licences contracts, or are they merely conditional licences? The status of an open source licence as a licence or contract dictates the types of remedies available if a licensee violates the conditions of the licence. If the conditions of an open source licence are mere contractual covenants, as the district court order stated, then injunctive relief is generally not available. If, however, violating the conditions places the activities of the licensee outside the scope of the licence, as the appellate court stated, the unlicensed activity is copyright infringement and far more likely to garner injunctive relief. Open source advocates are also concerned that the money damages for violation of open source licences under contract law may be more limited than those available under copyright law, which can include statutory as well as actual damages.

The Free Software Foundation has long taken the position that open source licences are licences rather than contracts – however, the two are likely not mutually exclusive. Most licensing contracts are both conditional licences and contracts. The Jacobsen case holds that violating the conditions of an open source licence can constitute copyright infringement, but not that open source licences are not contracts.

The flipside of this question – whether the conditions of a licence could be considered covenants whose performance might be ordered by a court – was not at issue. But this is the 'bet the company' issue for most corporate free software users. If providing source code were a contractual covenant, then failure to do so would be a breach of contract. However, even if it were a breach, specific performance of covenants is not generally available under contract law. Most companies' biggest fear when dealing with GPL is that they will be compelled to lay open proprietary software. In fact, the likely worst case is that they will be given a Hobson's choice: to lay it open and comply with GPL or to replace the GPL code.

On the question of whether the claim could sound in copyright, the appellate court reversed and remanded the case back to the district court to determine whether Jacobsen had established his claim to an injunction. The Federal Circuit opinion was hailed as a victory for open source licensing.

On 10 December 2009, on remand, the lower court issued an order granting and denying portions of motions for summary judgment by the parties. The court said: "Although it is undisputed that Plaintiff distributed the copied work on the Internet at no cost, there is also evidence in the record attributing a monetary value for the actual work performed by the contributors to the JMRI product", and that the record "may establish a monetary damages figure". This dispels the notion that no actual damages would be available for open source authors bringing claims of infringement, because of the royalty-free nature of open source licensing.

Jacobsen v Katzer was finally settled in 2010, obviating a second appeal. The settlement included an injunction by the district court against further infringing activities, and was not sealed. Most settlements are confidential, and an actual injunction by the court in a settlement is a bit unusual. But the injunction would have carried more precedential weight if it had been won in court rather than agreed to in settlement. (The district court had previously declined to issue an injunction.)

It was significant that this case involved the original Artistic Licence. If a permissive licence is enforceable, that lays substantial groundwork for enforceability of copyleft licences. In *Jacobsen*, the question of actual harm to the copyright owner was squarely at issue. For *Jacobsen* to prevail, the court had to decide that there was sufficient harm to give rise to a remedy. With a permissive licence, the harm is, essentially, failure to deliver notices. For a copyleft licence, the conditions are much more significant, and violating them would give rise to greater harm, and in turn, greater opportunities for relief.

8.4 *FSF v Linksys* (2008–2009)

This case was notable in that it was the first lawsuit filed for FSF as a plaintiff. Compliance issues regarding GPL software in Linksys consumer wireless routers had existed since Cisco acquired Linksys in 2003. (For details about the informal dispute that took place in the early 2000s, see the author's 2005 article "The Legend of Linksys" in *Linux Insider*.)[24] This dispute resulted in a fairly quick release of source code, but FSF continued to find compliance issues with the routers, and finally filed suit in 2008. The accused software included various GPL and LGPL components.[25] That suit was settled in 2009 with no substantial docket activity, via a settlement that, according to the press release describing it, was very similar to those in the BusyBox cases.[26]

8.5 *Artifex* cases (2008–present)

Artifex is the purveyor of the Ghostscript line of software, a popular pdf rendering tool. In 2008 and 2009, it filed some of the first lawsuits brought by a private company to enforce GPL. The first defendant was Premier Election Solutions, at the time a subsidiary of Diebold. That case was settled quickly and confidentially. In 2009, in an action related to its MuPDF software (a high-performance PDF rendering engine), it brought claims against Palm, Inc and various other defendants. Artifex's ongoing enforcement efforts help support its dual licensing business. Its latest case against Hancom, Inc advanced the state of the law on open source licensing. That settled at the trial court level, but an opinion in that case held that Artifex had the right to bring the claim based on breach of contract as well as copyright infringement, see *Artifex Software, Inc v Hancom, Inc*, No 16-cv-06982-JSC, 2017 US Dist 2017 WL 1477373 (ND Cal, 25 April 2017). In contrast, *Jacobsen v Katzer* held that an open source enforcement action could be brought based on copyright infringement, but did not foreclose the ability to plead the claim as breach of contract.

---

24 See www.linuxinsider.com/story/43996.html?wlc=1300413418.
25 See the complaint and FSF's press release here: www.fsf.org/blogs/licensing/2008-12-cisco-complaint.
26 See www.fsf.org/news/2009-05-cisco-settlement.html.

9. What to do if you receive a claim

Open source claims are not like other IP claims, and best practice for handling them is different. That is because the parties enforcing open source licences – mostly – have different motivations from traditional IP plaintiffs. As of 2019, most enforcement of open source licences still takes place via one of the advocacy organisations that seeks to serve the open source community. Advocacy organisations focus first on informal enforcement, and usually file a lawsuit only as a last resort. Advocates primarily want compliance, rather than damages or business disruption. Although the early enforcement actions of the SFLC sometimes included monetary settlements, these were often characterised as a contribution to offset the cost of litigation. Software Freedom Conservancy has followed this lead, and takes a leading role in enforcement today. Compared to the average IP dispute, the amounts involved are minimal. The correct response to an advocacy organisation is therefore a sincere, timely and robust compliance effort. If you receive an open source claim, it is crucial to act on it quickly. That may sound simple, but can be unexpectedly difficult, because not all open source claims are made through formal channels. Companies should train their technical employees to recognise complaints and take them seriously.

Companies should understand the distinction between compliance claims which can only be made by the copyright holder, and source code delivery demands which can only be made by a recipient of binary code. However, recipients whose demands for source code are unmet often complain to the authors, who do have the right to bring a claim.

Lately, strategic plaintiffs and profiteers have been emerging. Strategists generally prefer injunction and damages over compliance, because it serves their goal to disrupt the market for a competing product. Strategic plaintiffs may also make multiple claims, only one of which is copyright infringement. Those who receive such claims should consider the special legal issues that arise in open source claims. As a baseline, any lawsuit that involves open source licensing is likely to touch upon novel issues of law, as well as draw significant attention from the press and the free software community.

10. The future of open source

Open source software licensing is an area in which many significant legal issues – such as the scope of a 'work based on the program' under GPL, or what constitutes distribution of software – remain unresolved. Making business decisions on the use of open source can therefore be difficult: those decisions must often be informed by, at best, extrapolation of legal principles. But business is all about informed risk-taking, and businesses have signalled their willingness to accept many risks relating to open source software in order to take advantage of its benefits.

Enforcement of open source licences is still in its infancy, even if use of open source is not. As Amazon founder, Jeff Bezos, likes to say: "It's Day One". The legal principles of open source will doubtless change and grow with technology itself, and legal thinking on this topic will continue to develop as well. Open source software is here to stay, and it's a different and fascinating world.

Patents in Europe

Nicholas Fox

1. Overview of European Patent System

The European Patent Convention (EPC) establishes a system which makes it possible to obtain patent protection in 40 European countries[1] on the basis of a single patent application prosecuted in English, French or German.[2]

After an application has been filed, the European Patent Office (EPO) reviews an application to check that the application fulfils the formal requirements for a patent application including the form and content of the application, any required translation into one of the official languages of the EPO, the designation of the inventor, the appointment of a professional representative (if required) and the payment of the fees due.[3]

A European search report is drawn up in parallel with the formalities examination identifying the prior art which the EPO considers is relevant to the application.[4] This search report also includes an initial opinion on whether the claimed invention fulfils the requirements for grant.[5]

Applications are published normally 18 months from the filing date or priority date if priority is claimed.[6] The publication of the search report sets a six-month deadline for filing an examination request and responding to any issues identified in the initial opinion accompanying the search report.[7]

After a request for examination has been made and the examination fees paid, the EPO examines the application in the light of the search report taking into account the applicant's comments and any amendments filed.[8] Further rounds of correspondence take place as many times as the EPO consider

1 As of May 2016 the member states of the EPO were: Albania, Austria, Belgium, Bulgaria, Croatia, Cyprus, Czech Republic, Denmark, Estonia, Finland, France, Germany, Greece, Hungary, Iceland, Ireland, Italy, Latvia, Liechtenstein, Lithuania, Luxembourg, Macedonia, Malta, Monaco, Netherlands, Norway, Poland, Portugal, Romania, San Marino, Serbia, Slovakia, Slovenia, Spain, Sweden, Switzerland, Turkey and the United Kingdom. Additionally, the EPO has extension agreements with Bosnia and Herzegovina and Montenegro.

2 Article 14 EPC.

3 Article 90 EPC and Rule 57 EPC.

4 Article 92 EPC and Rule 61 EPC.

5 Rule 62 EPC.

6 Article 93 EPC.

7 Rule 70(1) and Rule 70a(1) EPC.

8 Article 94 EPC.

necessary until the application is found in order for grant or alternatively the EPO issues a rejection of the application.[9]

When an application is found in order for grant an applicant is asked to file translations of the allowed claims into the other two official languages of the EPO other than the language of proceedings.[10] The application can then proceed to grant. When granted, the granted European patent becomes a 'bundle' of individual national patents. To maintain the patents in certain countries, translations of the claims or the whole of the patent need to be filed with the various national authorities depending upon national requirements.[11]

After a European patent has been granted, it may be opposed by third parties who believe that that the application should not have been granted.[12] Any such oppositions must be filed within nine months from the date of grant of the patent[13] and can only oppose on the grounds that a patent should not have been granted because it does not fulfil the basic requirements for patentability discussed in section 2 below or the subject matter of the application was extended during prosecution beyond the scope of the application originally filed.[14]

Backlogs at the EPO are considerable and prosecution can be very slow. Typically, a patent may be granted four to five years from filing, although some cases take considerably longer to reach grant. Oppositions are normally decided in about three years or five to six years if a decision is appealed to a Board of Appeal.

The main official fees for filing and prosecuting a European patent application are as follows:

Due within one month from filing:

Filing fee	€210 (reduced to €120 if an application filed online)
Search fee	€1,300
Claims fees	€235 for the 16th and each claim up to the limit of 50
	€585 for the 51st and each subsequent claim

Due when requesting examination:

Designation fee	€585
Examination fee	€1,635

Due on grant:

Grant fee	€925
Publication fee	€75

9 Article 94(3) EPC.
10 Rule 71a(1) EPC.
11 Article 65 EPC.
12 EPO Oppositions are discussed in greater detail in the "Enforcement in Europe" chapter.
13 Article 99(1) EPC.
14 Article 100 EPC.

In addition, annual maintenance fees are due each year for the third year and successive years while the application is pending. These fees range from €470 for the third year through to €1,575 for the tenth and each successive year.

2. Basic requirements for patentability

The basic requirements for a patentable invention in Europe are set out in Article 52(1) of the EPC which states that:

European patents shall be granted for any inventions, in all fields of technology, provided that they are new, involve an inventive step and are susceptible of industrial application.

Further definition of what is meant by an invention susceptible of industrial application and the definitions of novelty and inventive step are then to be found in Articles 52 to 57 of the EPC. In addition to these four basic requirements, Article 83 of the EPC additionally requires that a patent application must disclose an invention such that it can be carried out by a person skilled in the art.

2.1 Excluded subject matter

There is no explicit definition of what constitutes an 'invention' in the EPC. Rather Article 52(2) of the EPC contains a non-exhaustive list of things which are not regarded as inventions as follows:

- discoveries, scientific theories and mathematical methods;
- aesthetic creations;
- schemes, rules and methods for performing mental acts, playing games or doing business, and programs for computers; and
- presentations of information.

Article 52(3) then proceeds to qualify this list stating that:

Paragraph 2 shall exclude the patentability of the subject-matter or activities referred to therein only to the extent to which a European patent application relates to such subject-matter or activities as such.

The meaning and scope of the exclusions and the term 'as such' is of great importance in the patenting of computer-implemented inventions and has generated significant amounts of jurisprudence at the EPO. This case law is discussed in detail in section 3 of this chapter.

In addition to the exclusions appearing in Article 52(2) of the EPC, Article 53 of the EPC prohibits patents from being granted:

- if their exploitation would be contrary to ordre public or morality;
- for plant or animal varieties or essentially biological processes for the production of plants or animals and methods of treatment of the human or animal body by surgery or therapy; and
- for diagnostic methods practised on the human or animal body.

Although of importance when considering the patentability of inventions in the life sciences in contrast to the exclusions in Article 52, these additional exclusions are of little relevance in the software and electronics field.

2.2 Novelty

Novelty is dealt with in Article 54 of the EPC, which provides that "an invention shall be considered to be new if it does not form part of the state of the art".[15] The state of the art comprises everything made available to the public (anywhere in the world) whether by written or oral description, by use or in any other way before the filing or priority date of the application.[16] Since the entry into force of the EPC 2000, for the assessment of novelty (but not obviousness) the state of the art includes the content of all European patent applications having an earlier priority date but published after the application in question (ie, co-pending patent applications).[17] A co-pending PCT application may also form part of the state of the art so long as it has been published in one of the official languages of the EPO or its translation into one of these languages has been filed with the EPO and published and the national fee has been paid.[18] Article 55 of the EPC provides a limited six-month 'grace period' for disclosures made in consequence of "an evident abuse in relation to the applicant or his legal predecessor" (eg, where the disclosure is made in breach of a duty of confidence owed to the inventor) or for disclosure of the invention at officially recognised international exhibitions.

The onus is on the party seeking to revoke the patent to prove that the disclosure or prior use was made available to the public before the priority date and further that the skilled person would have been able to put the prior art into practice in such a way as to carry out the invention. Interpretation of the disclosure is by reference to the knowledge of the skilled person in the field at the relevant date. Importantly, the purpose of the prior art disclosure is irrelevant for assessment of novelty and thus a disclosure in an unrelated technical field, which may be directed at a completely different technical problem, may still constitute an 'accidental' anticipation (even if the same disclosure would be irrelevant for assessment of inventive step).

The test for novelty is a stringent one. For a disclosure or prior use to anticipate a claim it must disclose all of the features of the claim (ie, only if the invention disclosed by the prior art would infringe the claim in question, if performed post-grant, will it deprive that claim of novelty). The test is not simply that the prior product or process was available to the public but that the

15 Article 54(1) EPC.

16 Article 54(2) EPC.

17 Article 54(3) EPC. Under the EPC 1973, an earlier filed European patent application was only considered to be prior art for the purposes of Article 54(3) EPC to the extent the application designated the same countries as the later application and only in respect of those countries.

18 Rule 165 EPC.

information conveyed by that product or process made the invention available. The limits of novelty are illustrated by the decision in *MOBIL/Friction reducing additives* G2/88 in which a paper proposal to use an additive as a rust inhibitor was not held to anticipate the use of the same additive as a lubricant even though the use of the additives in question for the known purpose would necessarily have achieved the new use as well. Further, for a claim to be anticipated, it must be inevitable that following the disclosure of the prior art something within the scope of that claim will result. The test of 'inevitability' is strictly applied.[19]

2.3 Obviousness

An invention is considered as involving an inventive step if, having regard to the state of the art, it is not obvious to a person skilled in the art.[20] For the purposes of considering inventive step and in contrast to assessment of novelty, later published European patent applications are not considered to form part of the state of the art.[21]

The European Patent Office adopts the so-called 'problem-and-solution approach' to the assessment of inventive step which involves the following steps:

- determining the 'closest prior art';
- establishing the 'objective technical problem' to be solved; and
- considering whether or not the claimed invention, starting from the closest prior art and the objective technical problem, would have been obvious to the skilled person.[22]

The 'closest prior art' is the one single prior art reference which discloses the combination of features which constitutes the most promising starting point for an obvious development leading to the claimed invention. Frequently this will be the prior art which requires the fewest structural and functional modifications to arrive at the claimed invention.[23]

Once the closest prior art has been identified, any advantages which flow from the differences between a claimed invention are then identified and these are used to establish the 'objective technical problem' which the claimed invention may be said to solve.

In establishing the objective technical problem, only differences which can be said to achieve a 'technical effect' or lend an invention 'technical character' are considered. Typically in the case of computer and software inventions such

19 T396/89 *Union Carbide/High strength polymers.*
20 Article 56 EPC.
21 *Id.*
22 Guidelines for Examination in the European Patent Office (2017), Part G – Chapter VII – 3, section 5.
23 T606/89 *Unilever/Detergent composition.*

technical character arises through software controlling an industrial process or processing data which represent physical entities or in the internal functioning of the computer itself or its interfaces which, for example, might affect the efficiency or security of a process, the management of computer resources or the rate of data transfer in a communications link.

Non-technical features and non-technical benefits such as purely economic benefits are ignored when formulating the 'objective technical problem'. This is because the EPO considers that deriving such benefits is not the solution to a technical problem and hence are considered incapable of supporting an arguable technical inventive step.[24]

This is achieved in the EPO by formulating the objective technical problem from the perspective of a person skilled in the art who is aware of the non-technical aspects of the invention. Thus, for example, when considering the computerisation of a novel business scheme, the formulation of the objective technical problem would be considered from the point of view of a developer or application programmer having the knowledge of the proposed business scheme to the extent that such a scheme was devoid of any technical benefits.

Having formulated the objective technical problem, the final stage of the 'problem-and-solution approach' involves assessing whether there is any teaching in the prior art as a whole that would have prompted the skilled person faced with the identified objective technical problem, to modify or adapt the closest prior art to arrive at the claimed invention. In making this final assessment the test applied is not merely whether a skilled person could have arrived at the claimed result but rather whether or not he or she would have done so in the hope of solving the objective technical problem or achieving some improvement or advantage.[25]

2.4 Sufficiency

Article 83 of the EPC sets out the requirement that "the application must disclose the invention in a manner sufficiently clear and complete for it to be carried out by a person skilled in the art" and this corresponds to the ground for revocation at Article 100(b) of the EPC. This 'sufficiency' requirement is an important means of maintaining the balance between, on the one hand, encouraging investment in innovation and rewarding invention and, on the other hand, ensuring that others can work the invention after the patent has expired. In combination with Article 84 of the EPC, the requirement for clarity and that the claims are "supported by the description", Article 83 also addresses the policy requirement that the patentee is only entitled to claim the contribution he or she has made to the art and taught in the patent. As stated

24 T931/95 *PBS Partnership/Pension Benefits* discussed in 3.3 below.
25 T2/83 *Rider/Simethionce Tablet.*

in *Genentech/Polypeptide expression* T292/85 there is a "general legal principle that the extent of the patent monopoly, as defined by the claims, should correspond to the technical contribution to the art in order for it to be supported or justified".

It is settled law that the disclosure of the application must be sufficient to enable the skilled person to perform the invention across the scope of the claim.[26] However, where the patentee has taught a principle of general application, teaching one way of performing the invention may be enough, even if the claim is of broad scope. Put at its simplest the claim scope needs to be commensurate with the technical contribution made.

As at the date of the application the skilled person, having read the application as a whole and in light of the common general knowledge, must be able to put the invention into practice without 'undue effort'.[27] This will be assessed on the facts of the case but a degree of trial and error will be allowed so long as this does not require an inventive step.

2.5 Industrial applicability

As indicated above, patents are only granted for inventions that are susceptible of industrial application, defined in Article 57 of the EPC to mean inventions that "can be made or used in any kind of industry, including agriculture". Relative to the other requirements of patentability set out in Article 52 of the EPC there is a paucity of decisions on industrial application. This is unsurprising as the term 'industry' is construed broadly and in most areas of technology the mere fact that a patent is worth applying for is of itself an indication that the invention has industrial value.

The European, US and UK law of industrial application was reviewed by the UK courts in *Eli Lilly v Human Genome Sciences* [2011] UKSC 51 (including an analysis of the following EPO case law: *Max-Planck* T870/04, *Johns Hopkins* T1329/04, Genentech T604/04, *ZymoGenetics* T898/05, *Bayer* T1452/06 and *Schering* T1165/06). A number of common principles were identified, including the propositions that industry must be construed broadly, that the industrial application must be derivable from the patent application (read with the common general knowledge), the need for a sound and concrete basis for recognising that the contribution could lead to practical application in industry and that "the patentee must make a full disclosure of his invention, including a practical use to which it can be put. It [is] not a hunting licence to find such a use".

---

26 eg, T409/91 *EXXON/Fuel Oils*.
27 T929/92 *Genentech/Human t-PA*.

2.6 Inequitable conduct

The list of grounds for opposition set out in Article 100 of the EPC and revocation in Article 138 of the EPC are exhaustive. Unlike in the United States and certain other jurisdictions, inequitable conduct is not a ground to revoke or render unenforceable a patent in Europe. The EPC does not deal with misrepresentation or fraud.

3. European Patent Office approach to excluded subject matter

3.1 *Travaux preparatoires*

The origins of the patentability exceptions in the EPC and in particular the origins of the exception for computer programs can be found in the *travaux preparatoires* for the EPC itself.[28] It is of course important to put the drafting of the EPC in its own historical context. Although the EPO opened its doors for business in 1978, the preparations and the drafting of the EPC began in the mid-1960s when computers and computer programs were still very much associated with the mathematics research departments within universities and not the ubiquitous consumer product that they are today. By way of example, at the time that the EPC was being drafted, early prototypes of such fundamentals of the computer sphere as the mouse were being developed at Stanford University, the ASCII coding system was just being agreed and B, the forerunner to the object-oriented programming language C was just coming into use.

It is against such a background that in 1965 the first draft set of proposed patentability exclusions was written. The initial draft which included exceptions for the patenting of matters such as scientific knowledge and theories, mere discoveries of substances occurring in nature and purely aesthetic creations made no reference to computer programs. At that time there was no consensus on whether or not computer programs could constitute inventions and the proposed draft merely reiterated categories of inventions which national practice throughout Europe consistently rejected as unpatentable subject matter.

Matters were then strongly influenced by the drafting of the Washington Agreement which brought the PCT into effect. The Washington Agreement did not include a list of matters which were to be excluded from patentability, Rather, rules 39 and 67 of the PCT listed certain inventions as matters which did not have to be searched or examined by the patent offices acting as international search and examination authorities. This list of exemptions

28 For a full discussion and analysis of the origins of the patentability exceptions see Justine Pila, "Article 52(2) of the Convention on the Grant of European Patents: What did the Framers Intend? A Study of the Travaux Preparatoires" (2005) 36 IIC: *International Review of Intellectual Property and Competition Law* 173–191.

included a reference to "computer programs to the extent that International Searching Authority (or International Preliminary Examination Authority) is not equipped to search prior art/carry out international preliminary examination concerning such programs".

Despite this limitation expressly making reference to the abilities of patent offices to undertake appropriate searches or conduct international examinations, the wording of Rules 39 and 67 was adopted as the list of excluded matter by the drafting committee. However this was done so with the inclusion of the 'as such' qualification of Article 52(3) applied across all excluded categories with a comment that in order for the boundaries of patentable subject matter to develop as technology developed, the final definition of patentable subject matter as such had ultimately to be decided by the judicial authorities.

3.2 *Vicom*

The scope of the exclusions from patentablity was first considered by the European Patent Office Boards of Appeal in *VICOM/Computer related invention* T208/84. The invention in *Vicom* was directed to a method of digitally processing images and the question facing the board was whether or not such methods were excluded from patentability on the grounds that they were mathematical methods as such. Considering the matter, the board initially noted that that there could be little doubt that any processing operation on an electric signal could be described in mathematical terms.

The board then established the existence of a 'technical effect' as the basis for distinguishing between a patentable mathematical method and an unpatentable mathematical method 'as such' stating:[29]

> *A basic difference between a mathematical method and a technical problem can be seen, however, in the fact that a mathematical method or mathematical algorithm is carried out on numbers ... and provides a result also in numerical form, the mathematical method or algorithm being only an abstract concept prescribing how to operate on the numbers. No direct technical result is produced by the method as such. In contrast thereto, if a mathematical method is used in a technical process, that process can be carried out on a physical entity ... and provides as its result a certain change in that entity. Generally speaking, an invention which would be patentable in accordance with conventional patentability criteria should not be excluded from protection by the mere fact that, for its implementation, modern technical means in the form of a computer program are used. Decisive is what technical contribution the invention as defined in the claim when considered as a whole makes to the known art.*

Following *Vicom*, through to the early 2000s the practice of the EPO in

29 *Vicom*, at paragraph 16.

relation to unpatentable subject matter was then fixed. The key to distinguishing between patentable and unpatentable subject matter was the existence of a 'technical effect'. That is to say, whether or not the claimed invention provided a technical solution to a technical problem. If it could be established that a claimed invention provided such an effect then the EPO would not reject a case on the grounds that it related to unpatentable subject matter under Article 52(2) and would proceed to examine a case for novelty and inventive step in the ordinary way.

3.3 *Pension Benefits*

Although the test in *Vicom* served the EPO very well for many years there were two issues with the test: a practical one and a theoretical one.

The practical issue was one of duplication. The EPO's problem-and-solution approach to inventive step[30] itself involved an assessment of whether or not a claimed invention provided a non-obvious technical solution to a technical problem. This meant that in the case of computer-implemented inventions the existence of a technical effect was tested for twice: once as part of the consideration of patentable subject matter and then once again as part of the test for inventive step.

The theoretical issues flowed from this duplicate test. When considering a patent application, the EPO always considers patentable subject matter before considering matters of novelty and inventive step. As part of the *Vicom* test, in order to establish whether or not a claimed invention related to patentable subject matter it was necessary to make a comparison with the prior art to establish what technical problem a claimed invention might be said to solve. However, unlike the case of the assessment of novelty and inventive step, the EPC does not contain a definition of the state of the art for the assessment of patentable subject matter.[31]

Changes to the EPO's approach began with PBS Partnership/Pension Benefits T931/95 which concerned a method of controlling a pension benefits program. The application described in detail a method of managing pension contributions and without any technical details went on to say that the management of the pension system could be implemented by a computer.

Rejecting the application, the Board of Appeal initially rejected the method claims included in the application on the grounds that they did not relate to technical subject matter, stating that:

> *All the features ... are steps of processing and producing information having purely administrative, actuarial and/or financial character ... the invention as claimed*

---

30 See 2.3 above.

31 The state of the art for the purposes of assessment of novelty is defined in Article 54(2) and (3) EPC. For the purposes of assessment of Inventive step only prior art under Article 54(2) is considered.

does not go beyond a method of doing business as such and, therefore, is excluded from patentability under Article 52(2)(c) in combination with Article 52(3) EPC; the claim does not define an invention within the meaning of Article 52(1) EPC.[32]

The board then when on to consider the apparatus claims, initially stating that:

In the Board's view a computer system suitably programmed for use in a particular field, even if that is the field of business and economy, has the character of a concrete apparatus in the sense of a physical entity, man-made for a utilitarian purpose and is thus an invention within the meaning of Article 52(1) EPC ... This means that, if a claim is directed to such an entity, the formal category of such a claim does in fact imply physical features of the claimed subject-matter which may qualify as technical features of the invention concerned and thus be relevant for its patentability. Therefore the Board concludes that an apparatus constituting a physical entity or concrete product suitable for performing or supporting an economic activity, is an invention within the mean of Article 52(1) EPC.[33]

However, subsequently when assessing inventive step, the board concluded that:

... the invention according to the application is an essentially economic one, i.e. lies in the field of economy, which, therefore, cannot contribute to inventive step. The regime of patentable subject-matter is only entered with programming of a computer system for carrying out the invention.

The assessment of inventive step has thus to be carried out from the point of view of a software developer or application programmer, as the appropriate person skilled in the art, having the knowledge of the concept and structure of the improved pension benefits system and of the underlying schemes of information processing as set out for example in the present method claims.[34]

The approach adopted in the *Pension Benefits* case solved the two issues with the previous *Vicom* test as it removed the double assessment of 'technical effect' for the apparatus claims. Instead, a simple test of concrete character – ie, whether or not a claimed apparatus was a physical man-made entity – was imposed for assessing patentable subject matter while the existence of a technical effect was retained for the assessment of inventive step.

3.4 *Hitachi*

The new approach established by *Pension Benefits* was then extended by the decision in *Hitachi/Auction* method T258/03. The invention in the *Hitachi* case concerned a computerised implementation of a Dutch auction. In such an auction, which is frequently used for trading financial products such as futures

32 *Pension Benefits*, reasons section 3.
33 *Ibid*, reasons section 5.
34 *Ibid*, reasons section 8.

and derivatives, the price paid by a successful bidder is determined by the timing of when a bid is received. This sensitivity to the timing of bids was addressed in Hitachi's invention by modifying the bidding process so that prices could be determined based on either the timing of a bid or a bid value included in a bidding instruction.

When assessing patentable subject matter, the board initially applied the test from *Pension Benefits* and concluded that the apparatus claims were not excluded under Article 52(2) as they comprised "clearly technical features such as a 'server computer', 'client computers' and 'a network'".[35]

The board then proceeded to consider the method claims and extended the concrete character test of *Pension Benefits* to cover such claims stating:

> *What matters having regard to the concept of 'invention' within the meaning of Article 52(1) EPC is the presence of technical character which may be implied by the physical features of an entity or the nature of an activity, or may be conferred to a non-technical activity by the use of technical means. In particular, the Board holds that the latter cannot be considered to be a non-invention 'as such' within the meaning of Article 52(2) and (3) EPC. Hence, in the Board's view, activities falling within the notion of a non-invention 'as such' would typically represent purely abstract concepts devoid of any technical implications ... A method involving technical means is an invention within the meaning of Article 52(1) EPC.*[36]

Having ruled that neither the apparatus claims nor the method claims lacked concrete character and hence were not purely abstract, the board then went on to consider novelty and obviousness taking into account only the 'technical features' in the claims. When doing so all the technical features of the claim were found to be present in the prior art and the only novel features of the claim were non-technical business features, and the case was therefore rejected on the grounds that the claimed invention lacked inventive step as the only problems which the invention could be said to have solved were solely within the business realm and the only technical aspects of the invention were limited to the mere automation of a novel business system.

3.5 *Duns Licensing*

The application of a concrete character test to both method and apparatus claims developed in *Hitachi* combined with an assessment of obviousness which considers only the obviousness of the technical aspects of the claims has been subsequently consistently followed by the EPO Boards of Appeal.

The EPO's current approach and the assessment of technical character as part of the assessment of inventive step is exemplified in the decision in *Duns*

---

35 *Hitachi*, reasons paragraph 3.7.
36 *Ibid*, reasons paragraph 4.5.

Licensing/Estimating sales activity T154/04 where the board held that in examining the patentability of an invention in respect of a claim, the claim must be construed to determine the technical features of the invention, ie, the features which contribute to the technical character of the invention. The board stated that it is legitimate to have a mix of technical and 'non-technical' features appearing in a claim, and even that the non-technical features may form a dominating part of the claimed subject matter. Assessment of inventive step is, however, based only on technical features, which thus have to be clearly defined in the claim. This is because non-technical features which do not interact with the technical subject matter of the claim are considered incapable of solving a technical problem which is equivalent to saying that non-technical features 'as such' do not provide a technical contribution to the prior art and thus may be ignored in assessing inventive step.

3.6 *Presidential Reference*

The modern approach of the EPO was endorsed by the Enlarged Board of Appeal in *Presidential Reference/Patentability of programs for computers* G3/08. This case arose through the President of the EPO making use of her power under Article 112(1)(b) of the EPC to refer questions of law where different opinions of the Boards of Appeal existed to the Enlarged Board of Appeal.

The main questions referred read as follows:

- Can a computer program only be excluded as a computer program as such if it is explicitly claimed as a computer program?
- Can a claim in the area of computer programs avoid exclusion under Article 52(2)(c) and (3) of the EPC merely by explicitly mentioning the use of a computer or a computer-readable data storage medium?
- Must a claimed feature cause a technical effect on a physical entity in the real world in order to contribute to the technical character of the claim?
- Does the activity of programming a computer necessarily involve technical considerations?

When the reference came up for consideration by the Enlarged Board, the reference was rejected as invalid on the grounds that contrary to the suggestion of the President there was no divergence of opinion in the jurisprudence but rather the development of the case law through decisions such as *Vicom, Hitachi* and *Pension Benefits* represented the natural development of the law and that the Boards of Appeal had been consistent in applying the law as it developed.

3.7 The modern approach

The decision of the Enlarged Board to reject the reference implicitly endorsed the application of the test stated in *Pension Benefits* and *Hitachi* as developed by *Duns Licensing*. The modern approach of the EPO is therefore to apply a low-level abstractness test for the assessment of patentable subject matter under Article 52(2). However, this is balanced by a requirement to demonstrate that an invention provides a technical solution to a technical problem when assessing inventive step under Article 56 of the EPC.

The modern approach to assessment of computer-implemented inventions is summarised in the EPO Guidelines for Examination. After noting that programs for computers are a form of 'computer-implemented invention' and that claims to such inventions can take the form of a method of operating said apparatus, the apparatus set up to execute the method, a readable medium carrying a program or the program itself, the Guidelines[37] state that:

> *The basic patentability considerations in respect of claims for computer programs are in principle the same as for other subject-matter. While 'programs for computers' are included among the items listed in Art. 52(2), if the claimed subject-matter has a technical character it is not excluded from patentability by the provisions of Art. 52(2) and (3). Moreover, a data processing operation controlled by a computer program can equally, in theory, be implemented by means of special circuits, and the execution of a program always involves physical effects, e.g. electrical currents. According to T 1173/97, such normal physical effects are not in themselves sufficient to lend a computer program technical character ... However, if a computer program is capable of bringing about, when running on a computer, a further technical effect going beyond these normal physical effects, it is not excluded from patentability. This further technical effect may be known in the prior art. A further technical effect which lends technical character to a computer program may be found e.g. in the control of an industrial process or in processing data which represent physical entities or in the internal functioning of the computer itself or its interfaces under the influence of the program and could, for example, affect the efficiency or security of a process, the management of computer resources required or the rate of data transfer in a communication link. As a consequence, a computer program may be considered as an invention within the meaning of Art. 52(1) if the program has the potential to bring about, when running on a computer, a further technical effect which goes beyond the normal physical interactions between the program and the computer. A patent may be granted on such a claim if all the requirements of the EPC are met ... Such claims should not contain program listings ... but should define all the features which assure patentability of the process which the program is intended to carry out when it is run ...*
>
> *Moreover, following T 769/92, the requirement for technical character may be satisfied if technical considerations are required to carry out the invention. Such technical considerations must be reflected in the claimed subject-matter.*
>
> *Any claimed subject-matter defining or using technical means is an invention*

37 Guidelines for Examination Part G – Chapter II – 3.6.

within the meaning of Art. 52(1) … Therefore the mere inclusion of a computer, a computer network, a readable medium carrying a program, etc. in a claim lends technical character to the claimed subject-matter.

If claimed subject-matter does not have a prima facie *technical character, it should be rejected under Art. 52(2) and (3). If the subject-matter passes this* prima facie *test for technicality, the examiner should then proceed to the questions of novelty and inventive step …*

3.8 Mathematical methods and mental acts

The exclusion of mathematical methods and mental acts from patentability follows from the general principle that purely abstract or intellectual methods are not patentable. Thus, for example, abstract mathematical methods such as a shortcut method of division would not be patentable. In contrast, methods of encrypting/decrypting or signing electronic communications may and routinely are regarded as a technical method, even if they are essentially based on mathematical methods.[38] This directly follows from the distinction between the substantive application of an excluded invention and the excluded subject matter as such which was made in *Vicom*.[39]

In general, the case law demonstrates that issues with the mathematical methods and mental acts exclusions can be avoided by appropriate claim drafting to ensure that the claims are directed towards something concrete which can be said to provide a technical solution to a technical problem.

One example of a case which was found to fall purely within the excluded subject matter was the invention considered in *General Electric Company/Nuclear Core loading* T914/02. In that case the claims of the invention were directed to a method for designing a nuclear core loading arrangement and the case was rejected on the grounds that the method specified neither use of means or measures of a technical nature, nor includes the provision of a physical entity as the resulting product and hence claimed a method which could be carried out mentally and was entirely lacking in technical character.

The possibility of avoiding the patentable exclusions by connecting an invention with a physical real-world object is demonstrated by *IBM/VSLI chip design* T453/91 which concerned both product claims to a VLSI-chip and a claim to method of designing the chip itself. In that case, whereas the product claims were held to be novel and inventive, the method claim was rejected because the claims only went as far as the step of designing such a chip and hence to the purely abstract matter of generating a 'design' which was wholly abstract and distinct from a real object.

A further example of the boundaries of patentable subject matter was

---

38 See, eg, T1326/06 *Giesecke/RSA Schlüsselpaarberechnung.*
39 Discussed in 3.2 above.

demonstrated in *Infineon/circuit simulation* T1227/05 which concerned a computer-implemented method with mathematical steps for simulating the performance of a circuit subject to 1/f noise. In that case the board concluded that the simulation of 1/f noise by feeding suitable random numbers into a circuit model had technical character by virtue of the fact that it was necessarily computer implemented. Turning then to which features might be considered for the purposes of assessing inventive step, the board was persuaded that the simulation of noise in a circuit was itself an adequately definite technical problem which could be said to be solved by the invention and hence all features addressing that issue could be considered when assessing obviousness.

A claim directed to a technical process in which a mathematical method is used restricted to a particular technical field is also considered not to be directed to a mathematical method as such. Thus, for example, the EPO Examination Guidelines[40] describe a method of encoding audio information in a communication system which aims to reduce distortion induced by channel noise as a potential example of a patentable invention. Similarly, in *RSA Schlüsselpaarberechnung* T1326/06 the Board of Appeal held a method of encrypting/decrypting or signing electronic communications to be a technical method, even though it was essentially based on a mathematical method.

Issues relating to the mental acts exclusion can also arise in the context of the automation of pre-existing processes as was demonstrated by a series of appeals involving IBM in the context of word processing. It should, however, be noted that the common theme running through all of these cases was that in each case the applicant was attempting to obtain a monopoly on the automation of a process previously undertaken manually, rather than a specific way or approach to such automation.

Thus, for example, in *IBM/Document abstracting* T22/85 the Board of Appeal had to decide upon the patentability of a method for automatically abstracting and storing an input document in an information storage and retrieval system and a corresponding method for retrieving a document from the system. Rejecting the case the board stated that it considered that the mere setting out of the sequence of steps necessary to perform the activity in terms of functions or functional means to be realised with the aid of conventional computer hardware elements did not import any technical considerations and could therefore neither lend a technical character to the activity nor to the claimed subject matter considered as a whole, any more than solving a mathematical equation could be regarded as a technical activity when a conventional calculation machine is used.

In *IBM/Text processing* T38/86 the board assessed the patentability of an invention which involved automatically replacing difficult expressions in texts

40 Guidelines for Examination Part G – Chapter II – 3.3.

with more straightforward ones. Again the case was rejected with the board ruling that the use of technical means for carrying out a method partly or entirely without human intervention, which, if performed by a human being, would require him or her to perform mental acts would be excluded from patentability if the invention did not involve a contribution to the art in a field not excluded from patentability. In such cases the implementation of the technical means to be used in those steps involved no more than the straightforward application of conventional techniques and had therefore to be considered obvious to a person skilled in the art. Similarly, if a claim to an apparatus did not specify any technical features beyond those already comprised in a claim pertaining to said method and furthermore did not define the apparatus in terms of its physical structure, but only in functional terms corresponding to the steps of that method then similarly the claim to an apparatus was also not allowable.

Similarly, in *IBM/Spell checking* T121/85 and in *IBM/Text editing* T65/85, where the cases related to the automatic detection and replacement of homophones, the Boards of Appeal concluded that the claimed word-processing systems failed to make a contribution in a technical field and hence were unallowable.

3.9 Business methods

The manner in which the EPO deals with patents in the business field is highly dependent upon a characterisation of where the EPO perceives the benefits and advantages of an invention seem to lie.

The mere fact that an invention arises in a business context is not fatal to an application. That was clearly established by the Board of Appeal in *Sohei/General Purpose Management System* T769/92. The invention in the *Sohei* case concerned a computer system for managing financial and inventory data. Although different types of data could be managed separately, data could be inputted into the system through a single 'transfer slip' displayed on the computer screen. The board accepted that although financial and inventory management would normally fall under the term 'doing business' and be excluded under Article 52(2) of the EPC and despite references to financial and inventory management appearing in the claims, the application was allowable. In reaching that conclusion, the board considered that the references to business data appearing in the claim were merely labels and considered the patentability of a generalised claim where different types of data were input via a single interface and then managed separately. Managing data in the manner claimed was considered to be a technical problem which was solved by the claimed invention and hence the application was allowable.

Where an invention involves both physical and business considerations, this is even more likely to find favour with the Boards of Appeal. Thus, for example, in *Pitney Bowes/System for processing mail* T767/99 a sorting machine which sorted mail on the basis of the timing of pick up and deliveries was considered to be allowable. The board accepted that the timing of deliveries was largely a business matter but considered that the claimed invention enabled improved efficiencies in the mail delivery business and thus constituted a technical invention in the field of mail sorting.

The situation in *Pitney Bowes* may be contrasted with the decision in *Comvik/Two identities* T641/00. In *Comvik*, the Board of Appeal rejected an application directed towards a novel SIM card having two different SIM identities which facilitated billing for mobile phone use on two different accounts. Considering the application, the board identified that the only benefits arising from the novel features in the claimed invention were purely business benefits and hence the claimed invention could not be said to provide a technical solution to a technical problem.

In *Nokia/Shopping with mobile device* T1670/07, the Board of Appeal summarised three arguments that applicants frequently advance unsuccessfully to try to argue in favour of patentability:

- The 'technical leakage fallacy', in which it is argued that the intrinsic technical nature of the implementation leaks back into the intrinsically non-technical nature of the problem;
- The 'broken technical chain fallacy' in which a chain of steps from providing information to its use in a technical process is broken by the intervention of a user: the possible final technical effect brought about by the action of a user cannot be used to establish an overall technical effect because it is conditional on the mental activities of the user; and
- The 'non-technical prejudice fallacy' in which non-technical aspects are invoked as a reason for not modifying the prior art, whereas these features cannot in fact contribute to inventive step.

In addition to the above, more recently in *TRILOGY/Software structure* T1755/10 and *EBAY/Image selection* T1834/10 the boards have added a further 'fallacy' which they refer to as 'the software implementation fallacy' which seeks to suggest that "modified software causes a modified behaviour of the computer and should for that very reason be considered as a technical implementation" which again has been rejected.

3.10 Scheme, rule or method for playing a game

Video games have given rise to a number of patent applications where the scope of the 'method for playing a game' exception has been ruled on by the Boards of Appeal.

In the case of video games, a fundamental distinction is made between, on the one hand, game rules, and on the other game design which improves playability or player interest. An example of a patent application being rejected on the grounds that it was directed solely to gaming rules is *IGT/Electronic Poker* T0336/07 where the board ruled that gaming rules "defining the regulatory framework agreed between [or with] players concerning conduct and conventions and conditions that are meaningful only in a gaming context" were not patentable. Similarly, in *BANDAI Game apparatus* T2127/09 an application directed to a novel form of the Tetris game was rejected because the only novel features in the game concerned the addition of a new gaming rule as to when blocks disappeared which was a matter entirely devoid of technical considerations.

In contrast, where innovations in the gaming field go beyond the mere rules of the game being played, the EPO will frequently allow applications to proceed to grant.

So, for example, in *KONAMI/Video game* T928/03 the Board of Appeal considered the patentability of the game mechanics of a video game involving the display of a ring-shaped guide mark at the foot of a player character to highlight a particular player. Considering the case, the board ruled that making a possibly concealed indicator clearly visible on a display screen contributed to the objective technical function of the display and that the functional quality was not affected by the fact that the visualised information was to be relied upon by a user playing a game. In reaching this conclusion the board emphasised the distinctions made in the earlier *Comvik* decision between technical and non-technical aspects of a claim. Whereas non-technical aspects of a claim could be utilised in the formulation of a technical problem, the actual contributions of any aspects having technical character had to be considered when assessing inventive step.

The EPO has also allowed applications based on the consideration of the underlying structure or narrative of a game. This occurred in *Labtronix Concept Inc/Auxiliary Game* T717/05. In the *Labtronix Concept Inc* case the Board of Appeal had to consider whether a gaming apparatus which periodically caused a player to access an auxiliary game was patentable. Ruling that the application was allowable the board expressed the view that game design in order to maintain a player's interest was a technical problem in that it was directly aligned with the purpose of the gaming apparatus itself. Thus the implementation of an auxiliary game and the manner in which progress towards that game was displayed to a user was relevant for the assessment of inventive step.

A similar result was achieved in *NINTENDO/Game machine and storage medium* T12/08 which concerned a video game where a player navigates around a gaming environment and randomly encounters other characters with whom the player interacts. The invention concerned a novel way of making the random appearance of the characters more unpredictable. Again, the board made the distinction between gaming rules and the mechanics of a game to make a game more interesting and ruled that the mechanism for making game play more unpredictable was considered to provide a technical solution to the problem of maintaining player interest.

3.11 Presentation of information

When interpreting the presentation of information exclusion, the EPO makes a clear distinction between the content of information and any technical benefits a particular type of presentation might achieve. The presentation of information solely defined by its content or semantic meaning is not patentable and this is the case regardless of the manner or form in which the information is presented. In contrast, the EPO accepts that if a particular arrangement or manner of presentation can be identified as having some kind of technical effect then that is potentially patentable.

Thus, for example, in *Nokia/Three-dimensional icons for graphical user interface* T1749/06 the board considered that an icon with alternating light and dark stripes conferring a three-dimensional appearance was potentially patentable since the claimed design of the icon could be said to solve the technical problem of how to provide an icon with a three-dimensional appearance. In particular it noted the distinction between 'what is displayed' and 'how something is displayed' commenting that the claimed effect was independent of the cogitative content of a specific icon.

A similar distinction was held in *IBM/On-line help facility* T887/92 which concerned the issue of how to make a help facility more user-friendly. Approving the application the board held that giving visual indications about the internal conditions of an apparatus or system was basically a technical problem. In that context, displaying only valid commands in a help panel had technical character because the content of the help panel reflected the status or condition of the system and hence a computer program which implemented such an invention would constitute technical means for carrying out a technical invention.

Where a particular manner of presentation is itself associated with a technical advantage then, again, such a presentation is a valid basis for an invention and this can extend to the form of a signal or a data carrier itself. Thus, for example, in *BBC/Television signal* T163/85 a colour television signal characterised by technical features of the system in which it occurred was considered to be patentable. It should be noted that following this decision, provided that a signal is defined by its technical features and not merely the specific information contained within the signal, then the EPO will not object to such a claim on the grounds that the signal *per se* is abstract or non-physical in character. In particular, this means that in contrast to the position in the United States there is no requirement to draft claims to exclude transitory media.[41]

3.12 Computer-readable media

Since the decision in *IBM/Computer program product* T1173/97 it has been the practice of the EPO to allow claims directed to computer-readable media storing computer interpretable instructions which when executed by a computer cause the computer to perform an otherwise patentable method or cause the computer to become configured as a patentable apparatus. In reaching this conclusion the board distinguished the effects that any computer program product had on a computer in that it had the potential to cause a computer to be configured or perform in a particular way and the further technical effect which actually arose when the underlying method or configuration was actually achieved.

The *IBM/Computer program product* decision was then later extended in *Microsoft/Clipboard formats* T424/03 were the board considered the patentability of claims for a computer program implementing an expanded clipboard for transferring between different programs such as is used in 'cut', 'paste' and 'copy' commands. Emphasising that a method implemented in a computer system represented a sequence of steps actually performed and achieving an effect, the board held that the claim to a computer program method could not be equated with a computer program as such. The board then went on to consider the allowability of a claim directed specifically for a computer program for implementing such a method and also considered such a claim to be allowable.

The practical result of these decisions is that the EPO will grant claims to computer-readable media and even to claims directed to computer programs themselves, provided that the underlying method or apparatus which is implemented by virtue of the program is itself patentable. It is the substance of the underlying invention rather than a particular form of the claim which is determinative. Provided the underlying invention can be found to give rise to a technical effect in the form of a technical solution to a technical problem, claims can be drafted to any aspect of the products or processes for implementing that invention.

4. Claim scope and infringement

The scope of protection conferred by a patent is of vital importance both for the assessment of a patent's validity and for determining whether or not a patent is infringed.

Strictly, assessment of infringement is irrelevant for the purposes of assessment of an invention by the EPO, as the EPO does not have any

---

41 The position in the United States is derived from the statement by David J Kappos, Subject Matter Eligibility of Computer Readable Media, 1351 OG 212 (23 February 2010) that "when the broadest reasonable interpretation of a claim covers a signal *per se*, the claim must be rejected under 35 USC § 101 as covering non-statutory subject matter".

jurisdiction over infringement matters. However, as a practical matter the drafting and prosecution of patent applications can only be understood in the context of an understanding of the law of infringement.

Infringement requires first that a prohibited act is carried out while the patent is in force and in the territory of the patent. Further, the product or process that is the subject of the act must fall within the scope of the claims.

The infringing acts, which differ depending on whether the claim is for a product or a process, are set out in Article 25 of the Community Patent Convention (CPC).[42] Article 25 defines the prohibited acts as:

- making, offering, putting on the market or using a product which is the subject matter of the patent, or importing or stocking the product for these purposes;
- using a process which is the subject-matter of the patent or, when the third party knows, or it is obvious in the circumstances, that the use of the process is prohibited without the consent of the proprietor of the patent, from offering the process for use within the territories of the Contracting States;
- offering, putting on the market, using, or importing or stocking for these purposes the product obtained directly by a process which is the subject matter of the patent.

In respect of infringement by offering (etc) the product obtained directly by a patented process, it is important to note that the requirement that the product is a direct result of the process has been strictly applied, severely curtailing the ability of patentees to take enforcement action where the process is performed abroad and then a product of that process is imported (ie, if the product has been further processed or altered in some way after performance of the claimed method then its importation and sale in a patent protected country will not be an infringement). For example, in the UK case of *Pioneer Electronics v Warner Music* [1995] RPC 487, the court held that the importation of compact discs manufactured using a patented master disc did not infringe a patent as it was only the manufactured discs and not the master disc which were imported into the United Kingdom.

As well as the primary acts of infringement identified above, Article 26 of the CPC provides for indirect infringement where a party supplies or offers to supply means relating to an essential element of the invention. However, unlike the primary acts of infringement for which knowledge or intention is irrelevant (save for offering a process for use), indirect infringement also requires the

42 Although all signatory states did not ratify the CPC and it never came into force, equivalent provisions are enacted in the laws of virtually all European states and the Convention itself remains widely influential in judicial decision making. Closely similar provisions are currently to be found in Article 6 of the draft Regulation 9224/11 which proposes provisions for the creation of unitary patent.

patentee to establish that the alleged infringer had the requisite knowledge (ie, that those means are suitable for putting and intended to put the invention into effect). For this requirement the objective knowledge of the reasonable person will suffice. The territorial requirements of indirect infringement also require close attention.

Of course, for there to be an infringement the relevant product or process must fall within the scope of the claims. The extent of protection conferred by the claims is determined by reference to Article 69 of the EPC and the Protocol on its interpretation. In essence this provides that the description and drawings shall be used to interpret the claims and that a balance must be struck between a very narrow 'literal' interpretation of the claims (that would not give fair protection to patentees) and a broad interpretation in which the claims serve only as a guideline (that would not give third parties a reasonable degree of certainty as to the extent of the patent).

5. Drafting and preparation of electronics and software patent applications

5.1 General matters

The description is the backbone of any patent application. The description must fulfil the formal requirements set out in the EPC and disclose the invention sufficiently clearly and completely so that the invention can be carried out by a person skilled in the art.[43] Additionally the description must provide support for the claims of the application.[44] When drafting an application it is also important to bear in mind that the application may require amendment and ensure that the description provides support for such amendment as it is not permissible to extend the content of an application beyond the content of the application as originally filed.[45]

The basic structure of a description of a European patent application is prescribed by Rule 42(1) of the EPC which requires that the description shall:

- specify the technical field to which the invention relates;
- indicate the background art which, as far as is known to the applicant, can be regarded as useful to understand the invention, draw up the European search report and examine the European patent application, and, preferably, cite the documents reflecting such art;
- disclose the invention, as claimed, in such terms that the technical problem, even if not expressly stated as such, and its solution can be understood, and state any advantageous effects of the invention with reference to the background art;

43 Article 83 EPC.
44 Article 84 EPC.
45 Article 123(2) EPC.

- briefly describe the figures in the drawings, if any;
- describe in detail at least one way of carrying out the invention claimed, using examples where appropriate and referring to the drawings, if any; and
- indicate explicitly, when it is not obvious from the description, or nature of the invention, the way in which the invention is industrially applicable.

Rule 42(2) of the EPC then goes on to provide that the description will normally be presented in the manner and order specified in Rule 42(1) of the EPC, unless a different ordering would lead to a better understanding or make the description more concise.

Perhaps, however, the most important aspect of drafting any patent application is the need to bear in mind the fact that a patent application is intended to be a persuasive document. During prosecution it should be such to persuade an examiner that a claimed invention is novel and inventive and hence suitable for being granted as a patent. Having been granted, a patent should hopefully on its face persuade competitors of its own validity and hence persuade competitors of the need to obtain a licence or avoid the scope of its claims. Finally, if a patent is ever enforced in the courts, the patent should be drafted so as to persuade a judge that the patent is valid and its claims are infringed.

5.2 Content of an application

The drafting of the introductory portion of a patent specification is particularly important in achieving these aims. Although formally the introduction is intended only to provide background to the invention and the background art, such an introduction provides the best opportunity to indicate the advantages that a claimed invention achieves and to present the invention as being a non-obvious departure from the prior art. In doing so the introduction can provide an indication of the 'technical effects' a claimed invention achieves.

Subsequent description of specific embodiments can then build on the groundwork established by the introductory portion of the specification. Here, in particular, it is important to address the specific issues which potentially arise in the realm of computer-implemented inventions, namely issue of patentable subject matter, basis for amendments and inventive step.

In the case of patentable subject matter, it is vitally important to ensure that any invention is described in concrete terms. This is best achieved by the description of any software-based invention in terms of processing undertaken by the computer itself rather than any human interaction with the computer. It is also important to avoid the abstract terminology which is sometimes beloved by computer programmers which suggest that an invention is somehow removed from reality. Although there is no problem referring to 'virtual memory' or 'virtual machines', it should be made clear that ultimately such virtual elements do have a physical reality as functional blocks of programming code or data storage in a physical machine.

Basis for amendments and inventive step largely revolve around describing the various aspects of an invention at an appropriate level of detail. Any new aspects of an invention can typically be described at three different levels of abstraction:

- structure;
- function; and
- advantage.

If an element is described solely in terms of structure, problems can arise when attempting to generalise a described structure to obtain claim language which covers alternative structures which achieve the same result. Conversely, describing an invention solely in terms of the advantages achieved results in a description being too abstract and claim language derived from such description is open to objection on the grounds that it claims a result to be achieved rather than the means for achieving such a result.

For most software inventions a balance is achieved by claiming an invention in terms of the functions which are to be performed by the various functional elements which make up an invention. The structural descriptions then provide a firm basis for such elements having a concrete existence, whereas references to the benefits and advantages which flow provide basis for arguments on inventive step.

In terms of structure, when describing specific embodiments, usually software inventions lend themselves to a top-down approach in which an invention is broken down into a number of functional modules the substructures and functions of which are then described in progressively more detail. Typically, such description involves the representation of an invention in terms of functional block diagrams and processing in terms of flow diagrams such as is illustrated in Figures 1 and 2 on the next page.

Figures 1 and 2. Two exemplary block diagrams showing the functional modules implementing a software invention with Figure 2 providing additional detail on the structures of the Re-gridding module of the system shown in Figure 1.

Figure 1. Overall structure of a number of interacting functional modules

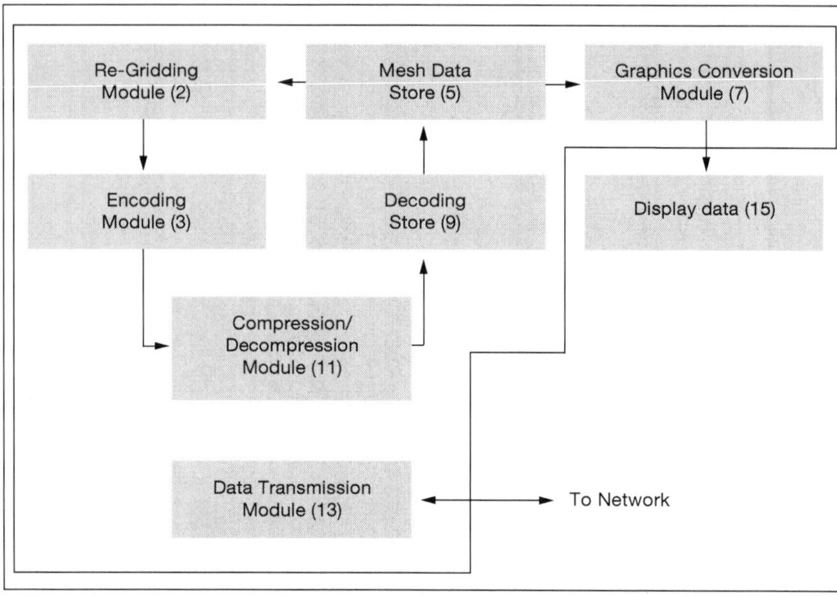

Figure 2. Modules in greater detail

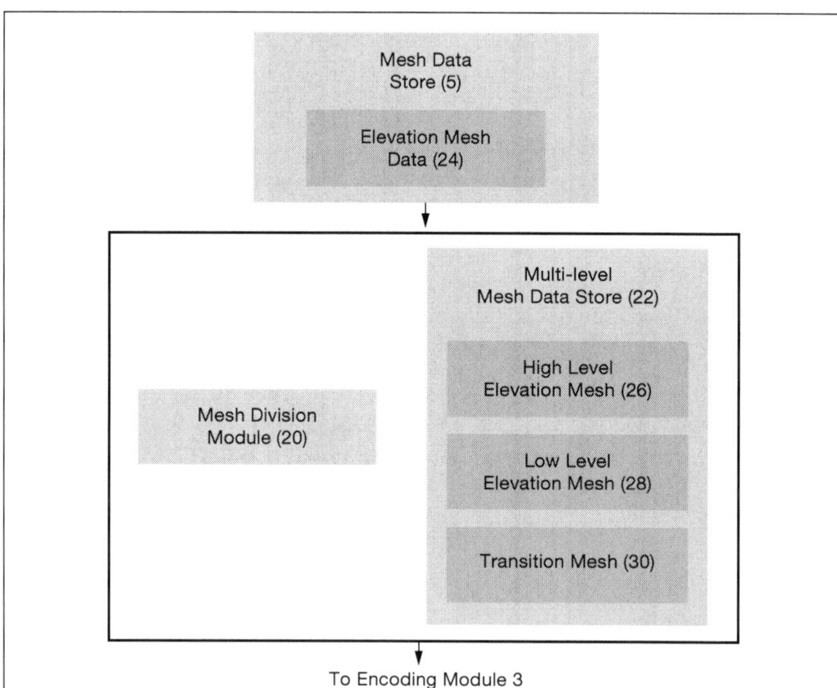

Figure 3. An exemplary flow diagram of the processing of a software invention

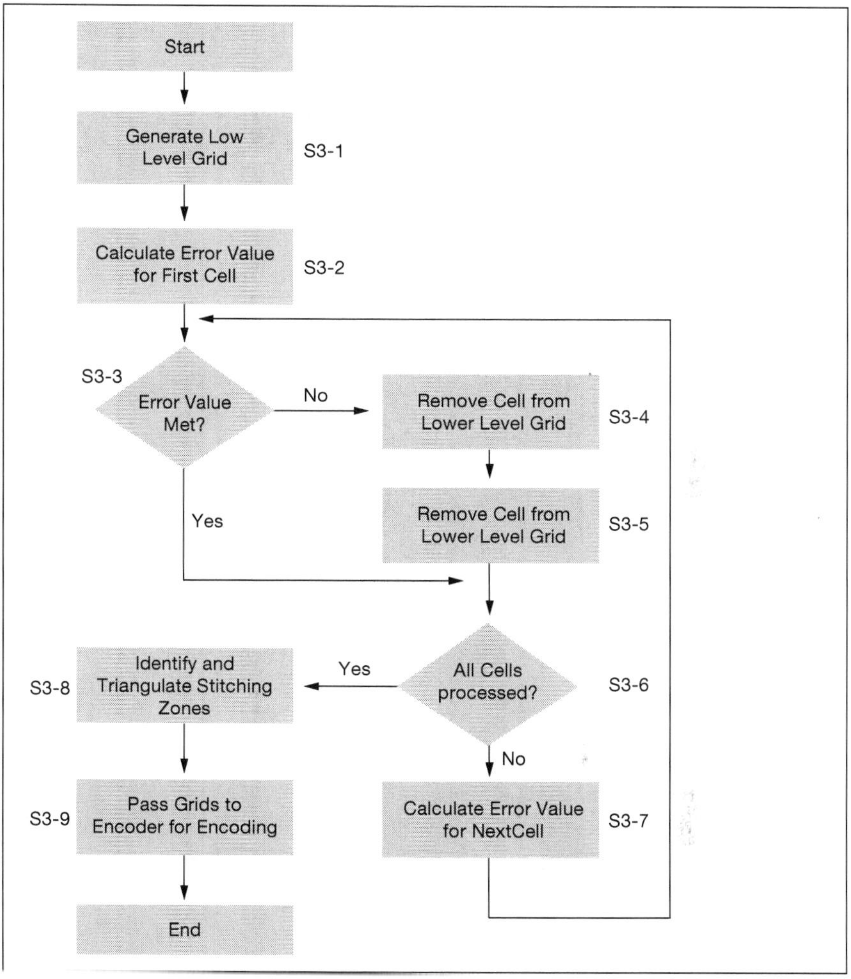

In the examples illustrated above, Figure 1 illustrates the overall structure of a number of interacting functional modules, whereas Figure 2 illustrates a selection of the modules in greater detail enabling the interaction and data processing of those modules to be discussed in greater detail. Processing undertaken by a software invention is normally best illustrated with reference to a flow diagram such as is shown as Figure 3 above. In the case of the above example, the flow diagram illustrates the processing of the re-gridding module of Figure 1, the structure of which is shown in greater detail in Figure 2.

Typically, most software inventions involve modifications to a limited number of software modules compared with the prior art. Having drafted a specific embodiment directed to novel implementations of such modules, it is important to consider and describe alternative ways of achieving the same effects to ensure that patent claims are supported across their entire scope. Such consideration is particularly important for software inventions given the flexibility inherent in software which means that patent claims which are too closely limited to a specific implementation can be easily designed around. It is also important to indicate in the description alternative aspects of the invention outside of the points of novelty to demonstrate the fact that such elements are not essential features of a claimed invention.

5.3 An illustrative example

The principles outlined above can be best understood with reference to a concrete example. WO2012/049471 is an example of a patent application for a computer-implemented invention developed by Visual Technology Services Limited and is the patent application in which the above illustrations originally appear.[46] The claimed invention concerns a particular method of encoding and compressing data representing elevation mesh arrays, that is data defining a regular uniform X-Y grid with a variable Z (height value). Such elevation mesh arrays are frequently used to represent surfaces generated based on LiDAR, Sonar, GIS, Geology and Geophysics data.

The invention in WO2012/049471 is based on two developments. First, the inventors appreciated that when encoding data representing an elevation mesh, traversing the regular X-Y grid in a regular pattern when encoding the mesh would generate data having large numbers of repeating elements making the data particularly susceptible to data compression.[47] Second, further data compression could be achieved by reducing the numbers of triangles used to represent a surface and hence which need to be represented, provided such a reduction was achieved without destroying the regularity present in the original grid structure.

The introduction to the application was then prepared with a view to emphasising the distinctions between the claimed invention and the background art. To that end, the description of the background art referred to conventional approaches to decimating triangle grid arrays which lead to the generation of irregular triangle grids and conventional approaches to encoding and compressing triangle grids and then identified the general aim of improving levels of data compression. From such an introduction it was apparent that modifying the decimation approach or the encoding or compression techniques could potentially achieve such a result. However,

---

46 The author would expressly like to thank Visual Technology Services Limited for their permission to use their patent application for the purposes of demonstrating the drafting of a software patent.

47 This is because the encoded data would involve representing the regular spacing of original X-Y grid.

Figures 4, 5 and 6. Exemplary illustrations of the processing of data to illustrate the functioning of a software invention

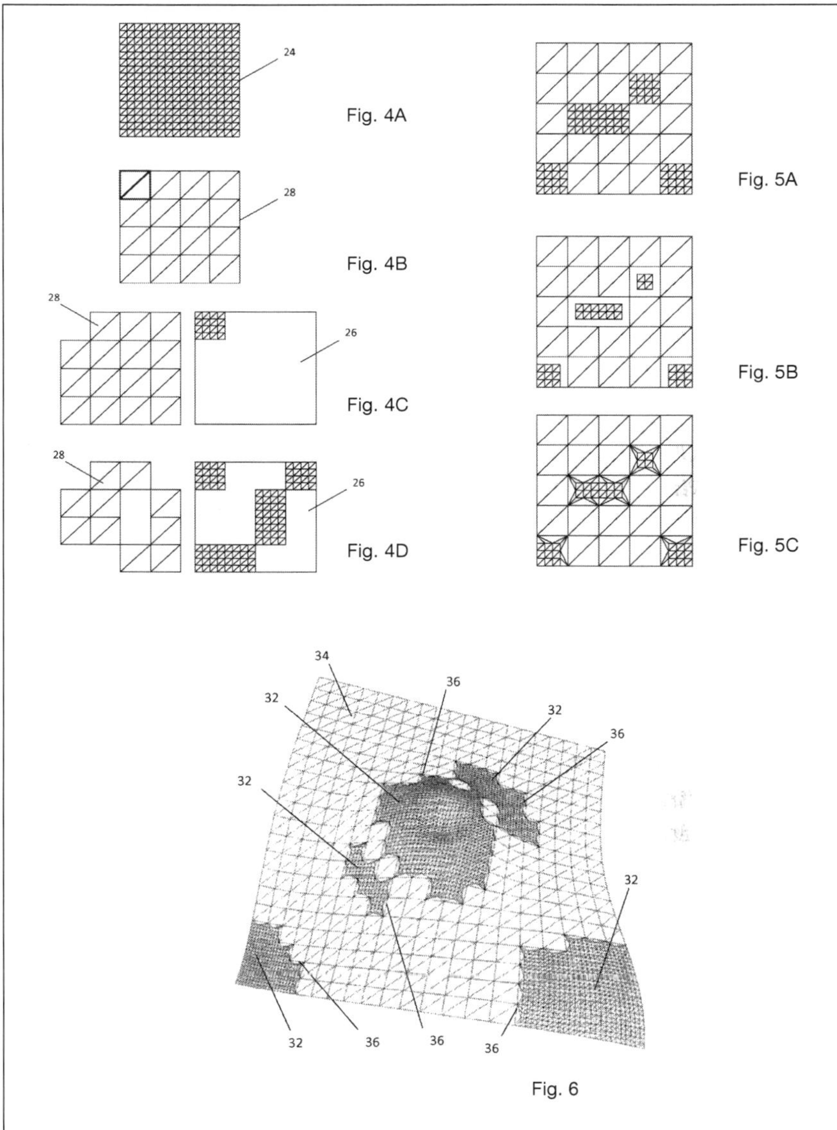

nowhere in the introduction was there any suggestion as to what modifications might be successful.

Turning then to the specific embodiment, the claimed invention was then described in terms of a block diagram incorporating a re-gridding module and an encoding module which enabled highly compressed representation of an elevation mesh to be achieved together with an outline of the processing which achieved the compression. By describing the invention in that way it should then be apparent which aspects of the described system were claimed to be inventive and what technical benefits were claimed to be achieved.

The processing undertaken by the different modules is then described referring to the illustrated flow diagrams. When such data involves the processing of data sets, it is frequently helpful to provide graphical illustrations of the data being processed. The following are examples taken from WO2012/049471.

In the case of the processing illustrated above, the illustrations of figures 4A–D and 5A–C correspond to the various steps in the flow diagram of Figure 3 reproduced above. Figure 6 then illustrates a grid after the processing undertaken by the re-gridding module. In the case of Figure 6, it should be apparent that the re-gridding module has generated a modified grid which retains a regular grid pattern but where different portions of the grid are represented by grids of different granularities.

The selected figures reproduced above illustrate a further aspect of drafting patent applications for software inventions in that, in the case of well-drafted applications, the nature of the processing and the invention should be evident from the drawings themselves. Thus, for example, in the case of the illustrations above, looking at Figures 1 to 6 it should be apparent that some kind of software module present within a larger system undertakes some kind of processing which converts data into the final structure illustrated in Figure 6. The detail of that processing and the benefits which arise from the conversion should then be evident from the description itself.

5.4 Claims

Drafting of patent claims is reliant upon a proper analysis of the essential features of an invention. The claims can then be drafted in a way which is clear and concise, including all essential features of an invention but without including any unnecessary limitations.

Under Rule 43(2) of the EPC, a European patent application is not normally allowed to include multiple independent claims in the same claim categories. Typically, this means that in the case of computer-implemented inventions an application will only include a single independent apparatus claim, a single independent method claim and a single independent claim to a computer-readable media. However, exceptions are permitted in the case of inventions involving a plurality of interrelated products (eg, transmitter and receiver), different uses of the same apparatus or alternative solutions to a particular problem where it is not possible to cover the alternatives in the same claim. The latter exception will, however, be construed narrowly and if at all possible multiple independent claims in the same category should be avoided as they are likely to lead to objections from the EPO.

The EPO prefers independent claims to be drafted in two-part form with a pre-characterising portion identifying the combination of features known from the prior art and a characterising portion identifying the elements introduced by the claimed invention.[48] The use of such a two-part form is frequently requested by examiners during prosecution. However, the two-part form is not obligatory and can be dispensed with if it is not appropriate for the invention being claimed.

Dependent claims should then be drafted providing fall-back positions in case the main claims are found to be invalid. The EPO charges excess claims fees of €225 for each claim in excess of 15 and €555 for each claim in excess of 50. In view of these fees it is generally most cost effective to limit claim numbers to around 15. If necessary this can be achieved by only including either dependent apparatus claims or dependent method claims, but not both, in the application on filing. If restrictions are required during prosecution the presence of the dependent claims will then normally provide the basis for making corresponding amendments to the other independent claims. This also remains the case in post-grant limitation proceedings.[49]

5.5 Conclusions

Frequently it is poor drafting rather than inherent lack of patentability which causes a patent application to fail.[50] Conversely, well-drafted patent applications are significantly easier to process through to grant at the EPO and result in stronger patent protection for an underlying invention.

Enforcing patents inevitably involves a comparison between patent claims and an alleged infringement the details of which will not have been known when a patent application is drafted. As a result, there is always room for argument about whether or not any particular infringement actually falls within the scope of a claim and considerable time, effort and expense is incurred considering such arguments during litigation.

Often many such problems can be avoided where sufficient time and analysis is spent drafting an application. If insufficient analysis is undertaken at an initial stage then the claims on filing may not adequately capture an invention. Similarly, unless a patent application is drafted with a view to demonstrating that a claimed invention provides an innovative technical solution to a technical problem and provides adequate support for subsequent amendments, prosecution is hampered and an applicant may be forced into making amendments which make enforcing a granted patent difficult.

---

48 Rule 43(1) EPC.

49 Articles 105a–c EPC.

50 A point well demonstrated in Keith Beresford's book, *Patenting Software Under the European Patent Convention*, Sweet & Maxwell, London (2000) which contains an exhaustive analysis of European patent applications where the computer program objection was considered by the Boards of Appeal prior to 2000.

Undoubtedly, there is a significant element of cost-benefit analysis required in the drafting of patent applications, particularly in the case of companies that file for large numbers of patents. If such patents are solely to be used in licensing negotiations where it is primarily the volume of patents which counts, then it is sufficient that applications are drafted merely to try to achieve the greatest number of patents granted. This can mean that quantity becomes more important that quality. However, if it is intended that a patent should be such to survive the scrutiny that is part of enforcement in contentious proceedings then extra time spent on the drafting and preparation of an application is likely to pay significant dividends later on.

The publisher acknowledges Scott Parker for his contribution to this chapter in the previous edition.

Enforcement in Europe

Nicholas Fox

1. The European patent litigation system

As has been explained in the "Patents in Europe" chapter, the European Patent Convention (EPC) establishes the European Patent Office (EPO) as the granting body for European patents. Once granted, a European patent is to be treated in each of the designated countries as having "the effect of and be subject to the same conditions as a national patent granted by that state".[1] Thus a European patent is often equated to a bundle of 'equivalent', or 'parallel', national patents, granting the patentee a monopoly right in respect of the invention in question in that particular designated EPC signatory state.[2,3]

Patents (whether national or European) have effect only within the borders of the designated country. Accordingly, if a company wants to 'clear the way' of a commercial rival's blocking patents to launch a new product in a number of European countries it may have to bring proceedings in those jurisdictions to revoke the patents registered there. Similarly, if a potentially infringing product is launched across Europe then infringement suits will need to be brought in the jurisdictions of the patents infringed. This provides litigants with the opportunity to 'forum shop', that is to choose to litigate in the court(s) that best suit that party's commercial objectives.

In practice, for reasons of cost and efficiency, patents are not litigated in every country protected in Europe. However, it is not uncommon for patents for important commercial products to be litigated in, say, three or four jurisdictions, on the basis that once the relative strengths of the parties' positions have been tested in these countries a settlement will be reached that will cover the others. The legal, procedural and strategic issues that guide the selection of the countries in which to bring proceedings are of course fundamental to successful litigation. Relevant factors may include:

- the specialist patent expertise of the court;

1 Article 2 EPC.
2 Note that the EPC signatory states are not limited to the European Union but also include, for example, Switzerland, Norway and Turkey. As at April 2019 there are 38 EPC signatory states and two extension states that recognise European patents upon request, namely Bosnia and Herzegovina, and Montenegro.
3 However, it should be noted that post-grant amendments at a national level may result in equivalent national designations of a European patent having different claims.

- the precedential value of a decision;
- the rigour of the procedure (including the ability to obtain disclosure and cross-examine witnesses);
- the value of being able to separate infringement and validity (as in the bifurcated systems of Germany and Austria);
- the availability of interim remedies (including injunctions);
- application of the law;
- the perception of how 'patent friendly' a particular court may be; and
- speed and cost (including the ability to recover legal costs from the other side if successful) (see the table below).

It is most common for patent litigation to be brought in one or more of the following countries, as the court systems in these are seen as being the most sophisticated in determining patent matters: Germany, the United Kingdom and the Netherlands.

The following table[4] summarises the volume of patent litigation in the larger jurisdictions in Europe.

Table 1. Patent litigation in Europe

	Average number of cases per year	Percentage involving European patents	Parallel actions	Proportion of pharmaceutical litigation
Germany	725	40%	55	Low
France	110	40%	5	Low
United Kingdom	80	80%	30	High
Italy	70	Unknown	Unknown	High
Netherlands	50	70%	10	Average

Direct comparisons with case numbers across different jurisdictions are difficult due to differences in national requirements as to what constitutes an

4 Figures taken from N Fox, "Brevets san Frontièrs: How Much Litigation Will Actually Take Place in the Unified Patent Court?" (2018) EIPR 85.

individual case.[5] However, by any measure, the number of patent actions brought in Germany clearly dominates the European litigation field.

Outside of the above major jurisdictions the volume of cases is low. Figures provided to the European Commission[6] indicated that, in 2005, 30–50 cases per annum were heard in Sweden, 25 infringement cases and 17 validity cases were heard in Austria, around 30 cases were heard in Belgium and Switzerland, 20 cases were heard in Portugal and Poland and around 10–15 cases were heard in Denmark. Other countries reported that very few or no patent infringement cases arose in their courts.

2. Harmonisation in Europe

2.1 Common foundations

The European Patent Convention (EPC) and the Community Patent Convention (CPC) have largely harmonised patent law throughout Europe. The EPC provides a uniform definition for the requirements for obtaining a patent[7] and the scope of protection that a patent confers[8] as well as defining the grounds on which a patent may be challenged.[9]

In contrast to the PC which was signed in 1973 and came into force in 1977, the CPC has never formally been ratified and has not come into force. However, in preparation for a community patent system every member state of the European Union has modified their patent law to be in conformity with the draft convention.

Direct patent infringement is defined by Article 25 of the CPC providing a patent proprietor with an exclusive right to make, offer to put on the market, import or stock products according to a patent or which are the direct product of a patented process.[10] Article 26 of the CPC then proceeds to extend patent protection to prevent the supply or offer to supply of 'essential elements' of an invention. A series of exemptions including exemptions for private non-commercial use, experimental purposes and infringement by visiting aircraft and ships is then provided for in Article 27 of the CPC.

2.2 European case law

As substantive patent law has been harmonised throughout Europe it should be unsurprising that the European national courts place significant weight on each other's case law and the case law of the EPO when interpreting the law.

5 For example, whether multiple parties and multiple patents can be combined in a single action or whether they must be filed as separate actions as well as whether or not actions for preliminary relief are counted as separate actions to the main litigation etc.

6 "Towards an Enhanced Patent Litigation System and a Community Patent – How to Take Discussions Further", Council document WD 11622/07 PI 135 (12 July 2007).

7 Articles 52–56 EPC.

8 Article 69 EPC.

9 Article 138 EPC.

10 Discussed in more detail in the "Patents in Europe" chapter.

Thus, for example, the UK Court of Appeal has stated:[11]

Advocates should recognise that where a point of patent law of general importance, such as the construction of a provision which by Treaty (either the EPC or the Community Patent Convention) is to be implemented by states parties to those conventions, has been decided by a court, particularly a higher court, of another member state, the decision matters here. For, despite the fact that there is no common ultimate patent court for Europe, it is of obvious importance to all the countries of the European Patent Union or the parties to the Community Patent Convention ('the CPC'), that as far as possible the same legal rules apply across all the countries where the provisions of the Conventions have been implemented. An important decision in one member state may well be of strong persuasive value in all the others, particularly where the judgment contains clear reasoning on the point.

In continental Europe which follows a civil law tradition, the acceptance of case law from other European jurisdictions and from the EPO generally passes without comment. This is in part because there is far less reliance on case law in such countries compared with common law countries such as the UK and Ireland. It is also because in the civil law systems any such case law is accepted at face value as expert consideration of the relevant law by a relevant expert tribunal to be used as such by the courts.

Acceptance of foreign case law is more problematic in the common law courts, which in the context of European patent litigation normally means the English courts. As the English courts apply a strict doctrine of precedent, foreign jurisprudence and EPO case law is only considered as being persuasive rather than binding authority. Although not binding authority, as noted by the Court of Appeal in *Grimme* quoted above, it is important to bring European case law to the Court's attention as in the absence of English authority it is highly likely that the court will follow the European lead.

A good example of this in the software field was the decision of Kitchin J in Astron Clinica's Application [2008] EWHC 85 (Pat) where having established that there were no relevant UK authorities on the matter, the judge decided to permit the grant of claims to an inventive computer program on a computer-readable medium citing support of an earlier decision of the German *Bundesgerichtshof* – the German Federal Supreme Court – *in Suche fehlerhafter Zeichenketten.*[12]

Similarly, in *Grimme* itself, the English Court of Appeal had to consider the correct application of Section 60(2) of the Patents Act 1977 on contributory infringement.[13] To do so, the judges consulted with their colleagues in Germany and the Netherlands on their national equivalents to Section 60(2) (and they

11 *Grimme Maschinenfabrik GmbH & Co KG v Derek Scott* [2010] EWCA Civ 1110.
12 Case No XZB 16/00; [2002] IIC 753.

looked at the language of the French equivalent). The German cases[14] were particularly influential in the court's decision that an infringement under Section 60(2) requires that the supplier of the means of carrying out a patented process knows, or it is obvious, that "the ultimate users will intend to do acts amounting to infringement".

However, where there is any relevant English case law, the English courts are obliged to follow that case law rather than European authorities. As is commented elsewhere in this book, this has led to significant departures between the English and European courts on some fundamental matters such as the patentability of computer software and inventive step.[15] Although there has been a softening of attitudes in the English courts – most notably in *Actavis UK Limited v Merck and Co Inc* where the Court of Appeal made it clear that where the Boards of Appeal of the EPO have formed a settled view of the application of an EPC provision the national courts should follow them[16] – these differences of approach remain.[17]

2.3 Exporting judgments

Most litigants in multi-national patent litigation are interested in using favourable judgments in one jurisdiction to leverage their position in another. However, as each of the European jurisdictions is entirely independent, the extent to which this is achieved is somewhat mixed.

Certainly, no formal estoppels arise merely because a party is successful in one court which would be binding on another. Rather, as was stated by the German Federal Supreme Court in its 2010 decision *Walzenformgebungsmaschine*[18] (*Cylinder-form shaping machine*), the courts feel obliged to "take into account decisions rendered by the EPO on the same questions and have to discuss their arguments; although not doing so, will not necessarily render the decision erroneous and appealable". Thus it is the case that the courts in one jurisdiction will consider seriously decisions in parallel cases. But ultimately the courts will make a decision based on the facts before them and, if necessary, will explain their reasons for reaching a different conclusion to another tribunal.

One example where a common approach was successful was the *Occlutech* series of contact lens cases in the English, German and Dutch courts. These

---

13 Under Section 60(2), a patent for an invention is infringed by a person who supplies or offers to supply the means, relating to an essential element of that invention, for putting an infringing invention into effect, when he or she knows or it is obvious that those means are suitable for putting, and intended to put, that invention into effect in the United Kingdom.

14 In particular, the *Bundesgerichtshof* case of *Deckenheizung* BGH X ZR 153/03.

15 Discussed in detail in the "Patents in Europe" and "United Kingdom" chapters.

16 [2008] EWCA Civ 444.

17 See, for example, P England, "Towards a Single, Pan-European Standard, Parts I, II and III" (2010) EIPR 195, 259 and 467 and N Fox and W Corbett, "UK and EPO Approaches to Excluded Subject-Matter and Inventive Step: Are *Aerotel* and *Pozzoli* Heading for the Rocks?" (2014) EIPR 27.

18 Case No Xa ZB 10/09, 15 April 2010.

cases concerned whether Occlutech infringed a patented medical device used for the occlusion of blood vessels. This consisted of a metal fabric formed of braided metal strands, clamped at the opposite ends of the device. Occlutech's allegedly infringing articles also used metal wire mesh, but with metal strands secured by welding at one end instead of both.

The English Court of Appeal[19] found that the skilled addressee would read the words 'opposite' and 'clamps' as meaning that the patent intended clamping at opposite ends of the device. To come to this conclusion, the court reiterated that a patent specification, while operating as a contextual aid to construction, is not necessarily determinative of the scope of the claims.[20] The language of the claims is important to decide whether they were intended to encompass all the ideas expressed in the teaching. The English Court of Appeal noted that the District Court of the Hague had also ruled in favour of Occlutech – clamping at both ends was an essential element of the claim; a decision later upheld by the Dutch Court of Appeal.

In Germany, the *Bundesgerichtshof* (the Federal Court of Justice) has also held that the devices in question did not infringe, either literally or as equivalents.[21] In doing so, and apparently echoing the English court's reasoning, this court held that the description is important only to the extent that it explains the subject matter of the claim. Where there is a contradiction between the description and the claim, the description cannot be used to correct the language of the claim.[22] Similarly, an infringement by equivalence would require the person skilled in the art to find the equivalent solution on the basis of the claim.

2.4 Procedural cooperation

There are also signs that judges are generally willing to cooperate across borders on procedural as well as substantive matters.

For instance, in the *Human Genome Sciences v Eli Lilly & Co* litigation, the English Court of Appeal asked an EPO Board of Appeal to expedite its appeal, and delayed the hearing date for the English appeal to ensure that the EPO Board would have a published decision before the English appeal was heard. The English court's reasoning for this was that if the Board of Appeal found the patent to be invalid, then it would be revoked across all the European states for which it is designated, and a costly English appeal hearing would be unnecessary. Conversely, if the Board of Appeal found the patent to be valid, its reasoning would be of interest to the English court.

Shortly after *Eli Lilly*, the Court of Appeal took similar steps in *Molnlycke*

---

19 *Occlutech GmbH v AGA Medical Corporation & Ors* [2010] EWCA Civ 702.

20 *Kirin-Amgen Inc v Hoechst Marion Roussel Ltd* [2005] RPC 9; *Virgin Atlantic Airways Ltd v Premium Aircraft Interiors UK Ltd* [2009] EWCA Civ 1062.

21 Case No X ZR 16/09.

22 See also the decision of the *Bundespatentgericht in Deutsche Telecom AG v The Patent GmbH*, BPaEG 253.

Health Care v BSN Medical Limited.[23] Jacob LJ could not determine on the papers in front of him whether a Swedish court had already been seised of the same issue as in the English proceedings. His solution was simply to contact the Swedish judge to ask.

An example in the electronics field arose in *HTC Europe Co Ltd v Apple Inc.*[24] In that case HTC succeeded in getting actions concerning the infringement and validity of various Apple patents expedited on the basis that "business needs to know where it stands" in relation to possible infringing activity in the telecommunications sector. However, HTC wanted to get a validity decision on one of the patents moved even earlier than this, because it thought an invalidity finding would persuade the Mannheim court to stay infringement proceedings on the equivalent German patent until validity could be heard in the *Bundespatentgericht* (the German Federal Patent Court). While Arnold J was prepared in principle to take the German proceedings into account, this particular attempt failed because of insufficient information before the court about the German procedural timetable.

2.5 Possibilities for conflict

Notwithstanding that the underlying law governing European patents is the same, the way in which this is applied by national judges can differ. Furthermore, the procedural differences[25] mean that under the current system conflicting decisions can arise in different member states.

The classic case of a different application of the law is *Improver Corporation v Remington Consumer Products.*[26] Here, the English House of Lords differed with both Dutch and German courts in its application of the Protocol to EPC, Article 69 to determine the scope of a patent claim, by finding that a claim covering the popular Epilady hair removing device was not infringed by the defendant's Smooth & Silky. In doing so the English judges took a stricter approach than that of the German and Dutch courts, who held that there is equivalence between the patentee's coiled spring mechanism for pulling out hair and the defendant's rubber tube with slits. No such 'doctrine of equivalence' is followed by the English courts.

A more recent example of diverging opinions between national courts arises from the 'patent wars' litigation relating to smartphone devices. Here, a large number of both standard essential and feature mobile phone patents is being litigated in parallel actions in numerous courts around the world. This gives rise to the possibility of different national courts giving inconsistent decisions, some of which have already become evident.

---

23 [2010] EWCA Civ 988.
24 [2011] EWHC 2396 (Pat).
25 See section 3, below.
26 [1990] FSR 181.

For example, three Apple patents have been the subject of validity decisions in both the District Court of the Hague (in a preliminary assessment only) and in the English Patents Court (first instance judgment); and a further validity-related injunction action has been heard before the Munich I Landsgericht. These decisions are shown in Table 2 where it can be seen that while the English court has found the '022 patent invalid, and the Hague court has also found it invalid on a preliminary assessment, the German Regional Court did not consider that it had a sufficiently high probability of it being invalid to stay an infringement action. As regards the '948 patent the English court differs with the Dutch court, and on '868 both agree that the patent is valid.

3. Key procedural differences

Differences of application and procedure between countries can have a large impact, giving rise to varying and inconsistent decisions between EPC countries. The fundamental dividing line is between those EPC signatory countries that adopt a civil law procedural approach and those that operate common law systems. While both systems are adversarial, the common law jurisdictions (ie, the United Kingdom and Ireland) typically use heavy disclosure of documents and long and detailed, oral cross-examination of witnesses at trial. Although there are exceptions, such as the sometimes lengthy hearing of experts and witnesses in Denmark, the civil approach of most other EPC countries places less reliance on general disclosure and oral evidence. In these countries, the emphasis is instead on conducting proceedings in writing. Further, the reliance on expert evidence (which is often a substantial and critical feature of UK and Dutch proceedings, for example) also varies in importance and application between jurisdictions. Furthermore, some jurisdictions, such as the Netherlands, use 'file wrapper' estoppel, taking into account amendments or even explanatory remarks made to a patent in the EPO, when deciding how to construe the patent.

Fact-finding measures also vary. For example, in France, the claimant may obtain an *ex parte 'saisie contrefaçon'* for the seizure of allegedly infringing material together with related documents, and later, during the proceedings, on the basis of the right of information. In Italy, the *descrizione* may allow the *ex parte* obtaining of a complete snapshot of a defendant's allegedly infringing activities, including insight into the accounts. In the Netherlands, disclosure orders and seizure orders are available.

These differences in evidential procedure can mean that, even when different national courts are dealing with the validity and/or infringement of equivalent European patents, each individual national court can be presented with a very different type, quality and amount of evidence relating to the same facts.

In Germany and Austria, proceedings are bifurcated between infringement

Table 2. Apple patents cases

European patent in dispute	**English Patents Court**	**District Court of the Hague (preliminary assessment only)**	**Munich I Landsgericht (consideration of whether high probability of invalidity only[27])**
EP 2 964 022 – Unlocking a device by performing gestures on an unlock image (user interface on a touch screen device that allows the user to unlock a device by dragging an image over the screen). Otherwise known as the 'slide-to-unlock' feature.	Invalid	Invalid	Valid
EP 2 098 948 – a touch event model (computer devices capable of responding to more than one touch at the same time).	Invalid	Valid	–
EP 2 059 868 – portable electronic device for photo management (more intuitive user interfaces for photo management).	Valid	Valid	Invalid

27 Such as to stay infringement proceedings pending a decision of the *Bundespatentgericht* in separate validity proceedings.

and validity hearings, so that different courts, and often different legal teams, will deal with these two issues in respect of the same patent. In certain circumstances, the final hearing of the infringement proceedings (which is typically many months before the validity hearing) may be stayed pending a validity decision. However, the standard to which the defendant needs to show the likelihood of invalidity of the allegedly infringed patent varies according to the regional division in which the infringement proceedings are pending.

Different courts also take different approaches to determining the level of damages awarded in the event of a patent being infringed as well as how that level of damages is impacted by the knowledge or otherwise of a potential infringer.[28]

The experience and expertise of judges differs between different countries too. In some jurisdictions where fewer patent cases are heard, judges are generalists. In others, they are patent specialists, often with scientific or technical backgrounds.

The differences between national procedures can, of course, result in great variations in the cost of litigation between EPC countries. A complex mix of tactical, timing and costs issues needs to be weighed up when deciding where it is most advantageous to litigate a European patent in a given set of circumstances. The following table summarises some of the key procedural differences between the courts in the most significant jurisdictions for patent litigation in Europe.

The most important features and differences for patent litigation in France, Germany, Italy, the Netherlands and the United Kingdom are summarised in Table 3.

Table 3. Most important features and differences in patent litigation in France, Germany, the Netherlands and the United Kingdom

	France	Germany	Italy	Netherlands	United Kingdom
Time to conclusion of proceedings on the merits	18 months to two years	Infringement: 10 to 15 months, dependent on venue. Revocation: 15 to 18 months	Two to three years (varying greatly depending on the Chamber)	Less than one year	10 months to one year

continued on next page

28 See, eg, N Fox et al, "Accounting for Differences: Damages and Profits in European Patent Infringement" (2015) EIPR 566.

	France	Germany	Italy	Netherlands	United Kingdom
Time to substantive appeal proceedings	18 months to two years	Infringement: up to 18 months Revocation: two to four years	One to two years (depending on the Chamber)	Approxi-mately one year	10 months
Time to obtain interim relief	*Inter partes*: five to six weeks *Ex parte*: A shorter timeframe, but only granted in exceptional circum-stances	A few days (*ex parte* grant) to some weeks or months (when oral hearing is held)	*Inter partes*: up to six to eight months *Ex parte*: a few weeks (but rarely granted)	*Ex parte*: immediate	As little as a few days
Bifurcation of validity and infringement proceedings?	No	Yes	No	No	No
Automatic stays pending EPO decisions?	At the court's discretion except in certain cases	No	No. But courts can adjust the timeline of the action in order to take into account EPO develop-ments	No. Stays are a matter of the court's discretion	No. Stays are a matter of the court's discretion
Disclosure	No general disclosure, but the *saisie contrefaçon* is available for gathering evidence of infringe-ment	No general disclosure, but pre-action orders for inspection or presentation of documents are available	No general disclosure, but a pre-trial *descrizione* search order is available	No general disclosure, but disclosure orders and search orders are available, both pre-trial and during proceedings	Yes, general and specific disclosure, but restricted in patent cases

continued on next page

	France	Germany	Italy	Netherlands	United Kingdom
Oral evidence	No, documentary evidence only	Emphasis mostly on documentary evidence but oral evidence does feature	Oral hearings are rare, but witness testimony is allowed. Discussion hearings take place to discuss expert evidence and experts can be examined by the court	Emphasis on documentary evidence but oral evidence does feature	Yes – detailed and lengthy cross-examination in the Patents Court. Very restricted in the Patents County Court
Expert evidence	The court can appoint an expert. The expert's findings are set out in a written report submitted to the court	Court experts are rarely appointed at first instance, but often appointed by the parties to support their respective cases. Experts can be examined by the court	One or more court experts are always appointed. Each party also appoints its own experts to liaise with the court expert	Usually an independent expert appointed by each party. Court-appointed experts are rare, as are joint experts. Experts can be examined by the court	Usually one or more independent experts appointed by each party. Court-appointed experts are rare, as are joint experts
Availability of cross-border injunctions	Not in practice, but possible in principle further to the *Solvay v Honeywell* decision of the CJEU	Dependent on the court, but, in principle, yes	Recognised in principle but not granted in practice to date	In some circumstances, such as the *Solvay v Honeywell* case	Not to date, but now possible in principle further to the *Solvay v Honeywell* decision of the CJEU
File wrapper estoppel doctrine	No	No	No	Yes	No

continued on next page

	France	Germany	Italy	Netherlands	United Kingdom
Specialist intellectual property (IP) courts	Yes	Yes	Yes	Yes	Yes
Judges with technical knowledge	No	Only in validity proceedings	No	Yes	Yes
Costs (€)	50k to 200k	50k to 250k	200k to 400k	50k to 200k	200k to 2m
Damages	Low	Medium	Low	Medium	High

4. EPO opposition

Despite the fragmented nature of the European patent system along national lines, there are three situations that provide an exception to the need to enforce or challenge patents at a national level. These are explained in this and the following two sections, beginning with EPO oppositions.

The EPC provides that for a period of nine months following grant of a European patent the validity of that patent may be challenged in opposition proceedings at the EPO.[29] The result of a successful challenge is that the patent is invalidated in all designated states (in which case the decision of the EPO, once the right to appeal to a technical board of appeal has been exhausted, is final[30] – ie, the patent cannot be 'resurrected' in proceedings before the national courts). The alternative outcomes are that the patent may be upheld in amended form (typically narrowed in scope), or it may be maintained as granted.[31] In either case the patent may then be challenged at a national level. In contrast, for infringement matters the national courts have exclusive jurisdiction. The EPO will not hear infringement proceedings.

An opposition may be filed by 'any person' (save for the patentee).[32] There is no test for standing. Indeed, an opposition may be brought by a firm of patent attorneys (allowing their client to remain anonymous). The notice of opposition must contain:

- a statement of the extent to which the European patent is opposed (ie, which claims are being challenged);
- a statement of the grounds on which the opposition is based; and

---

29 Article 99 EPC.

30 Save for the possibility of judicial review by the Enlarged Board of Appeal under Article 112a EPC where it is alleged that the outcome of the opposition was influenced by a fundamental procedural defect.

31 According to the EPO's annual report for 2009 in approximately 44% of oppositions the patent was revoked, in 30% of oppositions the patent was maintained in amended form and in 26% of cases the opposition was rejected.

32 G9/93 Opposition by patent proprietor.

- an indication of the facts and evidence presented in support of these grounds.[33]

As for the grounds on which the opposition may be based, these are set out in Article 100 of the EPC:

- that the subject matter of the patent is not patentable within the terms of Articles 52 to of the 57 EPC (ie, novelty, inventive step, exceptions to patentability, lack of industrial application and 'non inventions');
- that the invention is not disclosed clearly and completely enough for a person skilled in the art to carry it out; and
- that the subject matter extends beyond the content of the application as filed.

An opposition will first be examined for its admissibility on procedural grounds. Admissible oppositions will then be the subject of substantive examination. The opposition procedure is *inter partes* and both patentee and opponent(s) have the opportunity to submit their written arguments and to argue their case at oral proceedings. Although expert declarations are allowed, there is no scope for cross-examination of the other party's expert. That said, provided that notification is given, an accompanying person (typically a scientist or inventor) may play an important role at an oral hearing. Another procedural point worth highlighting is that a patent may be amended during opposition in order to meet a ground for opposition. Indeed, it is fairly common for the patentee to use a cascading series of auxiliary requests as a means to advance alternative positions as to the scope of the claims sought (in a way that does not feature to the same extent in most national court systems).

It is inherent in the system that national proceedings will run in parallel with EPO oppositions. Given that the typical EPO opposition may last around three years (five or six years if the decision is appealed), this is not uncommon. Different countries take different approaches to the question of whether national proceedings should be stayed pending resolution at the EPO. Although in certain countries and in certain circumstances this may arise (ie, the national court will exercise its discretion to stay the proceedings), the general trend is towards continuing with the national and EPO proceedings in parallel. The principal exception is in Germany where a nullity action cannot be commenced until the opposition period has lapsed or EPO opposition proceedings have been concluded.

A major reason why courts are reluctant to stay national proceedings pending the outcome of an opposition is the length of this process. However, there are encouraging signs that the EPO is trying to respond to the criticism it

33 Rule 76(2)(c) EPC.

has received over the duration of opposition proceedings and also that the EPO and national courts are prepared to engage with each other and adopt a more flexible approach to timing in order to avoid potentially unnecessary duplication, as described above. The flip-side of the pressure on parties to request acceleration (where appropriate) and the reluctance of national courts to stay proceedings is that it is difficult to achieve a tactical delay.

The primary advantages of the opposition procedure are certainty – if successful the patent will be knocked out across Europe – and cost, which is low compared with litigation in any one of the major patent litigation jurisdictions.[34] On the other hand, one needs to act quickly in order to file an opposition during the nine-month period following grant and one needs to be prepared to wait for the opposition procedure to run its course or initiate national proceedings in parallel and accept the risk of inconsistent decisions and unnecessary costs (eg, if a patent is upheld by a national court only to be revoked or narrowed by the EPO).

5. Cross-border injunctions and 'torpedo' actions

Some national courts (most notably the Dutch and Italian courts) have been prepared to interpret EC Regulation 44/2001 (Brussels I) in such a way as to permit them to decide patent issues outside their own borders (eg, to grant cross-border injunctions against a single defendant or group of related defendants). In particular, the so-called 'Spider in the web' approach was originally developed and used by the Dutch courts, under Article 6(1). Typically, an injunction was issued by a Dutch court against a group of defendants based, and infringing, in European countries outside the Netherlands, provided that their business activities were organised from a central office domiciled in the Netherlands (ie, where there was perceived to be sufficient connection between the claims that it would be expedient to determine these actions together to avoid irreconcilable judgments in parallel proceedings). This is so notwithstanding that the European rules on jurisdiction provide that patent validity (which is almost always raised as a defence to infringement) can only be determined by the national court in the country to which that part of the patent relates.[35]

However, following the European Court of Justice (ECJ) (now the Court of Justice of the European Union (CJEU)) rulings in *GAT v LuK* and *Roche v Primus*[36] in 2006, the practice of awarding cross-border injunctions became relatively rare. In *GAT v LuK* the ECJ held that the validity of a patent cannot be

34 Only rarely does the cost of an opposition exceed €100,000. By comparison, the cost of litigating a pharmaceutical patent in the UK High Court will cost at least £500,000 (with the cost of cases in other national courts to be added).

35 Article 22(4) of EC Regulation 44/2001.

36 Cases C-4/03 and C-539/03, respectively; both decided on 13 July 2006.

determined by the court in another jurisdiction and that this rule is mandatory and could not be derogated from. Indeed, where a court is seised of a claim that is primarily concerned with a matter over which the courts of another state have exclusive jurisdiction, the seised court is required to decline jurisdiction and as a consequence cross-border injunctions are not possible where validity is raised as a defence.

In *Roche v Primus* the ECJ held that where it is alleged that a European patent is infringed in multiple countries, infringement in all countries cannot be considered by a single court on the basis that there are multiple defendants within the same group (eg, sister companies) and the company domiciled in that country was responsible for determining a coordinated policy of carrying out the same allegedly infringing acts by the others. Nevertheless, in certain circumstances the Dutch courts demonstrated a continued willingness to grant cross-border injunctions, most notably leading to the decision in *Solvay SA v Honeywell Fluorine Products Europe BV*, which is discussed further below.[37]

In Italy, the historic willingness of the courts to seize jurisdiction for patent infringements across Europe combined with the slow pace of Italian litigation gave rise to the practice known as the 'Italian torpedo', in which a party fearing an infringement suit would file a claim for a declaration of non-infringement in Italy (or another slow jurisdiction such as Belgium). The intended effect of this is that if the patentee brings the same infringement dispute before the courts of another member state the court of that country will decline jurisdiction and stay the infringement proceedings under EC Regulation 44/2001. However, in recent years the force of the torpedo strategy has also diminished considerably, as a result of several factors including: courts in Germany and France ruling that in some circumstances a torpedo may be an abuse of law, the increasingly sceptical attitude of the Italian courts to the practice, the decision of several European courts (including in Italy) to decline jurisdiction on the grounds that no wrongful act had been committed in the member state of the court at issue and the *GAT v LuK* and *Roche v Primus* decisions mentioned above.

At the same time as the torpedo tactic has appeared to fall from favour in some countries, in the United Kingdom, where the prevailing view of practitioners is that such proceedings are considered an abuse by the courts, it has received some surprising support in the Court of Appeal. In *Research in Motion UK v Visto Corporation* [2008] EWCA Civ 153 Jacob LJ said:

> *... much ingenuity is expended on all this game playing. Despite the temptation to do otherwise, it is not easy to criticise the parties or their lawyers for this. They have to take the current system as it is and are entitled (and can only be expected) to jockey for what they conceive to be the best position from their or their client's point of view.*

37 District Court The Hague, the Netherlands, 22 December 2010, Case No 09-2275.

However, these comments were made *obiter* in a decision that concerned the application of Article 28 of EC Regulation 44/2001. The extent to which the court would follow this view in other cases is therefore uncertain.

The *Solvay v Honeywell* case referred to above concerns a number of parallel European patents for a method owned by Solvay. Solvay alleges that several of these patents are infringed by three Honeywell companies. One of these companies is Dutch and two are Belgian. In the course of the proceedings Solvay lodged an interim claim against the defendants in the main proceedings, seeking provisional relief in the form of a cross-border injunction against infringement for the duration of the main proceedings. Solvay argued that, because the Belgian defendants infringe in more than one country (and in the same countries as the Dutch defendant), all infringements should be dealt with in The Hague to avoid irreconcilable decisions between the courts of various jurisdictions (as under Article 6(1)).

Advocate General (AG) Cruz Villalón agreed with Solvay. In particular, the AG distinguished situations of the *Roche v Primus* kind from the present case because the infringements in *Roche* do not relate to the same acts carried out in the same member states, infringing the national parts of the same European patent. In *Solvay v Honeywell*, however, the same national patents were being infringed by the same products and the same parties, and so a collection of national decisions concerning infringements of each national patent could potentially result in conflicting decisions. The AG described this as "a more nuanced approach, carefully circumscribing the scope of the caselaw in *Roche*". The AG also dealt with the question of whether raising validity as an issue in interim proceedings required the court to pass jurisdiction back to the country in which each allegedly infringed patent is registered, by virtue of Article 22(4) and as indicated in *GAT v LuK*. Observing that part of the *raison d'être* of Article 22(4) is to avoid conflicting decisions, the AG nonetheless opined that courts seised of an interim application are not ruling on the merits or the existence of the infringement or on the validity of the patent; on the contrary, they normally confine themselves to consideration of whether the conditions for granting the interim measure sought have been fulfilled. Since there is no final decision on validity, there can be no conflicting decisions.

Following the AG's Opinion, the CJEU confirmed on 12 July 2012 that cross-border, interim injunctions are indeed permissible. In particular, the CJEU ruled that:

- Article 6(1) of Regulation 44/2001 must be interpreted as meaning that "a situation where two or more companies from different member states, ... are each separately accused of committing an infringement of the same national part of a European patent which is in force in yet another member state by virtue of their performance of reserved actions with regard to the same product, is capable of leading to 'irreconcilable judgments'". In such circumstances, cross-border interim relief is available; and

- in actions for infringement in which the invalidity of a European patent has been raised as a defence against the adoption of an *interim*, cross-border injunction, Article 22(4) Regulation 44/2001 does not apply. Such a defence cannot stand in the way of the adoption of a provisional measure concerning the cross-border prohibition against infringement.

Consequently, the cross-border interim injunction, much favoured by Dutch courts, is permissible to prevent "irreconcilable judgments" between EU national courts. However, where there is not a risk of "irreconcilable judgments" interim measures must be obtained in the courts of the country in which the infringement happens and where the European patent is registered (this will be the case, as exemplified by the facts of *Roche v Primus*, where each infringer's activities were strictly confined to a single, separate country and therefore informed separate national designations of the patent). The CJEU's ruling in *Solvay* therefore provides for wide forum shopping options for a claimant according to ease and speed of local injunction rules. In terms of time and cost, and until such time as the Unified Patent Litigation System (UPLS) is established (see the "Unified Patent Court" chapter), this is an attractive tool for claimants, because of its power to deal with a number of alleged infringements by multiple parties in one set of proceedings.

6. Other IPRs

6.1 Transnational rights

(a) Community design rights

The Community Designs Regulation[38] establishes Community-wide registered and unregistered design rights protecting the appearance of the whole or a part of a product resulting from the features of, in particular, the lines, contours, colours, shape, texture and/or materials of the product itself and/or its ornamentation.

The Community Designs Regulation provides for Community designs to be enforced in designated EU design courts.[39] Jurisdiction of such courts can be established either based on the domicile or country of establishment of a defendant, the domicile or country of establishment of a plaintiff, agreement between the parties or alternatively an act of infringement of a design in a particular jurisdiction.[40] Other than where litigation is based on the location of

38 Council Regulation (EC) No 6/2002.
39 Community Designs Regulation, Article 81.
40 Community Designs Regulation, Article 82.

infringement of a design, the regulation empowers a court to rule on infringement for the entire EU in a single action.

The already well-known example of a registered Community design being used in respect of alleged infringement of product shape is Apple's assertion of its Registered Design 000181607-0001 against a number of Samsung tablet devices in a number of European jurisdictions. In Germany, a number of preliminary injunctions were sought. Apple initially obtained a preliminary injunction against both German and Korean divisions of Samsung on a pan-EU basis, although the scope of this injunction was later reduced to Germany alone. However, the Dusseldorf Court of Appeal later found that it did have jurisdiction to issue interim pan-EU injunctions and did so in respect of the Tab 7.7 (but not the 10.1 and 8.9 Tabs as these were not found to infringe).[41]

In the Netherlands, Apple has failed to obtain a preliminary injunction against Samsung's Galaxy Tab 10.1. Furthermore, at the time of writing, in the first and only case where these issues have been tried on the merits in final proceedings, in the English Patents Court[42] Samsung sought and obtained a declaration of non-infringement of Apple's registered design in respect of the Tab 10.1, Tab 8.9 and Tab 7.7. Having pulled together a number of similarities and differences in the design features of the Samsung tablets and the Apple design, the court concluded that the Samsung products would have a different overall impression on the informed user.

(b) Community trademarks

An EU Community trademark system was established by the Community Trademarks Regulation (CTMR).[43] The Community trademark established by the CTMR is a unitary right governing the entire European Union. As with EU design rights, Community trademarks are enforced through the national courts and as was confirmed in the case of *DHL Express France SAS v Chronopost SA* (CJEU):[44] "the scope of the prohibition against further infringement or threatened infringement of a Community trademark ... extends, as a rule, to the entire area of the European Union". However, there is an important qualification to this general principle: "the acts or future acts of the defendant ... which do not affect the functions of the Community trademark, cannot therefore be prohibited". This means that if an act is not regarded as an infringement in a particular member state the scope of the injunction should be narrowed so as not to apply to that state. This would most commonly be the case where the allegedly infringing sign is not perceived as a trademark in a member state because of cultural differences such as language.

41 The jurisdiction is founded in Article 83(1) of Council Regulation (EC) No 6/2002 on Community Designs and applies to both registered and unregistered community designs.

42 *Samsung Electronics UK v Apple Inc* [2012] EWHC 1882 (Pat).

43 EU Regulation 207/2009

44 Case C-235/09, 12 April 2011.

6.2 Other rights

In addition to the unitary rights established by EU law, various national IP rights also exist and can be enforced via the European courts. Although in certain areas of law, in particular copyrights, semiconductor topographies and database rights there is a high level of harmonisation in law throughout Europe, differences in how such rights are enforced in different jurisdictions means that there is less scope for coordinated enforcement of such rights compared with patent litigation.

7. Border enforcement

7.1 Regulation 1383/2003/EC

Regulation 1383/2003/EC – the Counterfeit Goods Regulation – sets out the procedures for Customs authorities throughout the European Union to take action in respect of importation of goods believed to infringe particular IP rights. IP rights which the regulation covers include trademarks, design rights and patents. Pursuant to the Regulation, rights owners can record their rights with customs authorities for the purposes of having suspected counterfeit or pirated goods detained at EU ports of entry and for the rights owners to be notified of such detentions, for the purposes of taking action. Customs authorities can also initiate detentions of suspected counterfeit/pirated goods on their own account and thereafter contact rights owners.

In practice, while the procedures provided for by the Regulation are capable of working quite effectively to enable the detention of counterfeit products entering the European Union, the system is very far from perfect. In particular, the effectiveness and efficiency with which customs authorities in the different EU member states operate the system varies very significantly. The current system is also subject to the limitation that parallel imported goods from outside the community are not able to be detained by Customs, notwithstanding that dealings in those goods (eg, trademarked goods which have not been put on the market by the trademark proprietor or with its consent in the European Union) would amount to an infringement of the trademark proprietor's rights.

Following the European Commission's publication of its 'comprehensive' strategy for IP rights, the regime for Customs action against infringing IP rights (as with a number of the other matters dealt with in this chapter, eg, the Enforcement Directive) is in a state of flux. The Commission has published a new draft Regulation[45] which would replace the existing Regulation 1383/2003, if adopted. Changes to the existing system relevant to the electronics and

45 COM(711)285 dated 26 May 2011.

software sector which the new Regulation would bring (if adopted in its current form) include, in particular, extending the regime to enable detention of infringing parallel imported goods (ie, imports from outside the European Economic Area (EEA), introducing procedures to enable rights owners to call for Customs to carry out automatic destruction of small consignments of infringing goods without the need to revert to the rights owner on each occasion (this could potentially be very useful in view of the increasing use by infringers of multiple small consignments as a tactic to seek to minimise the risk of Customs detention) and changes to streamline procedures for destruction of goods in default of objection, as well as for the costs in relation to storage and destruction of small consignments of infringing goods to be borne by Customs authorities rather than rights owners. Most recently, developments in the new draft Regulation have been impacted by the decision of the CJEU in joined cases *Philips and Nokia*, in respect of goods in transit through the European Union from one non-EU country to another, as follows.

7.2 *Philips and Nokia*

The *Nokia* case involved Regulation 1383/2003 concerning customs actions against goods suspected of infringing certain IP rights and the measures to be taken against goods found to have infringed such rights. The *Philips* case, on the other hand, was referred in relation to the earlier Council Regulation (EC) No 3295/94 of 22 December 1994 (Regulation 3295/94), which was later replaced by Regulation 1383/2003. Both Regulations provide for initial action by Customs to detain potentially counterfeit goods followed by an opportunity for the rights holder to lodge an application for action. While there were a lot of similarities between the two cases, namely the detention by the customs authorities of goods in transit, there were some differences. *Nokia* involved trademarks and counterfeit phones, whereas *Philips* involved copyright, registered designs and counterfeit shavers. Furthermore, while *Nokia* focused on the lawfulness of the actions of the UK Customs authorities (which had suspended detention of the goods on the grounds that there was no 'actual' or 'real' infringement of the trademark claimed), *Philips* focused on whether or not it was essential to prove that the detained goods were to be traded in the European Union in order to reach a finding that the IP right allegedly infringed had indeed been infringed. *Philips* and *Nokia* were joined by the court because in both cases the references from the national courts posed questions about the interpretation of the EU legislation governing action by customs authorities against possible infringements of Community IP rights where goods were in the external transit procedure in accordance with Article 91(1)(a) of the Community Customs Code established by Council Regulation (EEC) 2913/92 of 12 October 1992, which "allows the movement from one point to another within the customs territory of the Community of ... non-Community goods, without such goods being subject to import duties and other charges or to commercial policy measures". According to case law, 'external transit' is based on a legal fiction, since the whole procedure unfolds as if the non-Community goods concerned had never entered the territory of a member state. The facts and background of the cases follow.

(a) ***Nokia***

In *Nokia*, Her Majesty's Commissioners of Revenue and Customs (HMRC) in the United Kingdom stopped a consignment of goods (comprised of mobile telephones, batteries, manuals, boxes and hands-free kits, each of which bore the trademark NOKIA) which had come from Hong Kong, destined for Colombia. HMRC sent Nokia samples of those goods and Nokia, after inspecting the samples, notified HMRC that the goods were counterfeit. HMRC then said it was uncertain how goods could be counterfeit within the meaning of Article 2(1)(a)(1) of Regulation 1383/2003 in the absence of any evidence that they might be diverted onto the EU market, and therefore concluded that, in the absence of such evidence, it was not lawful to deprive the owner of its goods. Nokia issued an application against HMRC, asking for the names and addresses of the consignor and the consignee together with any other relevant documents. The documents were sent, but Nokia could not identify the consignor or the consignee of the goods, concluding that they had both taken steps to hide their identity. Nokia commenced legal proceedings.

In his judgment in July 2009, Mr Justice Kitchin (as he then was) held that Regulation 1383/2003 did not entitle or require customs authorities to detain or seize counterfeit goods in transit where there was no evidence that the goods would be diverted on to the market in the member states because such goods were not counterfeit goods (under Article 2(1)(a)(i) of Regulation 1383/2003), without such evidence. The English court had given a literalistic approach to the interpretation of Regulation 1383/2003 which gave a result that Kitchin J recognised was not satisfactory. He said "I can only hope it provokes a review of the adequacy of the measures available to combat the international trade in fake goods by preventing their transhipment through Member States". On appeal, the Court of Appeal of England and Wales referred the following question to the Court of Justice for a preliminary ruling:

> *Are non-Community goods bearing a Community trademark which are subject to Customs supervision in a Member State and in transit from a non-Member State to another non-Member State capable of constituting 'counterfeit goods' within the meaning of Article 2(1)(a) of Regulation ... 1383/2003 if there is no evidence to suggest that those goods will be put on the market in the EC, either in conformity with a Customs procedure or by means of an illicit diversion?*

The hope among brand owners was that the CJEU would provide a more purposive approach, the purpose of Regulation 1383/2003 being to prevent counterfeit goods generally.

(b) *Philips*

In *Philips* the Antwerp investigations inspectorate of the Belgian Customs and Excise Administration detained a consignment of shavers from Shanghai which were suspected of infringing Philips' IP rights. The customs authorities sent Philips a photograph of the goods and informed it of the names of the companies that were involved. The customs declaration stated that the goods were declared under the temporary import arrangements without stating the country of destination. Philips brought an action before the Court of First Instance, Antwerp, seeking a ruling that its IP rights had been infringed and that, in accordance with Article 6(2)(b) of Regulation 3295/94 (later replaced by Regulation 1383/2003), the Court should use as its starting point the fiction that the goods seized had been manufactured in Belgium and should then apply Belgian law for the purposes of establishing the infringement (the 'production fiction'). The Court referred the following question to the Court of Justice for a preliminary ruling:

> *Does Article 6(2)(b) of Council Regulation ... 3295194 ... constitute a uniform rule of Community law which must be taken into account by the court of the member state which, in accordance with Article 7 of the Regulation, has been approached by the holder of an intellectual-property right, and does that rule imply that, in making its decision, the court may not take into account the temporary storage status/transit status and must apply the fiction that the goods were manufactured in that same member state, and must then decide, by applying the law of that member state, whether those goods infringe the intellectual property right in question?*

(c) *The CJEU decision*

The CJEU decided in *Philips* and *Nokia* that goods travelling through EU customs by the external transit procedure cannot automatically be considered counterfeit and therefore seized by Customs without sufficient grounds for suspecting that the goods are to be sold in the EU market. To do otherwise, the court says, would "hinder legitimate international trade transactions in goods transiting through the European Union". The decision means that seizure of counterfeit goods in transit through the European Union will not be achieved by brand owners simply identifying the goods as counterfeit to Customs. Instead, as matters stand, brand owners will likely need to provide evidence that there are sufficient grounds for suspecting that the allegedly counterfeit goods are likely to enter the EU market if no further action is taken, if they are to persuade customs officials to seize such goods in the first place.

The CJEU has given some guidance on the kind of evidence that could give rise to a legitimate suspicion, which does provide some help for brand owners on what they should be looking for if they are to have counterfeit goods in transit seized. Brand owners should look for evidence of the following, as appropriate indicators of sufficient grounds for suspecting allegedly counterfeit goods as likely to enter the EU market:

- the fact that the destination of the goods is not declared;
- the lack of precise or reliable information as to the identity or address of the manufacturer or consignor of the goods;
- a lack of cooperation with the customs authorities; and
- the discovery of documents or correspondence concerning the goods in question suggesting that there is liable to be a diversion of those goods to EU customers.

Such evidence may be able to provide sufficient indications that "one or more of the operators involved in the manufacture, consignment or distribution of the goods, while not having begun to direct the goods towards EU consumers, are about to do so or are disguising their commercial intentions", which would allow the customs authority to legitimately act.

In response to criticism of the *Nokia* and *Philips* decision, a number of amendments to the new draft Regulation 1382/2003 were agreed by the European Parliament in July 2012. Importantly, these include reversing the burden of proving the final destination of the goods to the holder of the goods, and the application of the IP laws of the member state in which the goods are found or where the application for detention is made.

8. Parallel importation issues[46]

Free movement of goods is one of the key principles of the European Single Market – allowing goods which are put on the market in one member state to move freely into other member states without being restricted by national laws. This principle is laid out in Articles 28 and 29 of the EC Treaty, which prohibit quantitative restrictions on imports and exports and measures having equivalent effect as between member states. This prohibition is subject to an exception set out in Article 30 of the EC Treaty to the effect that Articles 28 and 29 shall not preclude prohibitions or restrictions on imports, exports or goods in transit justified on the grounds of (*inter alia*) the protection of industrial and commercial property (which includes IP rights). However, the Article 30 exception is, in turn, restricted so as to ensure that IP rights (bearing in mind that these include national IP rights in member states such as national trademark registrations and are not limited to EU-wide unitary rights such as CTMs and RCDs), are not used to defeat the key principle of the free movement of goods. That restriction involves the application of the concepts of the 'specific subject matter' of the IP rights and the 'exhaustion of rights'.

The specific subject matter in respect of trademarks (to which many of the

46 The aim of this section is just to provide a brief overview of the underlying principles and concepts which are relevant to parallel importation issues in Europe. This is an extremely broad topic, encompassing, as it does, fundamental principles of the European Union and, specifically, both competition law and IP issues.

major European cases relate and which, in particular, are generally the rights in focus in the context of the parallel importation of pharmaceuticals) may be summarised as the right for a trademark owner to use/control the first marketing in the EEA of products bearing the trademark. The principle of exhaustion of rights, as set out under Article 7(1) of the Trademarks Directive, is that a trademark registration shall not entitle its proprietor to prohibit its use in relation to goods which have been put on the market in the Community under that trademark by the proprietor or with his or her consent. As will be apparent, the intention of this provision is that once goods bearing a registered trademark have been put on the market in the EEA by or with the consent of the trademark proprietor, then the proprietor's rights in respect of the trademark have been 'exhausted' and the goods are, thereafter, able to move around within the EEA freely without the trademark proprietor being able to assert its trademark rights to prevent parallel imports between EU member states. After a series of cases (the key ECJ decision being that in the *Davidoff v A&G Imports etc* joined cases C-414 to 416/99), it has been determined that a trademark owner's rights are not exhausted by marketing of products bearing its trademark outside the EEA and that the proprietor's consent to the importation of such goods into the EEA cannot generally be implied (eg, by the absence of express prohibition), nor inferred from the silence of the proprietor on the issue. In summary, therefore, trademark owners are generally able to object to parallel imports of trademarked products from outside the EEA, but may not object to parallel imports between member states once their rights to first-market the products in the Community have been exhausted.

The exhaustion principle set out in Article 7(1) of the Trademarks Directive is, however, qualified by Article 7(2) which provides that Article 7(1) shall not apply where there are 'legitimate reasons' for the proprietor to oppose further commercialisation of the goods, especially where the condition of the products is changed or impaired after they have been put on the market. What exactly amounts to 'legitimate reasons' which allow a trademark owner to oppose parallel importation into other member states of products which have already been put on the market in the EEA by it, or with its consent, has been the focus on the majority of the (extensive) case law in this connection, in relation to pharmaceuticals in particular.

The principle of exhaustion has most recently arisen in the CJEU case of *UsedSoft GmbH v Oracle International Corp*,[47] which concerns distribution rights in copies of software. Here, the question was whether Articles 4(2) and 5(1) of Directive 2009/24 on the legal protection of computer programs (dealing with the restricted acts in software and exceptions to those acts), prevented Oracle from opposing the resale of 'used' licences for copies of its databank software

47 Case C-128/11.

that had been downloaded from Oracle's website. The user right for the software comes with a right to store a copy of the program permanently on a server and to allow use by certain numbers of users by downloading to their work-stations. UsedSoft is a marketer of used software licences and acquired user licences to the databank software from customers of Oracle. Purchasers of these licences then download copies of the software from Oracle's website. All of which activity by UsedSoft prompted proceedings to be brought by Oracle. However, the CJEU has ruled that the right of distribution of a copy of a computer program which has been downloaded from the internet is exhausted if the right-holder has conferred the benefit of a right to use that copy for an unlimited period. Similarly, the second and subsequent acquirers of a user licence entailing the resale of a copy of a computer program downloaded from the copyright holder's website are able to rely on the exhaustion of the distribution right.

The publisher acknowledges Bas Berghuis van Woortman, Peter Meyer and Laura Orlando for their contribution to this chapter in the previous edition.

Unified Patent Court

Nicholas Fox

Michael Burdon
Simmons & Simmons

For over 40 years there has been movement towards the establishment of a unitary patent system in which a single unitary patent would have equal effect across Europe, driven by an aspiration to make patent litigation in Europe more predictable, quicker and cheaper, and ultimately to increase patent quality and boost European innovation and competitiveness. That aspiration had looked like becoming a reality with a new unitary patent and Unified Patent Court (UPC) being on track to launch in December 2017. However, the opening of the new court and with it the launch of the unitary patent has been delayed, first by the fallout from the decision of the United Kingdom to leave the European Union and then by a private challenge to the UPC system before the *Bundesfassungsgericht* (the German Constitutional Court) alleging that the proposed new court system is incompatible with German law. That challenge is ongoing and at present it is uncertain when or indeed if the UPC will come into existence. If the constitutional challenge is resolved and the court does come into existence, the UPC will represent a radical change to how patents are enforced in Europe.

1. Historical background

1.1 Community Patent Convention

The first step towards a unified patent system for Europe was the Community Patent Convention, which had at its heart the creation of a European Community patent. The Community Patent Convention was signed in 1975 but not ratified by all member states and so it never came into force. Several other initiatives followed in the wake of the Community Patent Convention, most notably the Council Regulation on the Community Patent and the Council's Common Political Approach. Ultimately these initiatives became mired in the same difficulties which led to the failure of the Community Patent Convention and were undone.

1.2 European Patent Litigation Agreement and Unified Patent Litigation System

In view of these failures, momentum gathered behind establishing an integrated European patent litigation system, rather than a Community Patent and to that end in the early 2000s a draft protocol for common rules of procedure for patent litigation and the establishment of a common European appeal court was proposed (the European Patent Litigation Agreement or EPLA). However, despite success in attracting participation from the European Commission, divisions resurfaced between those member states wanting a more integrated, EU-based system and those wanting to retain greater autonomy. As a result, the EPLA failed.

The EU Commission then proceeded, in 2008, to propose a Unified Patent Litigation System (UPLS) as a compromise. The UPLS plan reintroduced a single unitary patent covering all EU states. It also proposed a single system of European patent courts that would have jurisdiction for the infringement and validity of both existing European patents and for future unitary patents. The favoured approach was based on a set of largely decentralised first instance (Divisional) courts applying common procedural rules, a Central Division and a single Court of Appeal.

Notwithstanding that the UPLS had the support of the majority of stakeholders, its implementation gave rise to several fundamental issues of constitutional law concerning the relationship between the European Union and the European Patent Office[1] and the compatibility of the UPLS with EU law. These problems proved to be the downfall of the UPLS proposal as in March 2011 the CJEU ruled that the UPLS was not compatible with EU law on the grounds that member states cannot divest to an international body their powers over matters that require the implementation of EU law or deprive the CJEU of its powers to give preliminary rulings on matters of EU law.

1.3 Unified Patent Court

The Commission proposed a way around this constitutional deadlock in May 2011 by giving the Commission the right to bring proceedings against member states, should the new UPC (as it had been named by this stage) fail to refer EU law issues to the CJEU and not requiring the European Union to accede to the European Patent Convention. Importantly, this proposal would keep the new courts within the EU judicial system.

Although the Commission proposal dealt with the key constitutional issue, other fundamental issues remained, in particular the language regime for the new court and the location of the Central Division. However, one by one these issues were resolved. In late 2011 political agreement was reached on draft legislation outlining the language regime for a unitary patent and a unified patent court. Matters then moved on considerably when a compromise solution to the location of the Central Division emerged from a meeting of the European Council on 29 June 2012 by which the court would have its main seat in Paris,

1 A non-EU organisation established by the European Patent Convention.

but with further 'clusters' in London and Munich. The London cluster of the Central Division would take jurisdiction for matters including "life sciences, chemistry and human necessities". Munich would house administrative offices as well as being responsible for cases including advanced engineering, with all remaining matters, including those matters relating to software and electronics being the preserve of the Paris court.

Two regulations[2] establishing the framework for a Unitary Patent were then passed by the European Parliament on 11 December 2012 and signed into law on 17 December 2012 and an agreement for implementing a UPC was signed by 24 participating member states on 19 February 2013. This set the stage for the Unitary Patent and the UPC to come into existence four months after the Agreement had been ratified by 13 participating member states including the United Kingdom, Germany and France – the three participating member states where the greatest numbers of European patents were in force.

Austria was the first country to ratify the Unified Patent Court Agreement (UPCA) in August 2013. This was followed by the ratification by France in March 2014. Further ratifications followed. The future of the entire project was, however, then put in doubt again when the United Kingdom held a referendum and voted to leave the European Union. Since ratification of the UPCA by the United Kingdom was a mandatory requirement for the agreement to come into force, the court could not proceed without UK ratification. However, despite the outcome of the referendum, the UK government stated that it still intended to participate in the UPC system, drawing the distinction between the court being a multinational court permitted under EU law and membership of the EU itself. Although UK ratification was delayed by the vote and then again due to the calling of a general election in 2017, eventually the UK ratification process in the United Kingdom was completed in April 2018.

With the ratification of the UPCA by the United Kingdom, France and 14 other participating member states, the only remaining barrier to the UPCA coming into force is ratification by Germany. Although participation in the court was approved by votes in the German parliament in March 2017, ratification has not yet been completed due to a private challenge before the *Bundesverfassungsgericht* arguing that the UPC system is contrary to German law. In February 2018, that case was put on the list of cases the *Bundesverfassungsgericht* hoped to decide in 2018. Until the case is decided, ratification by Germany cannot be completed and the opening of the court is therefore on hold. Decisions before the *Bundesverfassungsgericht* can take a considerable time before a judgment is issued and hence any start date for the new court is unknown.

---

2 EU Regulations 1257/2012 and 1260/2012, referred to herein as the Unitary Patent Regulation and the Languages Regulation respectively.

2. The Unified Patent Court

2.1 Structure of the UPC

The UPC will be made up of a Court of First Instance; a Court of Appeal; and a Registry.[3] The Court of First Instance will include a Central Division (based in Paris, with sections in London and Munich),[4] which will have jurisdiction to hear invalidity cases, declarations of non-infringement, and in certain circumstances infringement cases. London will deal with patents in Sections A and C of the International Patent Classification (Human Necessities, Chemistry and Metallurgy) and Munich will deal with the patents in Section F (Mechanical Engineering, Lighting, Heating, Weapons, Blasting), leaving Paris dealing with the remainder (including Physics, Electricity and Paper, which encompasses electronics and telecoms).[5] In addition, there will be a number of local and regional divisions located throughout the contracting member states.[6] These local and regional divisions will have jurisdiction to hear infringement cases (together with any counterclaim for revocation), interim measures (including injunctions), provisional protection, and prior use issues.

Figure 1. UPC structure

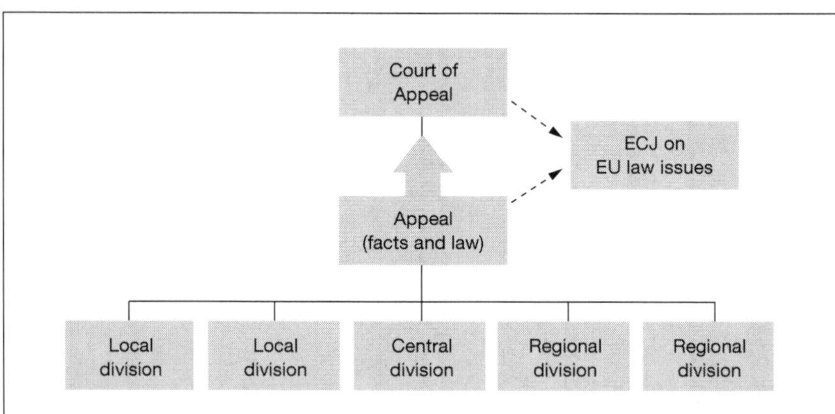

Broadly, infringement actions must be brought before:

- the local or regional division where the actual or threatened infringement occurs;[7] or
- the local/regional division of the contracting member state where one

3 UPCA, Article 6(1).
4 UPCA, Article 7(2).
5 UPCA, Annex II.
6 UPCA, Article 7(3) and (5).
7 UPCA, Article 33(1)(a).

defendant has its residence or principal place of business, or place of business[8] or, when the defendant(s) is domiciled outside the territory of the contracting member states, the Central Division.

This means that, where there are infringing products on sale across a number of contracting members and/or a number of defendants, the patentee has considerable scope to choose where to sue the defendant for infringement. Actions for revocation or non-infringement, on the other hand, must be brought before the Central Division[9] unless an action for infringement between the same parties referring to the same patent already has been brought before a local or regional division. When a defendant to an infringement action counterclaims for revocation of the patent, the local or regional division hearing the infringement action has the discretion to either:

- hear the whole case before a panel of judges (three in total), including a technical judge, allocated from a Pool of Judges which comprises legally and technically qualified judges (for each field of technology) of the Court of First Instance;[10]
- refer the counterclaim to the Central Division and suspend or proceed with the action for infringement;[11] or
- with the agreement of the parties, refer the whole case to the Central Division.[12]

2.2 Jurisdiction[13]

When it is established the UPC will be a new supranational court that will have jurisdiction not only over new unitary patents (see section 3 below) but also over existing European patents granted by the European Patent Office (EPO) which have not during the transitional period been opted out of its jurisdiction. This transitional period will run for the first seven years of the UPC's existence, extendible for a further seven years on the basis of a consultation after the first five years. During this transitional period:

- Actions for infringement or revocation of national patents granted by the EPO (ie, not a Unitary Patent) may still be brought before national courts unless an action has already been brought before the UPC.[14]
- A patentee can opt out of a European national patent from the UPC's

8 UPCA, Article 33(1)(b).
9 UPCA, Article 33(4).
10 UPCA, Article 33(3)(a).
11 UPCA, Article 33(3)(b).
12 UPCA, Article 33(3)(c).
13 In addition, the UPC will have a limited jurisdiction over acts of infringement of European patents granted by the European Patent Office which occur outside of the participating member states as is discussed in P Véron, "Extent of the Long-Arm Jurisdiction Conferred on the Unified Patent Court by Art. 71(b)(3) of the Brussels I Regulation as Amended by Regulation 542/2014 of May 15, 2014: Turkish Delight and a bit of Swiss Chocolate for the Unified Patent Court" (2015) 37 EIPR 588–596.
14 UPCA, Article 83(1).

jurisdiction, unless an action has already been brought before the UPC. Then only national courts will be able to hear revocation, infringement and declaratory applications relating to that patent.[15] The opt-out will last for the life of the patent unless withdrawn. The opt-out can be withdrawn at any time up until the moment that the patent is litigated.[16]

Once the transitional period has finished the UPC will have exclusive jurisdiction over all Unitary Patents and all European national patents granted by the EPO designating a member state which has joined the UPC, provided that such patents have not been opted out of the jurisdiction of the court during the earlier transitional period.

2.3 Language regime

The language regime of the UPC is relatively complex.

Article 49(6) of the UPCA provides that proceedings before the Central Division will use the language of the patent. This will be English, French or German. In practice, English is likely to dominate as approximately 80% of European patents granted by the EPO are in English, with only around 15% being in German and around 5% being in French.

Articles 49(1) and (2) provide that local/regional divisions where infringement actions can be initiated must identify the languages which they are prepared to work in and a plaintiff can chose between any of the identified languages.[17] The languages offered by the local/regional divisions will be a mixture of local official languages of the member states where the divisions are established and the EPO languages: English, French and German. A number of the proposed local/regional divisions have identified which languages they will use. In most cases the courts have identified that they will offer claimants the choice of using a local language or English.[18] The proposed Swedish/Baltic regional division has declared that all proceedings held there will be in English alone.

Although in general a claimant has a free choice amongst the languages offered by a local/regional division, there is one important restriction. If a defendant is domiciled and infringing in only one jurisdiction then a plaintiff must bring proceedings in the local official language for the local division where infringement is occurring.[19]

Judges may use any language permitted by the division in their proceedings[20] providing translations into the language of proceedings as

15 UPCA, Article 83(3).
16 UPCA, Article 83(4).
17 UPC Rules of Procedure, Rule 14(2).
18 For example, the Danish, Belgian and Finnish Divisions.
19 UPC Rules of Procedure, Rule 14(2)(b).
20 UPC Rules of Procedure, Rule 14(3)(c).

appropriate. The rules also provide for the language of the patent to become the language of proceedings either if agreed or by order following a request.[21]

2.4 Forum shopping and conflicts within the Unified Patent Court

The structure of the UPC means that it will frequently be the case that a plaintiff has a choice as to whether to start an action in a particular local or regional division or the Central Division. This is because competence of a division can potentially be established on the basis of limited infringement occurring in one jurisdiction (eg, a test purchase) even if the bulk of infringement occurs elsewhere. Although the procedures of the UPC are uniform across the entire court, the selection of the division where proceedings are brought will be important as it determines the physical location where hearings will occur, the choice of the language of proceedings and most importantly the selection of judges who might hear a case.

2.5 Internal conflicts

The Rules regarding internal conflicts within the UPC are relatively clear. Article 33(4) of the UPCA provides that if an action for infringement between two parties is pending in a local or regional division, any revocation action between the same parties relating to the same patent may only be brought as a counterclaim in that same local or regional division. Thus, the option of bringing a revocation action directly to the Central Division is excluded as soon as infringement proceedings are initiated in a local or regional division.

If a revocation action between two parties is pending, an infringement action may be brought before any competent local or regional division or as a counterclaim before the Central Division. If infringement proceedings are initiated before a local or regional division and a counterclaim for invalidity is then brought in those proceedings, the corresponding revocation action pending before the Central Division is stayed until a decision is made on whether or not the local/regional division will hear the entire case or have the revocation action transferred to the Central Division.[22]

If a declaration of non-infringement between two parties is pending in the Central Division and an infringement action is brought before a local or regional division, the rules[23] adopt a simple time-based cut off. If the infringement action is brought within three months from the date when the declaration of non-infringement was sought, the declaration of non-infringement is stayed. Otherwise, both actions proceed and the judges consult and agree future prosecution (potentially including staying proceedings).

21 UPC Rules of Procedure, Rules 321–323.
22 UPCA, Article 33(5) and UPC Rules of Procedure, Rule 75(3).
23 UPC Rules of Procedure, Rule 76.

Finally, where more than one action concerning the same patent is pending before different panels at any time the panels may, after hearing the parties, order that the actions should be heard together, in the interests of the administration of justice and avoiding inconsistent judgments, provided the division has jurisdiction over the disputes.[24]

2.6 Jurisdictional conflicts with the national courts

Jurisdictional disputes between the UPC and the national courts are potentially more problematic.

The Recast Brussels Regulation (Regulation (EU) 1215/2012) recognises the UPC as a "common court".[25] This means that the ordinary rules on *Lis Pendens* for settling jurisdictional disputes between European courts should apply.

These provide that where proceedings involving the same cause of action between the same parties are brought in the courts of different member states, any court other than the court first seised is obliged to stay its proceedings of its own motion until the jurisdiction of the court first seised is established.[26] The courts then contact each other to determine who was first seised and the second court declines jurisdiction[27] with the date when documents instituting proceedings are lodged with the court determining when a court is seised.[28] In situations concerning the same parties and the same cause of action, a stay of proceedings is mandatory. In contrast, where non-identical but related actions are pending in the courts of different member states, any court other than the court first seised may optionally stay its proceedings.[29] Actions are deemed to be related where they are so closely connected that it is expedient to hear and determine them together to avoid the risk of irreconcilable judgments resulting from separate proceedings.[30]

The provisions of the Brussels Regulation have led to extensive case law in the patent field particularly regarding the extent to which an application for a declaration for non-infringement in one jurisdiction (typically a jurisdiction where such actions are slow (eg, Italy or Belgium) should require a court elsewhere in Europe to stay later-filed infringement proceedings.

It is likely that similar issues may arise in relation to parallel litigation in the UPC and national courts as the courts will share jurisdiction over national patents granted by the EPO during the transitional period. Article 34 of the UPCA provides that decisions of the UPC cover the territory of all contracting member states where the European patent has effect. Therefore, a proprietor

---

24 UPC Rules of Procedure, Rule 340.
25 Brussels Regulation, Articles 71a and 71b.
26 Brussels Regulation, Article 29(1).
27 Brussels Regulation, Article 29(2) and (3).
28 Brussels Regulation, Article 32.
29 Brussels Regulation, Article 30(1).
30 Brussels Regulation, Article 30(3).

faced with a declaration of non-infringement in a slow-moving jurisdiction will not be able to avoid that action by excluding that country from the infringement action.

3. The Unitary Patent

Currently when a patent is granted by the EPO, a patentee has to choose in which of the 38 countries covered by the European Patent Convention to validate the patent. The arrival of the Unitary Patent will provide the patentee with the additional option: to designate the patent as having 'unitary effect', meaning that it will obtain an indivisible Unitary Patent that provides uniform protection and has equal effect in all participating member states.[31] A patentee will retain the option to obtain a bundle of national patents in the countries participating in the Unitary Patent as an alternative to obtaining a Unitary Patent. For countries not participating in the Unitary Patent, the patentee will be able to validate a bundle of national patents in the normal way. The decision to validate invoices balances the cost and benefit of broad geographical protection. The cost of renewal fees for a Unitary Patent will equal the cost of renewing national patents in Germany, France, the United Kingdom and the Netherlands.

A Unitary Patent will cover all the territories of those EU countries that have signed up to and ratified the UPCA at the date of registration of that Unitary Patent.[32] Although 25 EU member states have signed up to the Unitary Patent System (all EU member states other than Croatia, Poland and Spain) a Unitary Patent registered at the outset will not cover all 25 territories because some have not yet ratified the agreement. This means that Unitary Patents granted in the first few years of the system are likely to have a different territorial scope to those Unitary Patents granted later on.

To obtain a Unitary Patent, in a transitional period lasting at least six years, and extendable to 12 years, a translation of the patent must be filed with the EPO.[33] This translation must be in English if the patent was prosecuted in German or French or in any EU language other than English if the patent was prosecuted in English. The transitional period will end when it is determined that machine translation facilities are sufficiently good that no translations are required.[34] The translation must be filed within one month of the patent application on which the Unitary Patent is based being granted by the EPO.

A Unitary Patent, like any patent, is an object of property. The relevant law (eg, formalities for transfers, default co-ownership rules etc) for a Unitary Patent will be determined by the nationality of the first-named European applicant. If

31 Unitary Patent Regulation, Article 5.
32 Unitary Patent Regulation, Article 18(2).
33 Languages Regulation, Article 6.
34 Languages Regulation, Article 6.

there are no European applicants, German law will apply.[35] This default provision, together with the relatively large numbers of German originating European patents, will mean that in approximately 80% of cases the law and formalities applicable to the ownership and transfer of German patents will apply.

4. Litigation in the Unified Patent Court

4.1 Overview

The procedure in the UPC is a hybrid of procedures in several major EU countries.

Notable features of the UPC procedure as set out in the Rules of Procedure are:

- The procedure is a front-loaded procedure which will mean that litigants will need to get their arguments, evidence and strategy settled up front.
- In common with most European patent litigation such as occurs in Germany and the Netherlands, there is a strong emphasis on written arguments rather than the oral presentation.
- There is, however, a possibility of cross-examination of experts and witnesses and for orders obliging litigants to provide information regarding the functioning of allegedly infringing products.[36]
- The rules provide for a fixed timetable for proceedings with a view to having cases completed within 12 months. A hearing date is fixed early in the proceedings[37] and there are very limited opportunities to deviate from the normal timetable.
- The rules provide for strong judicial case management. All amendments to a case require the permission of the court. At the same time there is considerable flexibility as to how a judge may apply the procedures to a particular case.

4.2 Stages of an action

Actions at the UPC will have three main stages:[38]

- a written procedure;
- an interim procedure; and
- an oral hearing.

(a) Written procedure

Proceedings are commenced by the plaintiff filing a Statement of Claim. The

35 Unitary Patent Regulation, Article 7.
36 UPC Rules of Procedure, Rules 190–198.
37 UPC Rules of Procedure, Rule 28.
38 UPC Rules of Procedure, Rule 10.

Rules of Procedure set out the information that must be provided in pleadings and require that a plaintiff sets out their case in full in their Statement of Claim.

More specifically, in addition to providing formal details about the case (eg, the identity of the parties, the identity of the patent in suit etc), a Statement of Claim is required to include:

- an indication of the facts relied on;
- the evidence relied on where available, and an indication of any further evidence which will be offered in support;
- in a claim for patent infringement, the reasons why the facts relied on constitute an infringement of the patent claims including arguments of law and where appropriate an explanation of the proposed claim interpretation and in the case of a claim for revocation, the reasons why the facts relied upon cause a patent to be invalid; and
- an indication of any order the claimant will seek during the interim procedure.[39]

The case is then served on the defendant. Service is undertaken by the Registry by being served at the defendant's statutory seat, central administration or principal place of business within the contracting member states. Revocation/declaration of non-infringement actions may be served at the address for service on the EPO or national Patent Register. A defendant can refuse service if proceedings are not in a language which the defendant understands and can also require a plaintiff to provide them with a translation of the pleadings into a language which they understand.[40]

During this stage the action will be assigned to a panel of a division. The panel will appoint one of the judges as the judge-rapporteur who will act as the presiding judge.

(b) *Interim procedure*[41]

This is essentially the case management stage of the proceedings where the preparations needed for the oral hearing are made after the written pleadings have been served. During this stage the judge-rapporteur may hold interim conferences in order to identify the main issues and clarify each party's position. The judge-rapporteur will establish a schedule for the proceedings and issue case management orders such as service of further pleadings or evidence, appointment of experts and the matters to be the subject of oral evidence. The judge-rapporteur will also discuss with the parties whether it is possible to settle the dispute or for it to be referred to mediation or arbitration.

39 UPC Rules of Procedure, Rules 13 and 45.
40 UPC Rules of Procedure, Rules 270–279.
41 UPC Rules of Procedure, Rules 101–106.

A key decision will be whether a claim for infringement should be heard together with a counterclaim for revocation.[42] This decision will made by the judge-rapporteur having considered the pleadings and having given the parties an opportunity to be heard. Such a hearing may potentially be by telephone or by video conference.

If a revocation case is transferred to be heard in the Central Division separately from an infringement case, the judge-rapporteur can stay the infringement case until the Central Division rules on the revocation case. The judge-rapporteur is obliged to stay the case if the court considers that there is a high likelihood of the patent being revoked.

(c) Oral procedure[43]

The oral hearing takes place before a panel of judges. The parties make their submissions and, if permitted, any cross-examination of witnesses and experts takes place. The oral hearing is limited to specific issues identified by the judge-rapporteur and will normally be concluded within one day,[44] although there may be an earlier hearing to deal with witness evidence.[45] After the oral hearing the judge-rapporteur will make a decision on the merits in writing within six weeks. Decisions and orders are directly enforceable from date of service in each participating member state, with the enforcement procedures of the member state where enforcement takes place applying. First instance decisions may be appealed to the Court of Appeal within two months of the final decision.

4.3 Intervention[46]

The Rules of Procedure provide some limited opportunities for third parties to intervene in proceedings before the court.

The possibility of intervening is limited to interveners who can establish a legal interest in the result of an action submitted to the court. In order to do so the intervener must establish that they have an interest in the whole or part of a claim, order or remedy sought by one of the parties.

An application for intervention is only permissible if it is made before the closure of the written procedure and is made by filing a written Statement of Intervention. The judge-rapporteur may invite a third party to intervene either on its own motion after hearing the parties or on a reasoned request by a party. Alternatively, a party can force the intervention of a third party it wants to be bound by a decision. Failure to intervene may prevent a third party arguing that an action was wrongly decided or improperly conducted.

---

42 UPC Rules of Procedure, Rule 37.
43 UPC Rules of Procedure, Rules 108–117.
44 UPC Rules of Procedure, Rule 113.
45 UPC Rules of Procedure, Rule 115.
46 UPC Rules of Procedure, Rules 313–317.

If an application to intervene is filed, the court will consider the application and make an appropriate order either adding the party to the proceedings or refusing the application to intervene. It is not possible to appeal against an order refusing an application to intervene.

4.4 Damages[47]

If a decision on the merits finds that there has been infringement of a valid patent, then damages may be recovered to put the injured party in the position they would have been in had the infringement not occurred. The successful party needs to lodge an application for separate proceedings for the determination of damages within one month from the final decision on the merits on both infringement and validity. The application may include a request for an order to lay open books and must indicate the recompense sought (damages, licence fees, profits) and, if relevant, calculations concerning lost profits or the infringer's profits. An interim award of damages may be made to cover, as a minimum, the expected costs of the damages procedure.[48]

4.5 Provisional and protective measures[49]

Patentees may apply to the UPC for provisional measures including injunctions to prevent imminent infringement or to stop ongoing infringement, orders for the preservation of evidence, inspection and freezing of assets. The court can, in its discretion, order provisional measures with or without hearing the alleged infringer. Protective letters may be filed by a prospective defendant at the UPC in anticipation of an application for an injunction being lodged by a patentee. Whether any protective letters have been filed is one of the factors that the court will take into account when deciding whether to grant an injunction without hearing the defendant or whether an oral hearing is required. Consequently, protective letters have the potential to be an important strategic tool for prospective defendants.

The Rules of Procedure[50] expressly provide for the court to order a *Saisie*, an order to authorise the seizure of evidence from an alleged infringer, similar to the process used in the French courts. Such an order may be requested prior to initiating proceedings. The request must identify the exact location of evidence to be preserved, reasoned evidence why preservation is required, and set out the facts and evidence relied upon. The court has the option to inform a prospective defendant of the existence of a request for a *Saisie*. Grant of a *Saisie* is at the court's discretion and the court is obliged when considering that discretion to

47 UPC Rules of Procedure, Rules 125–144.

48 See N Fox *et al*, "Accounting for Differences: Damages and Profits in European Patent Infringement" (2015) 37 EIPR 566–574 for a discussion on national practices on the assessment of patent damages in Europe and the likely assessment of damages for patent infringement in the Unified Patent Court.

49 UPC Rules of Procedure, Rules 190–200 and 205–213.

50 UPC Rules of Procedure, Rules 192–198.

consider the urgency of the request, whether the request is well founded and the probability of the destruction or unavailability of evidence.

Similar procedures are provided in the Rules of Procedure[51] permitting the court to order an alleged infringer to make products and premises available for inspection or for the court to order the freezing of an alleged infringer's bank account where there is a danger that an alleged infringer may try to move assets outside the jurisdiction.

The process for applying for provisional measures such as a preliminary injunction is similar to that in a main action.[52] The process is initiated as a written procedure by pleadings which identify the requested measures and the reasons why such measures are necessary along with the facts and evidence in support. Such applications may be made on an *ex parte* basis. The court may request a defendant to respond and/or summons the applicant or both the applicant and the defendant to the Oral Hearing. Protective letters can be filed.[53] When deciding whether or not to issue preliminary measures, the court will take into account: whether the patent has been previously upheld in opposition/other proceedings; the urgency of the action; and the likely validity of the patent and evidence of infringement.

4.6 Recoverable costs[54]

The rules of the UPC provide that a successful party is entitled to recover reasonable and proportionate costs. A ceiling is set on recoverable costs, which is subject to the value of the claim (see Figure 2). The maximum award of recoverable costs is €2 million for a case valued at more than €50 million. For particularly complex cases the ceiling may be raised but cannot exceed a maximum of €5 million. The court has discretion, on request by a party, to lower the ceiling if the amount of recoverable costs would threaten the economic viability of the requesting party, particularly if that party is a university, non-profit organisation, public research organisation or natural person.

4.7 Court fees[55]

Court fees before the UPC include fixed fees and value-based fees. Infringement actions and declarations of non-infringement will have fixed court fees of €11,000 plus an additional value-based fee on a rising scale which applies to claims with a value of more than €500,000. The maximum value-based fee is €325,000 for actions with a value of more than €50 million. The fixed fee for revocation actions is €20,000. Value-based fees will not apply to revocation actions or counterclaims.

---

51 UPC Rules of Procedure, Rules 199 and 200.
52 UPC Rules of Procedure, Rules 205–213.
53 UPC Rules of Procedure, Rule 207.
54 UPC Rules of Procedure, Rules 150–157.
55 UPC Rules of Procedure, Rule 370.

Figure 2. Recoverable costs

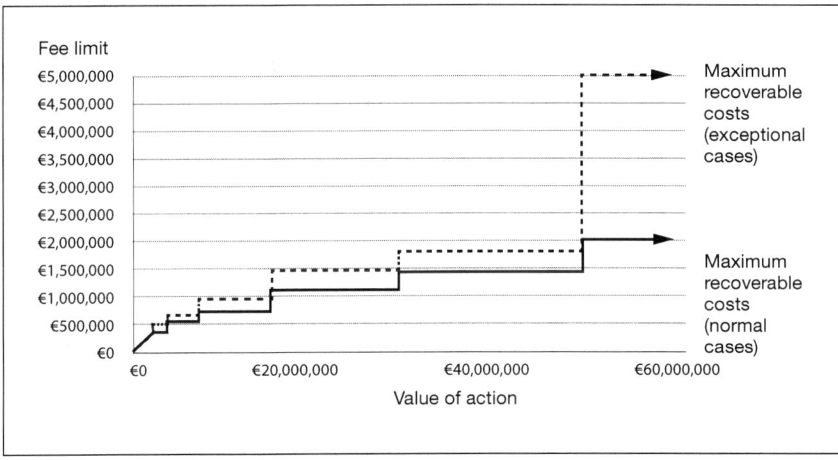

Figure 3 shows that the court fees at the UPC for lower-value cases are broadly comparable to those for a German action. Fees for cases between €10 million and €50 million are lower at the UPC. For cases in excess of €50 million in value the total fees at the UPC are slightly higher, with total fees of €336,000 payable. The fees seem high compared with the United Kingdom, where the maximum court fee is about €12,500. For high-value cases, however, particularly where infringement occurs in multiple jurisdictions, the differences between UPC court fees and fees for initiating national proceedings are unlikely to be a significant factor driving the choice of forum.

Figure 3. Court fees

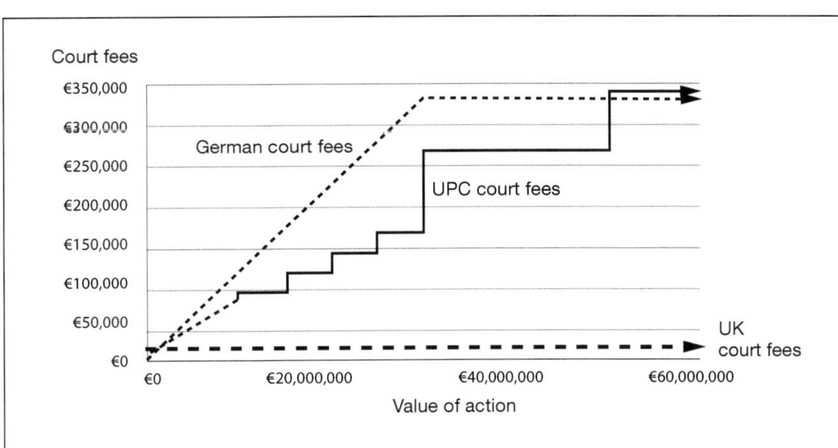

4.8 Valuing a case

Case value is used to determine not only the ceiling of recoverable costs at the UPC but also the court fees payable. Although the Preparatory Committee has issued some guidance on the application of the rules on recoverable costs and court fees, there is likely to be an element of variation as to how the guidelines apply in individual cases. In any event, only reasonable and proportionate costs may be recovered. The cost ceilings are a safeguard against undue recovery not an indication of the amount that will be recoverable. In case of partial success or in exceptional circumstances, the court may order the parties to bear their own costs or apply a different apportionment of cost based on equity.

As the above graphs show, there is interplay between the cost ceiling and the court fees, with the key values at the €30 million and €50 million thresholds:

- Up to €30 million – for infringement proceedings valued over €16 million up to and including €30 million the ceiling for recoverable costs is €1.2 million (€1.5 million in exceptional cases) and the total court fee is €161,000 (€11,000 fixed fee plus €150,000 value-based fee).
- Up to €50 million – for infringement proceedings valued in excess of €30 million and up to and including €50 million the ceiling for recoverable costs is €1.5 million (€1.875 million in exceptional cases) and the total court fee is €261,000 (€11,000 fixed fee plus €250,000).
- Over €50 million – for infringement proceedings valued in excess of €50 million the ceiling for recoverable costs is €2 million (€5 million in exceptional cases) and the total court fee is €336,000 (€11,000 fixed fee plus €325,000 value-based fee).

Moving from the 'up to €30 million' band to the 'up to €50 million' band therefore means an additional €100,000 is payable in court fees but this also raises the ceiling on recoverable costs by €300,000 to €1.5 million (by €375,000 to €1.875 million in exceptional cases). Moving to the 'over €50 million' band means paying an additional €75,000 in court fees but also raises the ceiling by €500,000 to €2 million and potentially by €3.5 million to a maximum of €5 million in exceptional cases. With this in mind, it seems that cases valued at around €30 million and around €50 million will be the most likely to result in disputes over valuation.

4.9 Appeals[56]

Decisions of the Court of First Instance in the UPC can be appealed to the UPC Court of Appeal. Permission to appeal is required unless the appeal is from:

- a final decision of the Court of First Instance;
- a decision terminating proceedings;

56 UPC Rules of Procedure, Rules 220–243.

- a decision changing the language of the proceedings to language of the patent;
- an order to produce or preserve evidence or to provide information about third parties or to inspect premises; or
- a decision granting a freezing order or an interim injunction.

Permission to appeal may be given by the Court of First Instance within 15 days of the decision. If the Court of First Instance refuses to grant permission, an application for permission can be made to the Court of Appeal. When considering whether to grant permission, the court is obliged to consider the objective that the Rules of Procedure should be applied consistently across all divisions of the court.

Filing an appeal does not automatically have suspensive effect. An appellant must lodge a request that the implementation of a decision is suspended pending the appeal. Such applications are then considered by the Court of Appeal. It is only possible to request that a decision is suspended when it is one of the decisions (listed above) for which permission to appeal is not required. Applications for suspensive effect for appeals requiring permission to appeal which will typically involve appeals of procedural matters are not permitted.

If an appellant appeals a final decision of the Court of First Instance or a decision terminating a case, a Statement of Appeal (ie, formal notice of appeal) must be filed within two months of date of service of the decision being appealed. Formal Grounds of Appeal (ie, formal reasons and evidence in support) have to be filed within four months of the date of service of the decision being appealed. In the case of appeals other than appeals against a final decision of the Court of First Instance or a decision terminating a case (ie, appeals where an appellant requires the permission of the court – typically appeals on procedural matters), a Statement of Appeal and Grounds of Appeal are due within 15 days of service of the order embodying the decision being appealed.

A Statement of Response, which may include a Statement of Cross Appeal, is then due from the other parties to the proceedings. The Statement of Response is due within three months in the case of appeals against a final decision of the Court of First Instance or a decision terminating a case and within 15 days of service of Grounds of appeal in all other cases.

The remainder of the appeals procedure is then similar to the procedure at first instance. In an interim stage a judge-rapporteur appointed to the case fixes the date for an oral hearing and can make any order to put the appeal in order so it is ready for the hearing. A judgment is issued following the Oral Hearing. The Court of Appeal may deal with all matters or refer them back to the Court of First Instance for reconsideration.

4.10 Rehearing and re-establishment of rights

The Rules of Procedure contain two sets of provisions to address the possibility of errors occurring during a case.

(a) *Re-establishment of rights*[57]

Re-establishment of rights provides a means by which litigants can reinstate court proceedings in the event of a failure to observe a time limit. The party requesting re-establishment is required to establish that such a failure was "despite all due care having been taken by the party" and "was outside his control". An application for re-establishment of rights must be filed within one month of the removal of the cause of the non-observance of the time limit and within six months from the original deadline which was missed. The omitted act also needs to be completed by the same deadline. If such an application is filed, the other parties to the litigation are invited to comment and then the court rules on whether to allow the request.

(b) *Rehearing*[58]

The rehearing procedure is a procedure of last resort which exists to provide litigants with a means to reopen a final decision of the UPC where a fundamental procedural error has occurred.

Any application for a rehearing is required to be filed:

- within two months of the discovery of a fundamental defect or the service of final decision;
- within two months from the issuance of a final court decision that a criminal act impacting a UPC court decision has occurred; and
- in any event within 10 years.

Such an application can only be based on very limited grounds including allegations that:

- an excluded judge took part in the proceedings;
- a fundamental violation of the right to be heard occurred;
- the court failed to decide on a relevant request; or
- a breach of human rights occurred.

Further litigants are obliged to have raised such objections at the original hearing unless it was impossible to do so. If successful, the original decision of the court where the error occurred will be overturned and set aside and the original litigation will be resumed.

---

57 UPC Rules of Procedure, Rule 320.
58 UPC Rules of Procedure, Rules 245–255.

France

Pierre-Olivier Ally
Bitfury Group
Marianne Schaffner
Reed Smith LLP

The French electronics and software market has been constantly growing since 2010, increasing by 30% in 2014, a further 12% in 2015, and a further 12% between 2016 and 2017. Between 2016 and 2017, the sector has created 10,000 jobs even if there are some issues in recruitment and talent retention. Innovation remains an absolute priority with a growing share of the R&D. France boasts world leaders in software, with its top three software companies: Dassault Systèmes, Criteo and Ubisoft reporting turnovers of €3.280 billion, €2.036 billion and €1.732 billion respectively in 2017.[1]

France is an attractive forum for IP litigation. Particularly noteworthy is the *saisie-contrefaçon* (seizure order) – a powerful tool for collecting evidence before filing an action on the merits. This is an area in which France has led the way, with similar procedures being extended to the rest of the European Union through the EU Directive on the Enforcement of IP Rights (2004/48/EC). However, compared to France, with its long history of the use of such orders, similar orders are something of a novelty in other European countries; and other countries can be reluctant to grant such orders, which are far easier to obtain in France.

If the Unified Patent Court proceeds, Paris will play a key role with the section of the central division of the court of first instance responsible for hearing disputes involving patents falling within the physics and electricity International Patent Classes, including inventions relating telecommunications, computers and electronics having its seat in Paris.

1. National patent procedure

1.1 Filing

Patent protection in France can be obtained via either a national French patent application or a European patent application filed with the European Patent

1 For more information on the French electronics and software market, see Ernst & Young, "*Top 250 des éditeurs de logiciels français*", 8th edition, October 2018. A compilation of statistics regarding the top 250 French software editors and creators is also available at www.ey.com.

Office (EPO). If an applicant chooses to use the Patent Cooperation Treaty (PCT) system, a French patent can be granted only via the EPO, as France has declared the national route to be closed for PCT-based applications.

An applicant is free to apply via one or both routes. If an application for a European patent is submitted at the same time as a French national patent application, any granted French patent will cease to exist upon grant of the corresponding European patent following expiration of the opposition period, if the two patents are identical in scope. If not, then the French and European patents will coexist.

France has adopted on 22 May 2019 and published in the Official Journal on 24 May 2019, the so-called Loi PACTE that creates a provisional patent application, similar to the US provisional patent application. The French provisional patent application will confer a priority right which will allow the applicant to complete and finalise its patent application.

1.2 Novelty-based examination system

The French Intellectual Property Office (INPI) currently performs only novelty-based examination and does not assess inventive step. Prior art searching of new applications is outsourced to the EPO, which issues a search report that subsequently forms the basis of the examination by INPI. The applicant is given an opportunity to amend the claims if INPI raises objections. In most cases, the applicant must convince INPI that the invention is novel. If INPI refuses to grant the patent, an appeal can be lodged before the Paris Court of Appeal within one month of issuance of INPI's decision.

However, the Loi PACTE provides under Article 122 that the INPI will have to assess inventive step. This new provision will enter into force in one year and will have a significant impact on French patent examination.

French law also provides for utility models, which must meet the same patentability requirements as a French patent. However, no substantive examination is performed. Rather, a search is conducted on the application and a search report is issued; the application is subsequently granted. A utility model confers protection for six years, but the Loi PACTE has prorogated the protection to 10 years. While a utility model was not convertible into a patent, it will be according to the Loi PACTE. A patent application can be converted into a utility model. In practice, utility model protection is rarely used or enforced in France. The changes introduced by the Loi PACTE aim to make the utility model more attractive, as is the case in Germany.

1.3 Timescales

The search report is issued within approximately nine months of filing. A French patent is issued within approximately two years of filing. INPI offers no accelerated prosecution procedure.

It is worth underlining that the Loi PACTE provides for the creation of an opposition procedure before the INPI but the conditions and the rules of procedure have not yet been defined.

1.4 Outline of costs

The main official fees for filing and prosecuting a patent application are as follows:

- filing fee – €36;
- search fee – €500; and
- grant fee – €84 (plus €40 for each additional claim in excess of €10).

The annuities for the second to fifth year amount to €38; between the sixth and 20th year, annuities range from €76 to €790.

2. Patents for computer-implemented inventions

2.1 Examination of software patents

Article L611-10(2)(c) of the Intellectual Property Code (IPC) states that software is not patentable subject matter and is thus excluded from patentable inventions:

> 2. *The following in particular shall not be regarded as inventions within the meaning of the first paragraph of this Article*
>
> ...
>
> c) *schemes, rules and methods for performing mental acts, playing games or doing business, and programs for computers.*

However, Article L611-10(3) provides that Article L611-10(2)(c) only applies to the extent to which the patent application or the patent relates to software *per se*:

> 3. *The provisions of (2) of this Article shall exclude patentability of the items referred to in these provisions only to the extent to which the patent application or the patent relates to such subject matter or activities as such.*

Consequently, software is considered to be patentable if it involves or is involved within the use of a technical device.

In the *Mobil Oil* case[2] the French Supreme Court clearly stated that although software is not patentable on its own, it can be patented if it is part of a larger invention, provided that the other patentability conditions are met. In *Schlumberger*[3] the Paris Court of Appeal ruled that the mere fact that software is involved in an invention as a necessary step in order to obtain a technical result does not bar the invention from patentability, provided that the patented invention consists of an industrial process.

---

2 *Mobil Oil v Dir INPI*, French Supreme Court (Commercial Chamber), 28 May 1975.
3 *Schlumberger v Dir INPI*, Paris Court of Appeal, 15 June 1981.

Further case law has since confirmed this principle, ruling that: "While computer programs are in principle excluded from patentability, a process may not be deprived of patentability on the sole ground that one or more of its steps are carried out by a computer to be controlled by a program".[4]

INPI and the Paris First Instance Court[5] examine whether the software invention has technical effect in order to determine whether it is patentable. In other words, the computer program must include functions which solve a technical problem, and implementation by a computer must produce a further technical effect beyond the usual physical interactions (eg, electrical current) between the software and hardware through which the program is executed.

The patentability of computer-implemented inventions is dependent on the specific circumstances of each case. For instance, French courts have considered the following inventions to be patentable (without having then necessarily validated the patent with regard to novelty or inventive step):

- a process allowing mobile phone users to exchange text or audio messages anonymously, as the process was likely to produce a technical effect materialised by the transmission of the message in anonymous form, regardless of whether these functions could be provided by software provided that the latter not be claimed as such;[6]
- a method of using the telephone network to produce sounds connected with simultaneous browsing of a webpage, as each step of the method required physical means, such as a voice server, an internet server, a computer terminal, and a telephone and involved technical considerations such as the sound emission on the telephone which was triggered by a navigation action;[7]
- a computer system for collecting money for issuing commercial benefit tickets, on the grounds that a structure that constitutes a computer system is patentable regardless that an intended result was directed to providing a commercial advantage and marketing a store;[8] and
- a remote managing system for industrial maintenance, as it provided technical means of information enabling it to control the operation of the industrial equipment and to monitor the maintenance company's operations.[9]

In contrast, the following inventions have been found not to be patentable:

- a cash collection system, which only described a mechanism imposed by the operations to be carried out directly induced by an associated data structure and the theoretical constraints imposed on anyone who

4 *Cafetel v Index Multimedia*, Paris First Instance Court, 20 July 2006.
5 In France, the Paris First Instance Court has exclusive jurisdiction over patent matters.
6 *Freever v Atchik*, Paris First Instance Court, 10 June 2005.
7 *Cafetel v Index Multimedia*, Paris First Instance Court, 20 July 2006.
8 *Infomil v Atos*, Paris First Instance Court, 20 November 2007.
9 *Kone v Mr Azpitarte*, Paris Court of Appeal, 5 June 2009.

wanted to put merchants referenced in one list in contact with buyers referenced in another;[10]

- a trademark filing process consisting in making a referral of brands, a selection of products or services, the validation of these entries and the sending of requests remotely, as these means were used for non-technical purposes, such as the collection of legal and administrative information necessary for the filing of a trademark, with a view to solving problems of a purely economic nature and therefore related only to the exercise of intellectual and economic activities and not to an invention;[11]
- a process allowing the display, on a television screen, of a program guide and detailed information on programs and channels, as the object of the invention was only to enable a user to access information and the contribution of the invention resided only in the content thus made visible and its presentation in two successive windows;[12]
- a process for switching multimedia sessions from a mobile device to a device on a local network, as (i) 'computer program product' claims are only a language device which should not allow patents to be granted in breach of Article 52(2) of the European Patent Convention, and (ii) no further technical effect was produced by the claim directed to the 'recording medium on which the program is stored';[13]
- a simplified application management system directed to adapting an abstract template to customise an application, as the presence of a further technical effect – the way the user was indeed assisted – was not demonstrated;[14] and
- a search engine for extracting information and more particularly displaying the results of a search request as well as browsing databases and entering requests into databases, as it was not sufficiently technically characterised and did not go beyond the mere steps implemented by a librarian in the manual search for a book, thus constituted a purely intellectual process.[15]

If an invention relates only to a computer program, a business method[16] or a purely abstract concept,[17] patentability will be denied. If no technical effect can be found in the description, or if technical means are not described and explained with the aim of reaching a technical effect, INPI will reject the application.

---

10 *Syrdrec v Carte Bleue*, Toulouse First Instance Court, 1 August 2005.
11 *Cabinet Malemont et Ordipat v Dir INPI*, Paris Court of Appeal, 9 September 2005.
12 *Starsight Telecast v Kudelski*, Paris First Instance Court, 7 June 2013.
13 *Orange v Free*, Paris First Instance Court, 18 June 2015.
14 *Xaga v Ewalia*, Paris First Instance Court, 18 November 2016.
15 *Dassault Systèmes v Sinequa*, Paris Court of Appeal, 16 December 2016.
16 See, for example, *Cotranex v Dir INPI*, Paris Court of Appeal, 15 March 2006.
17 *Exalead v Sinequa*, Paris First Instance Court, 19 March 2010.

2.2 Relationship with EPO law

In substance, the French approach closely reflects the EPO's approach. However, if an invention only provides for some technical means without involving any real technical effect, INPI will reject a patent application directly on patentability grounds, rather than declaring that the invention lacks inventive step, as is existing EPO practice under its *Hitachi*[18] case law.

Thus, for example, in *Sagem*[19] the Paris Court of Appeal declared a decision of INPI well-founded, which had refused to grant a patent for a method of electronically ordering products at a sales outlet on the grounds that the technical effect requirement had not been met, and that the invention amounted to a business method. This decision was the first issued by a French court of appeal on the patentability of business methods.

However, in the *Infomil v Catalina*[20] and *Infomil v Atos*[21] rulings, French courts held that an invention aimed at delivering coupons to customers was patentable, provided that what was claimed was a device characterised by a precise structure and defined by its features, which constituted its technical aspect.

For most inventions related to software, it is advisable to apply for a European patent designating France via the EPO, rather than directly via INPI, as the standards applied during French examination are not as clear as those applied by the EPO, and the possibility of a borderline case being rejected may be slightly higher if the French national system is used.

3. Infringement

3.1 Claim interpretation

Article 69 of the European Patent Convention and Article L613-2 of the IPC provide that patent claims must be construed in light of the descriptions and illustrations.

Under French law, the doctrine of equivalents allows for patent protection even where there is no literal infringement of the claims. The French conception of the doctrine centres on the general means of the patented invention. Whether the invention will benefit from this protection depends on whether the function of these means is novel. In case of novelty of function, the doctrine of equivalents applies. However, if the function is already found in the prior art, the protection afforded by the patent will be limited to the form of the new claimed means.[22]

---

18 *Hitachi*, EPO, 21 April 2004 (T-258/03).

19 *Sagem v Dir INPI*, Paris Court of Appeal, 10 January 2003.

20 *Infomil v Catalina*, Paris First Instance Court, 21 June 2002; confirmed by *Infomil v Catalina*, Paris Court of Appeal, 28 April 2004.

21 *Infomil v Atos*, Paris First Instance Court, 20 November 2007.

22 See *Fagor Brandt v Whirlpool*, Paris Court of Appeal, 13 February 2009.

Case law holds that there is equivalence when the devices take on different forms, fulfil the same function and are aimed at achieving the same result.[23] However, French courts also hold that the interpretation, especially when the wording is ambiguous or unclear, should not lead to extension of the scope of a claim.[24]

3.2 Acts of infringement

Article L613-3 of the IPC[25] defines 'infringing acts' as follows:

In the absence of the consent of the patent proprietor, the following acts are prohibited:

a) Making, offering, putting on the market or using a product which is the subject matter of the patent, or importing or stocking a product for such purposes;

b) Using a process which is the subject matter of the patent or, when the third party knows, or it is obvious in the circumstances, that the use of the process is prohibited without the consent of the owner of the patent, offering the process for use on French territory;

c) Offering, putting on the market or using the product obtained directly by a process which is the subject matter of the patent or importing or stocking for such purposes.

3.3 Contributory infringement

Article L613-4 of the IPC[26] states that contributory infringement exists if the following four elements are met:

- The means of implementing the invention are related and suitable to implement as an essential element of the invention;
- Such means have been delivered and/or offered to be delivered in France, to be implemented in France;
- The third party which supplied said means knew or the circumstances must have made it obvious that these were suited and intended for effecting the invention in France; and
- Such means are not staple commercial articles (except where the supplier induces the person to whom it delivers to commit infringing acts, in which case this last requirement does not apply).

23 See *Passat v Perga*, Paris Court of Appeal, 27 March 2002.

24 *Nettobus* v *Alvan Blanch, Onet* and *Prodim*, Paris Court of Appeal, 10 May 1994.

25 See also Article L615-1 of the IPC.

26 *"1. It shall also be prohibited, save consent by the owner of the patent, to supply or offer to supply, on French territory, to a person other than a person entitled to work the patented invention, the means of implementing, on that territory, the invention with respect to an essential element thereof where the third party knows, or it is obvious from the circumstances, that such means are suited and intended for putting the invention into effect.*
2. Paragraph 1 shall not apply where the means of implementation are staple commercial articles, except where the third party induces the person supplied to commit acts prohibited by Article L. 613-3.
3. Persons carrying out the acts referred to in items (a), (b) and (c) of Article L. 613-5 shall not be deemed persons entitled to work the invention within the meaning of paragraph 1."

3.4 Network inventions

(a) Implementation by a single party: all steps required to be carried out in France

According to French law, acts may constitute infringement of a French patent only if they are committed on French territory. Thus, network inventions (ie, inventions involving a combination of software and/or hardware) of which at least one step or component is established outside of France cannot be protected on the sole basis of a French patent, because the infringing acts are not entirely committed in France.

(b) Implementation by more than one party: contributory infringement if all acts are carried out in France

When different steps or components of a network invention are implemented, on French territory, by more than one party, no infringement legally occurs because none of these parties individually implement the entire invention.

In such instance, however, the contributory infringement doctrine of Article L613-4 allows the rights holder to bring a legal action against the party supplying, on French territory, an essential[27] element of the invention (eg, the software necessary for the invention) for it to be implemented on French territory.[28]

However, since Article L613-4 of the IPC requires the essential means to be delivered and/or offered to be delivered in France, and also that the invention be then implemented in France, the delivery or an offer to deliver, in France, the essential means of a network invention does not amount to contributory infringement of a French patent, if the invention is then, at least partially, implemented abroad.

With regard to acts of infringement of a network invention which are partially or entirely committed abroad at the instigation of a French party, the concept of contributory infringement could also be used to try to pursue such acts on the basis that they produce their most important effects on the French territory, but there is as yet no case law in this respect.

4. Copyright protection

4.1 Copyright protection for software programs

Special protection for software under copyright law has been adopted by Law 85-660 (3 July 1985). Previously, courts ruled that software was governed by Law 57-298 (11 March 1957), which protects any work under the common copyright regime.

---

27 See, for example, *Applimo v Atlantic*, French Supreme Court (Cass Com), 28 April 2004.

28 *L'Air Liquide v Yara*, Paris First Instance Court, 29 October 2008.

Additionally, EU Directive 91/250/EEC on the legal protection of computer programs (14 May 1991), providing for uniform protection of software across Europe, was implemented in France by Law 94-361 (10 May 1994).

French law offers no definition of 'software', although "software including its preparatory design material" is expressly referred to as a protectable work in Article L112-2 (13) of the IPC.

Software protection may include both internal specifications (ie, the functions to be carried out by the software and its architecture) and external specifications (ie, the software interfaces), and extends to prototypes, as well as object and source codes.

Software must be original in order to be protected under copyright law. 'Originality' is not defined by the law; rather, it has been determined by the courts. Software will be considered to be an 'original' work if it bears the imprint of the author's personality, consisting of "a personalized effort going beyond the mere implementation of an automatic and constrictive logic".[29] It is also specified that 'originality' is not equivalent to the concept of 'novelty' used in patent law.[30] In particular, in a software infringement action the plaintiff has the burden of proof of establishing the originality of his or her software.[31]

The imprint of an author's personality may exist in the creative expression, regardless of any assessment of its perceived artistic or economic value,[32] or in the software composition and its architecture, regardless of its novelty.[33]

Software may be protected:

- regardless of type, form and language of expression, technical or economical merit or purpose;[34]
- in the absence of registration, to the extent that the work is protected "by the mere fact of its creation".[35] However, it is advisable to place the software under escrow to evidence its creation date (in a sealed envelope[36] provided by INPI, at a notary's office or at the French Agency for Programme Protection);[37] and
- regardless of the absence of any mention of copyright, which is not mandatory but may nonetheless be useful for informational and preventive purposes.

29 Two decisions of the French Supreme Court (Full Court) of 7 March 1986, *Babolat Maillot Witt v Mr Jean X*, (No 83/10477) and *Williams Electronics v Jeutel* (No 85/91465), applied by the French Supreme Court (Commercial Chamber) in *Alix v Codix*, 17 October 2012.

30 See *Microsoft*, French Supreme Court (Commercial Chamber), 14 November 2013.

31 See *Microsoft*, French Supreme Court (Commercial Chamber), 14 November 2013 and *Alix* v *Codix*, Montpellier Court of Appeal, 6 May 2014.

32 *Mr X* and *Mr Y*, French Supreme Court (Criminal Chamber), 9 September 2003.

33 *SARL Vidéo informatique par numérisation et compression d'images VINCI v Périphériques et matériels de contrôle*, Paris Court of Appeal, 7 May 1999, No 1997/18104.

34 Article L112-1 of the IPC.

35 Article L111-1 of the IPC.

36 *Enveloppe Soleau.*

37 See the Agency for Programs Protection website: www.app.asso.fr/.

4.2 Other subject matter protected by copyright

Under French law, industrial designs which fulfil the originality requirement are protected under the common copyright regime. The 'unity of the arts' doctrine provides that French copyright law does not differentiate between designs that are industrial and those which are artistic in nature.

Similarly, following the European Court of Justice (ECJ) decision of 22 December 2010,[38] graphical user interfaces are not protected by the specific copyright deriving from EU Directive 91/250, but rather by the common copyright regime. As such, their protection is subject to the condition of originality.

4.3 Limitation of copyright protection

Ideas, principles and algorithms[39] which underlie any element of a computer program are not protected by copyright, but may still be protected under the laws of unfair competition or parasitism.[40] As a general principle of copyright law, works are protected in their form or expression, excluding the ideas and principles on which they are based.

The functionalities of software are not afforded copyright protection, as they are treated as ideas.[41]

Although the look and feel of original software is not excluded from copyright protection under the common copyright regime, it may be easier to seek protection through industrial design, or unfair competition and parasitism.

4.4 Infringement

Any act committed in violation of the author's rights may constitute infringement. With respect to software, Article L122-6 of the IPC expressly provides that the violation of the following authors' rights constitutes infringement:[42]

1° the permanent or temporary reproduction of software by any means and in any form, in part or in whole;

2° the translation, adaptation, arrangement or any other alteration of software and the reproduction of the results thereof;

3° the placing on the market, whether for reward or free of charge, including rental, of the software or of copies of the software by any process.

---

38 See *Bezpe nostní softwarová asociace – Svaz softwarové ochrany v Ministerstvo kultury*, ECJ, 22 December 2010, C-393/09.

39 *Auriol v Widmer*, Paris Court of Appeal, 23 January 1995: "An algorithm is a simple succession of operations reflecting functionalities and therefore it is not capable of being protected" and *Microsoft*, French Supreme Court (Commercial Chamber), 14 November 2013.

40 See section 7.3 below.

41 *Perrin v Microsoft*, French Supreme Court (First Civil Chamber), 13 December 2005.

42 Article L335-3(2) of the IPC (which incorporates by reference Article L122-6).

4.5 Permitted acts

A user who has been authorised by the owner to use the software (an authorised user) benefits from special rights compared to those of the common copyright regime. Specifically, an authorised user has the right to reproduce, translate, adapt, or arrange the software (including to correct errors), since such acts are necessary for use of the software.[43] However, the rights holder may, by agreement, reserve its right to correct errors and specify special conditions under which the user is entitled to enjoy these rights.[44]

An authorised user is entitled to make a back-up copy (not to be confused with the 'private copy' authorised by Article L211-3 of the IPC) of the software, if this is necessary for the purpose of maintaining use of the software.[45]

An authorised user is also entitled to observe, examine or test the functioning of the software for the purpose of identifying the ideas and principles which underlie any element of the software, a practice often referred to as 'reverse-engineering'. The authorised user may conduct this review in the context of normal use of the software, while loading, displaying, running, transmitting or storing the software.[46]

Additionally, an authorised user has the right to decompile the software – that is, to change the language in which the software is written into something that is more understandable to the lay person. This can be achieved by taking the language of the original program (object code) and changing it into a source code that can be better understood. The user is entitled to reproduce or translate the form of its code if such operations are essential for the purpose of obtaining information that is necessary to achieve the interoperability of the software with other software.[47]

Software decompilation may be carried out only:

- by the user entitled to use the software or on his or her behalf by a person authorised to do so;
- for the single purpose of enabling the interoperability of multiple software products and only in connection with parts of the software that are necessary to achieve interoperability; and
- if the information necessary to achieve interoperability has not been easily and quickly made available to the user or the person acting on his or her behalf.

In any case, the rights granted to the user must be carried out in a manner that does not modify the software or harm the normal exploitation of the owner's rights.[48]

---

43 Article L122-6-1(I)(1) of the IPC.
44 Article L122-6-1(I)(2) of the IPC.
45 Article L122-6-1(II) of the IPC.
46 Article L122-6-1(III) of the IPC.
47 Article L122-6-1(IV) of the IPC.
48 Article L122-6-1 of the IPC.

4.6 Prohibited acts

Since 1957, French Law has provided for a 'private copy' exception under which the reproduction or representation of a work or performance does not require consent of its rights holders if such reproduction is strictly limited to the copyist's private, personal and non-commercial use.

However, Article L122-5 2° of the IPC provides that this exception does not apply to copies of a software other than the back-up copies made under Article L122-6-1(II) of the IPC. A user is therefore not entitled to make a private copy of any software.

With the advent and democratisation of blank high fidelity digital storage media, and the subsequent dramatic increase in the number of private copies made, the Law of 3 July 1985 introduced private copying remuneration (PCR) to compensate for the loss suffered by copyright holders. Since then, a percentage of the purchase price of blank media or equipment intended for copying music and images (such as blank CDs and DVDs, digital audio players and USB memory sticks) is thus directed to a levy payment towards creators, publishers, performers and producers.

Of the collected sums, 75% are paid directly to the creators, publishers and producers of the works copied, thereby contributing to the process of creation, while the remaining 25% contribute to funding cultural initiatives. The calculation of the remuneration is based on the declared storage or recording capacity of each media, expressed in gigabytes (GB).

Payment distribution has been assigned to an independent administrative authority, known as the Commission de la copie privée, which defines the tariffs and type of recording and storage media subject to the levy and the methods of declaration and collection. Although the levy is excluded for recording media acquired for professional ends, almost all consumer recording and storage media now fall within its scope, the latest addition being Network Personal Video Recorder (nPVR). PCR does not, however, apply to back-up copies of software.

5. Design right protection

Industrial designs protect the appearance of a product.[49] As such, design protection is not appropriate for software *per se*, since software products have no form of external appearance and such protection is expressly excluded by Article L511-1 of the IPC. However, multiple graphical interfaces or audiovisual effects of a software product can be protected through the French industrial designs regime, since these are distinct elements.[50]

In order to be protected, a design must meet the novelty and individual

---

49 Article L511-1 of the IPC.

50 Some examples of graphical interfaces registered as industrial designs at INPI are available on its online database under the following registration numbers: 20120014, 044225 and 977636.

character requirements.[51] Pursuant to Article L511-3 of the IPC, a design will be considered as being novel: "if, on the date of the filing of the application for registration or on the claimed date of priority, no identical design or model has been disclosed". Moreover, a design will be considered to possess individual character: "if the overall visual impression it produces on the informed observer differs from that produced by any design or model disclosed before".

Property rights over an industrial design are acquired through registration of that design, which is granted to the author or his or her successor.[52] On registration, the applicant is deemed to be the beneficiary of such protection.[53]

The registration of an industrial design takes effect from the filing date of the application for a five-year period, which may be extended by additional five-year periods up to a maximum of 25 years.[54]

6. Confidential information

In France, trade secrets are now protected by the New Law on the protection of trade secrets of 30 July 2018 implementing the EU Directive 2016/943 of 8 June 2016. 'Trade secret' is defined by Article L.151-1 of the French Commercial Code which requires that three cumulative criteria be met:

- the information must be secret, that is, not generally known or readily available to persons in the relevant industry;
- the information must have a real or possible commercial value; and
- the persons having control over the information must have taken reasonable steps to keep it secret.

Pursuant to Article L.151-4 of the French Commercial Code, a trade secret acquired without the consent of the trade secret holder is unlawfully acquired, whenever carried out by unauthorised access to, appropriation of, or copying or, more generally, any other conduct which, under the circumstances, is considered dishonest and contrary to commercial practices.

The system does not introduce any specific criminal remedy, only the civil liability of the infringer being incurred (Article L.152-1). While not making fraudulent appropriation of trade secrets a specific criminal offence, other existing offences can, for example, be relied on in the following situations: theft, handling stolen goods, breach of trust, disclosure of manufacturing secrets (*secret de fabrique*), and unlawful intrusion into a computer system.

Among the civil remedies, the court can order, subject to a penalty fine for non-compliance, any proportionate measure to prevent or put an end to a trade secret misappropriation as well as an award of damages, assessed on the basis of:

---

51 Article L511-2 of the IPC.
52 Article L511-9 of the IPC.
53 *Ibid.*
54 Article L513-1 of the IPC.

- the adverse economic consequences of the misappropriation of the trade secret;
- the moral prejudice; and
- the profits made by the author of the misappropriation, including savings in terms of intellectual, tangible and promotional investments as a result of the infringement.

Alternatively, Article L.152-5 of the French Commercial Code specifies that the court may, upon the motion of the author of the misappropriation, grant a financial compensation provided that he or she proves that:

- he or she did not know that the secret had been unlawfully obtained;
- prohibition or injunction measures would cause him or her disproportionate harm; and
- a financial compensation appears to be 'reasonably satisfactory'. The amount of such compensation is capped and correlated with the amount that the author of the misappropriation would have had paid in seeking the consent to use the trade secret.

Furthermore, Article L621-1 of the IPC contains a specific reference to the Labour Code, which applies only to disclosure of confidential information by a director or an employee. Consequently, the only type of trade secret that is protected under the code is manufacturing secrets. Protection is thus limited to the disclosure or attempted disclosure of manufacturing secrets. Such disclosure is subject to a fine of up to €30,000[55] and up to two years' imprisonment.[56]

7. Other IP rights

7.1 Database rights

Pursuant to Article L112-3 of the IPC (which implemented EU Directive 96/9/EC on the legal protection of databases into French law):

> *The authors of translations, adaptations, transformations or arrangements of works of the mind shall enjoy the protection afforded by this Code, without prejudice to the rights of the author of the original work. The same shall apply to the authors of anthologies or collections of miscellaneous works or data, such as databases, which, by reason of the selection or the arrangement of their contents, constitute intellectual creations.*
>
> *Database means a collection of independent works, data or other materials, arranged in a systematic or methodical way, and capable of being individually accessed by electronic or any other means.*

---

55 This amount is multiplied by five for legal persons.
56 Article L621-1 of the IPC.

Therefore, if an arrangement of works is considered to be original, it will benefit from copyright protection, regardless of whether the collected/compiled works are protected as original works themselves.

Pursuant to Articles L341-1 and L341-3 of the IPC, without prejudice to copyright protection or any other rights over the database or any of its components, the producer of a database has the right to prevent third parties from extracting and/or re-using the whole or a substantial part, evaluated qualitatively and/or quantitatively, of the database.

In order to benefit from the foregoing regime, it is necessary to show that there has been a qualitatively and/or quantitatively substantial investment – be it financial, material or human – in either obtaining, verifying or presenting the contents of the database.[57] Therefore, a database can benefit from both the common copyright regime and *sui generis* regime (a special form of protection regime outside the known framework).

A database protected under Articles L341-1 and following of the IPC will benefit from a shorter protection term compared to the common copyright regime term. *Sui generis* protection is granted for a 15-year period[58] from the date of completion of production of the database, or for a 15-year period[59] from the date on which the database is first made available to the public if it is made available before the expiry of the former period.[60] However, where a protected database is the subject of a new substantial investment, its protection is extended for a further 15-year period.[61]

7.2 Semiconductor topographies

Topographies of semiconductor products have been protected under French law since 1987 following the adoption of Law 87-890, which implemented EU Directive 87/54/EEC.

A final or intermediary topography of a semiconductor will receive protection pursuant to Article L622-1 of the IPC if the following conditions are met:

- the topography is the result of its creator's own intellectual effort; and
- it is not commonplace in the semiconductor industry.

The creator[62] (or his or her successor) of the semiconductor topography must file an application within two years of first commercialisation and within

57 Article L341-1 of the IPC.

58 Fifteen years from 1 January of the calendar year following completion of the production of the database.

59 Fifteen years from 1 January of the calendar year following the date on which the database was first made available to the public.

60 Article L342-5 of the IPC.

61 *Ibid.* Fifteen years from 1 January of the calendar year following that in which this new investment was made.

15 years of the topography being first fixed or encoded, if it has never been exploited.[63]

From the filing date or the date of first commercial exploitation (if earlier),[64] third parties are prohibited from reproducing the protected topography and commercial exploitation or importation to that end of such a reproduction or of any semiconductor product incorporating it.[65]

7.3 Protection against unfair competition and parasitism

In addition to the foregoing forms of protection for software products in France, tort law[66] offers complementary protection.

An unfair competition and parasitism action can be initiated at the same time as an infringement action. However, this action is admissible only if it is based on a set of facts different from that on which the infringement action is based.

The cause of action for unfair competition and parasitism is based on Articles 1240 (formerly 1382) and 1241 (formerly 1383) of the French Civil Code, which provide for the main French law principles applicable to tort law. The elements of the cause of action of unfair competition and parasitism are:

- a wrongful act;
- loss or damage; and
- a causal link between the wrongful act and the damages suffered by the plaintiff.

Parasitism is a specific form of unfair competition, which French courts have defined as the unfair appropriation of a third party's intellectual work.[67]

For instance, in an action for software copyright infringement and unfair and parasitic competition against the plaintiff's former business partner, the French Supreme Court ruled that the defendant's behaviour had been parasitic in nature: "the contentious software was designed and developed by [the

---

62 Article L622-3 of the IPC. Furthermore, Article L622-2 of the IPC provides that:
"The following shall be eligible for the benefits of this Chapter:
a) Creators who are nationals of a State party to the Agreement Establishing the World Trade Organization or who have either their habitual residence or a real and effective industrial or commercial establishment in such country, and their successors in title;
b) Persons meeting the aforesaid conditions of nationality, residence or establishment, who in a Member State engage in the first commercial exploitation anywhere in the world of a topography not protected by this Chapter, for which exploitation they have received exclusive authorization from the entitled person for the whole of the European Community or the European Economic Area.
c) Persons other than those referred to in the foregoing paragraph shall be eligible for the benefits of this Chapter subject to evidence of reciprocity with the countries of which they are nationals or in which they are established."

63 Article L622-1 of the IPC.

64 Article L622-6 of the IPC which further provides that "any registration relating to a topography that has not been commercially exploited within a period of 15 years from the date on which it was fixed or encoded for the first time shall cease to have effect".

65 Article L622-5 of the IPC.

66 Articles 1240 (formerly 1382) and 1241 (formerly 1383) of the French Civil Code.

67 See *Perrin v Microsoft*, French Supreme Court (First Civil Chamber), 13 December 2005.

defendant] using the latest research work [of the plaintiff], and ... this misappropriation of know-how, made possible by the exposed contractual relationship between the parties, allowed the defendant to achieve significant savings at the expense of the plaintiff".[68]

8. Ownership, inventorship and compensation

8.1 Patents

The right to a patent belongs to the inventor.[69] An applicant is deemed to be the inventor.

Article L611-7 of the IPC provides a specific regime in relation to inventions created by employees. There are two types of employee invention:

- those made during the course of normal work activities; and
- those made outside the course of normal work activities, attributable or non-attributable to the employer.

Inventions made during the course of work activities belong to the employer, provided that the employee had an inventive mission. Consequently, the employer is required to provide the employee with an additional compensation as established by any collective agreements, company agreements and the individual's employment contract. Such additional compensation is usually triggered by the filing of a patent and/or exploitation thereof. Various criteria may be taken into account for the purposes of evaluating such compensation, including the general context of the research, the difficulties of developing and perfecting the invention, the importance of the inventor's personal contribution and/or the degree of industrial and commercial interest of the invention to the employer and the number of inventors having contributed to the conception of the invention. Any dispute relating to an inventor's additional compensation is submitted to the CNIS (*Commission Nationale des Inventions de Salariés*) joint conciliation board, the proposal of which shall be deemed to constitute an agreement between the parties if, within one month of its notification, neither of them has submitted the case to the Paris First Instance Court.[70]

Inventions attributable to the employer made outside the course of work activities include any invention made in the employer's field of activity or created with the help of the employer's resources. Such inventions may be claimed by the employer against payment of a 'fair price' which, failing agreement between the parties, shall similarly be assessed by the CNIS joint

68 *Ibid.*
69 Article L611-6 of the IPC.
70 Article L615-21 of the IPC.

conciliation board or the First Instance Court. Assessment of the 'fair price' is based "on both the initial contributions of either of them and the industrial and commercial interest of the invention".[71]

All other inventions belong to the employee.

Inventions made by inventors enjoying the status of civil servant or public official, as is often the case of public research universities' inventors, have a right to receive a €3,000 'patent bonus' as well as a yearly 'profit-sharing bonus' which amounts to 50% (for revenue up to €68,659) and 20% (for revenue exceeding the previous cap) of the revenue derived from the invention.[72]

Inventions made by corporate officers, trainees, consultants or any other non-employees belong to them, absent any assignment to the contrary.

8.2 Copyright

Under French law, the mere existence of a service contract under which the author created the work does not grant the author's contractual partner any IP rights in the software.[73] Except in the case of collective works, the author is deemed to have created the work. Collaborative works (ie, any work created by more than one person)[74] are jointly owned by their authors, who may exercise their rights by mutual agreement. Where the contribution of each author differs, the authors are entitled to exploit their personal contributions separately, provided that they do not harm the exploitation of the work as a whole.[75]

Any transfer of copyright must result from an express and written assignment agreement, which shall identify each of the assigned rights and the field, scope, purpose, territory and duration of the exploitation.[76]

Pursuant to Article L113-2(3) of the IPC, a 'collective work' is defined as:

(i) a work created at the initiative of a natural or legal person (ii) who edits it, publishes it and discloses it under his/her direction and name (iii) and in which the personal contributions of the various authors who participated in its production are merged in the overall work for which they were conceived, without it being possible to attribute to each author a separate right in the work as created.

Unlike the common copyright regime, in which the mere existence of an employment contract creates no rights for the benefit of the employer, the rights relating to software and its documentation created by one or more employees in the course of their duties or following the instructions of their employer are owned by the employer.[77] No additional compensation for the employee is provided for by the law relating to such creation.

---

71 Article L611-7 of the IPC.
72 Article R611-14-1 of the IPC.
73 Article L111-1 of the IPC.
74 Article L113-2 of the IPC.
75 Article L113-3 of the IPC.
76 Article L131-3(1) of the IPC.
77 Article L113-9 of the IPC.

Unless proved otherwise, the collective work is owned by the party under whose name it has been disclosed.[78]

9. Counterfeiting

9.1 Criminal liability

Under French law,[79] the unauthorised reproduction, use or commercialisation (or any other infringing act)[80] of software products that are protected under copyright law is a criminal offence. This offence is punishable by a fine of up to €300,000 and up to three years' imprisonment.[81] In the case of a second or subsequent offence, the foregoing penalties may be doubled.[82] They may also be accompanied by additional penalties, such as closure of the business, recall of the infringing product, seizure of revenues generated by the offence and destruction of infringing goods.[83]

When an offence is committed by a company, the fine may be quintupled.[84]

Moreover, the rights holder may also apply to the customs authorities in order to have the counterfeiting goods detained.[85]

9.2 Illegal downloads

In 2006, the French Parliament adopted a law to prevent illegal downloads of copyrighted material on the internet.[86] One of the adopted measures enables a rights holder to ask the court, in summary proceedings, to order the necessary measures to prevent or prohibit the use of a software product which is mainly designed for illegal downloading of copyrighted material.

9.3 HADOPI

The High Authority for the Dissemination of Works and the Protection of Rights on the Internet (HADOPI) was created in 2009 in order to stop illegal peer-to-peer downloading.

Pursuant to these provisions, a different infringement regime applies to illegal downloading of copyrighted material from the internet. This particular type of infringement is not enforced via regular legal channels. HADOPI first sends a notice to the illegal downloader. If, six months after the notice, the infringing acts have not ceased, HADOPI sends a second notice. Finally, if the infringement has not ceased within one year of the second notice, the infringer

---

78 Article L113-5 of the IPC.
79 See Articles L335-1 and following of the IPC. With regard to patents, see Articles L615-12 and following.
80 See section 3.3 above.
81 Article L335-2 of the IPC.
82 Article L335-9 of the IPC.
83 Articles L335-5, L335-6 and L335-8 of the IPC and Article 131-39 of the French Penal Code.
84 Article L131-38 of the French Penal Code.
85 See EU Regulation 1383/2003 of 22 July 2003 and Article L335-10 of the IPC.
86 See Articles L336-1 and following of the of the IPC.

may be served with a €1,500 fine. Since 2013, the complementary penalty consisting in having one's internet connection withdrawn for up to a year is no longer available to courts.[87]

As of 14 September 2017, HADOPI had sent more than 9.6 million warning emails, 686,318 letters and transferred 1,308 cases to the courts. Since it began operating, only 189 criminal penalties have been issued, for fines ranging mainly from €50 to €1,000.[88] Recently, a decree of 9 March 2017, provided for compensation for Internet Service Providers (ISPs) due to the costs of the system of HADOPI's 'graduated response'. However, this decree has yet to enter into force.[89]

It is worth noting that HADOPI might extend its power to be the body in charge of controlling the licence agreements between the owners of copyrights and the online sharing platforms as provided by Article 17 of the EU Directive on Copyright in the Digital Single Market which was implemented into French law on 24 July 2019.

10. Enforcement

10.1 Obtaining information

Although French law does not provide for a common-law style discovery procedure, it nonetheless offers a wide range of tools for gathering evidence. Rights holders are allowed to collect:

- evidence of infringement, even if no infringement action has been initiated, on the basis of an *ex parte* procedure called *saisie-contrefaçon*; and
- information as to the extent of the infringement through the exercise of a 'right of information', when the rights holder has sufficient evidence to bring a legal action.

(a) Saisie-contrefaçon

The *saisie-contrefaçon* has been part of the French system since 1992. This procedure enables the rights holder to have, upon grant of an order issued in the vast majority of cases on an *ex parte* basis by the President of the First Instance Court, the alleged infringement recorded by a bailiff both to enter any place where the infringement might be observed and to seize the items of evidence of the infringement. Since 2007,[90] identical provisions have governed the *saisie-contrefaçon* for patents,[91] trademarks,[92] industrial designs[93] and semiconductor

---

87 Article L335-7 of the IPC.

88 HADOPI Annual Activity Report for 2016–2017.

89 Decree No 2017-313 of 9 March 2017 concerning the modalities for compensating identifiable and specific additional costs of services provided by the electronic communications operators at the request of the High Authority for the Dissemination of Works and the Protection of Rights on the Internet.

90 Law 2007-1544, 29 October 2007, implementing the EU Directive on the Enforcement of IP Rights (2004/48/EC).

91 Article L615-5(2) of the IPC.

topographies.[94] However, special provisions apply to software.[95] A combination of these special rules and the rules of the common *saisie-contrefaçon* regime applicable to copyright[96] apply to alleged infringements concerning software.

Pursuant to Article L332-4 of the IPC, two combinable types of *saisie-contrefaçon* are available to rights holders:

- a detailed description (*saisie descriptive*) with or without taking samples of the allegedly infringing software or database; and
- a physical seizure (*saisie réelle*) of:
  - the allegedly infringing software or database as well as any related documents (invoices, marketing and technical documents, copies of accounting books etc); and
  - the equipment and tools used to produce or distribute the allegedly infringing software or database.

The *saisie-contrefaçon* application must identify:

- the rights holder and the IP right on the basis of which the application is made;
- the party that is the subject of the *saisie-contrefaçon*;
- the subject matter of the *saisie-contrefaçon*, ie, the specific measures sought;
- the name of the bailiff who will perform the *saisie-contrefaçon* and, where applicable, the names of the experts who will assist him or her; and
- the place(s) where the *saisie-contrefaçon* will be carried out.

Even if the purpose of a *saisie-contrefaçon* is to collect evidence of the infringement, the applicant will nonetheless have to provide some *prima facie* evidence which is likely to show that infringement is at least plausible,[97] in line with the requirements of EU Directive 2004/48/EC.

If the rights holder's application complies with all relevant requirements, the President of the First Instance Court may issue an order authorising the rights holder to execute a *saisie-contrefaçon*, usually at the alleged infringer's premises, but also in that of any third party.

The *saisie-contrefaçon* is carried out by a bailiff who can be assisted by experts but without the presence of the rights holder or any of its employees.

The seizing party must bring an infringement action (civil or criminal) within 20 working days or 31 calendar days, whichever is longer from the end of the *saisie-contrefaçon*. Otherwise, the *saisie-contrefaçon* will be void.

---

92 Article L716-7 of the IPC.
93 Article L521-4 of the IPC.
94 Article L622-7 of the IPC.
95 Article L332-4 of the IPC.
96 Articles L332-1 to L332-3 of the IPC.
97 *SFR, Huawei and Bouygues v High Point*, Paris Court of Appeal, 28 January 2014; *Duplo v MDI Digital* Technology, Paris First Instance Court Interim Relief Judge, 8 September 2017.

(b) Right of information

Since 2007,[98] identical provisions have governed the right of information for copyright,[99] patents,[100] trademarks,[101] industrial designs[102] and semiconductor topographies.[103]

The purpose of the right of information is to enable the plaintiff to:

- establish the origin of the distribution channels of the infringing goods, allowing, if necessary, extension of the action for infringement against other parties involved; and
- evaluate damages.

Therefore, if a right of information is granted, the court may order the defendant or any other party involved to produce documents or information relating to:

- the name and address of the producer, manufacturer, distributor, supplier and other prior possessors of the infringing goods or services, as well as the name and address of the wholesaler or retailer; and
- information regarding the quantity of products that have been produced, commercialised, delivered, received or ordered, as well as the price obtained for the products and services in question.

A right of information may be requested either as part of a pending infringement action, by submitting a request before the pre-trial judge, or autonomously, once a decision on the merits has been issued.[104] The request can equally be directed against the alleged or condemned infringer, as well as against a third party.

10.2 Interim relief

Since 2007,[105] except for copyright, identical provisions have governed interim injunctions for patents,[106] trademarks,[107] industrial designs[108] and semiconductor topographies.[109]

(a) Copyright

Copyright, including software protected by the specific copyright regime, is

---

98 Law 2007-1544 of 29 October 2007.
99 Article L331-1-2 of the IPC.
100 Article L615-5-2 of the IPC.
101 Article L716-7-1 of the IPC.
102 Article L521-5 of the IPC.
103 Article L622-7 of the IPC.
104 *New Wave v Alltoys*, ECJ, 18 January 2017, C-427/15.
105 Law 2007-1544, 29 October 2007.
106 Article L615-3 of the IPC.
107 Article L716-6 of the IPC.
108 Article L521-6 of the IPC.
109 Article L622-7 of the IPC.

governed by the common interim injunction regime of the Code of Civil Procedure. The applicant has the following options:

- in urgent cases, file an application for interim relief where such application may not be seriously challenged or where the facts of the case justify the grant of interim relief;[110] or
- file an application based on evidence of an imminent loss and/or the existence of a manifestly unlawful disturbance resulting from the infringement.[111]

For instance, having found similarities between software products which were likely to trigger a likelihood of confusion, a court issued an interim injunction based on the existence of a 'manifestly unlawful disturbance' against the defendant which offered the infringing software for downloading.[112]

(b) *Other IP rights*

IP rights other than copyright are subject to the following unified regime:

- *Any person having the right to sue for infringement may bring summary proceedings before the competent civil court in order to have ordered, if necessary with penalty fines, any measure intended to prevent an imminent infringement or to terminate the continuation of allegedly infringing acts, against the alleged infringer or intermediary agents whose services the latter uses;*
- *Following an* ex parte *application, the competent Civil Court may also order any urgent measure, when circumstances require that such measures should not be taken in the presence of both parties, in particular when any delay might cause irreparable harm to the plaintiff;*
- *Ruling on* inter partes *or* ex parte *proceedings, the Court may only order the requested measures if evidence reasonably available to the applicant makes it likely that his or her rights shall be infringed or that such infringement is imminent.*[113]

A lower degree of evidence of infringement is required for the purpose of obtaining an interim injunction compared to the level required for the purpose of an action on the merits, to the extent that:

- the evidence is limited to the evidence that is 'reasonably available'; and
- it is evidenced that the infringement (or threat thereof) is 'likely'.

---

110 Article 808 of the French Code of Civil Procedure: "In all cases of urgency, the President of the Court of First Instance may order, in summary proceedings, all measures that do not encounter any serious challenge".

111 Article 809 of the French Code of Civil Procedure: "Even if serious challenges are raised, the President of the First Instance Court may order, in summary proceedings, measures ... to avoid an imminent loss or to terminate a manifestly unlawful disturbance".

112 *Conex v Tracing Server*, Paris Commercial Court of First Instance, 24 October 2002.

113 Articles L615-3 (patents), L521-6 (industrial designs), L716-6 (trademarks), L343-2 (databases) and L622-7 (semiconductor topographies) of the IPC.

According to recent case law, only *prima facie* and serious challenges made with regard to the validity of a patent make infringement implausible and preclude interim relief from being granted.[114]

Various measures may be ordered by the judge:

- prohibiting further infringement, in conjunction with penalties;
- making the continuation of the infringing acts subject to the lodging by the plaintiff of a guarantee to cover possible indemnification of potential loss suffered by the defendant if the infringement is subsequently considered to be ill-founded; and
- granting provisional damages if the existence of the plaintiff's loss 'is not seriously challengeable'.

(c) Duty to initiate proceedings following the order

The plaintiff must bring an infringement action (civil or criminal), or file a complaint with the public prosecutor, within 20 working days or 31 calendar days, whichever is the longer, of issue of the order; otherwise, the order will be void upon the defendant's simple request.

10.3 Unjustified threats

The use of warning letters, which is common in France, is not regulated or prohibited by the French IPC or Civil Code. French law does not provide for a specific action for unjustified threats of patent infringement proceedings.

However, a party that is subject to these types of threat can bring suit based on general French tort law, in particular for unfair competition.[115]

A company or person can make a claim for unfair competition if, for instance, warning letters are sent to its clients containing information regarding a potential or impending suit for infringement.[116] Therefore, reasonable care should be exercised in asserting that a third party has infringed a patent before any court ruling on the merits to that effect.[117]

The recipient of unjustified threats can also apply to the courts for a declaration of non-infringement of a patent, by which the judge issues a ruling recognising the non-infringing nature of the goods and/or methods. The application cannot be filed before the patentee is invited to respond as to the opposability of the patent with regard to the use in dispute.[118]

In the case of infringement proceedings, the defendant can also assert that the proceedings were frivolous. If the judge rules that the patentee's claims are in fact frivolous, the claimant can be ordered to pay a maximum fine of

114 *Gilead v Mylan*, Paris First Instance Interim Relief Judge, 5 September 2017.
115 See section 7.3 above.
116 See, for example, Paris Court of Appeal, 3 November 1993.
117 See, for example, *KIS v Games Diffusion International*, Paris First Instance Court, 16 May 2000.
118 Article L615-9 of the IPC.

€10,000[119] without prejudice to the defendant's right to claim damages for abusive proceedings pursuant to Article 700 of the Code of Civil Procedure. It is difficult to establish that the claim is frivolous, since malice or bad faith must be evidenced by the defendant. Hence, the chances of success of such motions are relatively low.

10.4 Remedies

In most infringement cases, an injunction will be ordered to enjoin the infringer from selling or manufacturing the infringing products. Recall, forfeiture and/or destruction of the infringing products, as well as publication of the decision, may also be ordered.[120]

According to a general principle of civil law,[121] damages are awarded for the sole purpose of compensating the loss suffered by the plaintiff. Therefore, under French law, no punitive damages may be awarded.

However, since 2007,[122] "the profits made by the infringer" may be taken into account in the assessment of the suffered loss:

- *For the purpose of evaluating damages, the court takes into account the negative economic consequences, including loss of profits which the injured party has suffered, any profits made by the infringer and the moral prejudice caused to the rights holder by the infringement.*
- *However, the court may, as an alternative and on the request of the injured party, set the damages as a lump sum that cannot be inferior to the amount of royalties or fees which would have been due if the infringer had requested authorisation to use the IP right in question.*[123]

Hence, the plaintiff has a choice between two methods of damages calculation:

- a detailed calculation, including the loss suffered, the profit made by the infringer and the moral prejudice depending on the sovereign assessment of the court; or
- a lump sum (only if requested by the plaintiff).

11. Hot topics

In one year, France has amended its legal IP system to make it more efficient and to protect better innovation in France.

This includes:

- the New Law on protection of trade secrets of 30 July 2018 implementing the EU Directive 2016/943 of 8 June 2016;

119 Article 32-1 of the French Code of Civil Procedure.
120 Article L335-10 of the IPC (copyright).
121 Article 1240 (formerly 1382) of the French Civil Code.
122 Law 2007-1544, 29 October 2007.
123 Articles L331-1-3 (copyright), L615-7 (patents) and L716-13 (trademarks) of the IPC.

- the Loi PACTE of 22 May 2019 which significantly changed the French patent system by introducing an inventiveness assessment, creating a French provisional patent application as well as an opposition procedure; the regulatory framework of these changes shall be defined in the coming months; and
- the New Law of 24 July 2019 implementing the EU Directive on Copyright in the Digital Single Market.

The advent of the blockchain technology, a digitised, decentralised, public ledger of transaction, has been acknowledged by many as having a potential impact on the proof of ownership, and how licences and assignments of IP rights are recorded and managed, as well as on monetisation thereof.

The tamper-proof nature of the blockchain technology would also make its use possible to provide evidence of creations or inventions and give a definite date to an idea or project, much in the same way as the *Soleau* envelope service provided by the INPI currently makes it possible.

Furthermore, the number of cryptocurrency and blockchain-related patent applications being filed and published, which nearly doubled in the United States in 2017,[124] reflects the interest that these technologies generate, as well as the impact they could have on intellectual property in the future.

The number of patent applications in the field of artificial intelligence (AI) is also soaring, with a nearly threefold increase worldwide between 2012 (708 applications) and 2016 (2,888 applications) and a more than 500% increase in the US in the relevant 'AI data-processing systems' class.[125] The various legislative bodies around the globe will also soon have to grapple with issues related to the ownership of AI-generated copyrighted works and patentable inventions, as this is currently not specifically dealt with by existing laws. Should we create the concept of the skilled machine in the art to assess the novelty and inventiveness? How should we treat an infringement committed by AI and who should be considered to be liable for that infringement?

Lastly, quantum computing, which is based on the use of quantum-mechanical phenomena, such as superposition and entanglement, to perform computing operations, is also set to significantly impact the area of software and electronics, with patent applications in this field being forecast to rise by 430%, from less than 100 publications in 2014, up to more than 400 publications projected through 2017.[126]

---

124 See www.coindesk.com/rate-blockchain-patent-applications-nearly-doubled-2017/.

125 *Trends and Priority Shifts in Artificial Intelligence Technology Invention: A global patent analysis*, Fuji and Managi, 2017, available at www.rieti.go.jp/jp/publications/dp/17e066.pdf.

126 See https://patinformatics.com/wp-content/uploads/2017/10/Quantum-Computing_Full_Report_Final_opt.pdf.

Germany

Erik Gelke
Leinen & Derichs Anwaltssozietät
Martin Köhler
Hoyng Rokh Monegier

Germany is the most active jurisdiction for IP infringement litigation in Europe. This applies in particular to patent litigation where the German bifurcation procedure, in which infringement and nullity are tried separately, is regarded as 'patentee friendly' and provides for quick and cost-effective patent enforcement. The courts in Dusseldorf and Mannheim have been involved in many major global patent litigation actions in the field of electronics and software over the last few years, developing a considerable body of case law.

1. National patent procedure

1.1 Filing

A German national patent application can be filed at the Patent and Trademark Offices (PTO) in Munich (headquarters), Jena or Berlin. Alternatively, an application can be filed online using software which can be downloaded from the office's website. Patent applications can be filed in any language. However, if the application documents are not in German, a translation must be filed with the PTO within three months.[1] Germany is also a member of the Patent Cooperation Treaty and national patent applications can also stem from pending international patent applications.

1.2 An examination system[2]

Patent applications are subjected to full substantive examination by the PTO with patent examiners checking if an application has technical character, novelty, inventive step and industrial applicability.

After the application documents have been filed and the filing fee paid the patent examiner performs a preliminary examination. This is an initial check of the content of the application to confirm that the formal requirements for a patent application have been met and also to identify any obvious obstacles to patentability.

---

1 Section 35(1), sentence 1 of the Patents Act.
2 For further information, see www.dpma.de/english/patents/patent_protection/index.html.

Substantive examination begins when an applicant files a request for examination and pays the examination fee. The examination request can be filed at any time up to seven years from the filing date.[3] If no examination request is filed the application is deemed withdrawn.[4] An applicant has the option of filing a separate search request asking the PTO to undertake a search of the application and provide the applicant with copies of the prior art before requesting substantive examination.

If the substantive examination confirms that the application meets the requirements for a patent the application is granted.

Anyone can oppose the grant of a patent within three months of publication of the patent specification. A notice of opposition is required and should state the reasons why an opponent considers that grant of a patent would be unlawful. If an opposition is filed and the opposition fee paid, a panel of PTO patent examiners re-examine the patent and either maintain the patent as granted or in an amended form; or revoke the patent.[5]

Appeals against decisions in opposition procedures can be lodged with the Federal Patents Court.

After the expiry of the opposition period, validity of a patent can still be challenged in a nullity action brought before the Federal Patents Court.

1.3 Timescales

Patent applications are published 18 months after filing or priority date if priority is claimed. Apart from that, applicants can influence timescales. If applicants wish to prolong the application procedure they can delay filing their request for examination for up to seven years from the application date. Conversely, if the applicant files for examination promptly, such as within four months from filing application, the application procedure can be completed within 24 to 30 months.

Opposition proceedings and appeal proceedings both usually take between 18 and 24 months. As the opposition procedure is designed as a post-grant procedure, a patent can be enforced while the opposition is pending and until there is a final and binding revocation. However, if an infringement court finds a high likelihood that a patent will ultimately be finally revoked in opposition proceedings, the infringement court may stay the infringement proceedings until the opposition is decided.

3 Section 44(2), sentence 1 of the Patents Act.
4 Section 58(3) of the Patents Act.
5 Sections 59 and 60 of the Patents Act.

1.4 Outline of costs

The main official fees for filing and prosecuting a patent application are:

Application fee	€60
Excess claims fees	€30 for each additional claim over 10
Search fee	€250
Examination fee	€350 (reduced to €150 if a search request has been filed)

The application fee and excess claims fees are reduced to €40 and €20 per claim if an application is filed online for 10 or more claims.

Annual maintenance fees are payable on pending applications and on granted patents. These range from €70 for the third year through to €1,940 for the 20th year of a patent.

2. Examination of software patents

The patentability of software may be raised at any time during examination proceedings, in post-grant opposition proceedings or during nullity proceedings. The case law of the Federal Patents Court and of the Federal Court of Justice on the patentability of software is characterised by a number of case-specific decisions from which some general guidelines may be derived.

According to Section 1(1) of the Patent Act, in order to be patentable an invention must have technical character. Technical character is considered to be present, if an invention at least partially involves some technical aspects. Examples of an invention's technical character having been established include:

- the embedding of a program in a technical device;
- processing, storage and transfer of data via technical means; and
- where a processing system's elements interact directly with particular data.[6]

The mere use of a device or its components may be sufficient to fulfil the requirement of technical character, even if this only represents a partial technical aspect of an invention.[7]

Subject matter excluded from patentability according to Section 1(3) and (4) – includes programs 'as such'. German practice in this respect is essentially identical to the approach adopted by the EPO. Under established German case law, programs are only patentable if they are designed to solve a specific technical problem.[8] According to this concept, the mere content of a program –

6 Federal Court of Justice of 20 January 2009, X ZB 22/07 (*Equipment for selecting medical examination methods*); of 22 April 2010, Xa ZB 20/08 (*Dynamic document generation*).

7 Federal Court of Justice of 20 January 2009, X ZB 22/07 (*Equipment for selecting medical examination methods*); of 26 October 2010, X ZR 47 (*Display of topographic information*); of 24 February 2011, X ZR 121/09 (*Website display*); of 22 April 2010, Xa ZB 20/08 (*Dynamic document generation*).

8 Federal Court of Justice (BGH), Decision of 17 October 2001 (*Incorrect strings*); decision of 22 April 2004, X ZB 22/07 (*Online banking*).

the program 'as such' (as excluded from patentability in Section 1(3) and (4) of the Patents Act) – does not serve a technical function.

Consequently, patent applications are examined to rule out programs which do not solve a concrete technical problem. The question is whether a patent's teaching serves to solve a concrete technical problem which goes beyond mere data processing. This is the case, for instance, if an external device (eg, a computer tomograph) is controlled by a computer program, requiring consideration of conditions outside the data-processing unit as part of the claimed teaching.[9]

The resolution of a concrete technical problem can be shown if device components are modified or addressed in a new way or if a program is determined by external technical aspects. Similarly, the adaption of a computer program to the technical features of a computer may be deemed to be a technical solution. In other words, because a computer is a technical device, software tailored to its characteristics may have technical character.[10]

The technical character of a claim has been denied in cases in which the teaching did not go beyond mere data processing; for example:

- software designed to offer interactive help;[11] or
- a method of graphically displaying all websites visited by a user during a session to create an easier-to-use overview.[12]

Finally, as with the EPO approach, the technical character of a teaching will be considered as part of the examination of inventive step.[13] When assessing the presence of an inventive step over the prior art, the courts will take into consideration only those elements involved in the solution of a technical problem by technical means; that is, only such features that present a technical character.[14]

3. Patent infringement

3.1 Claim interpretation

The scope of protection of a patent covers literal infringements as well as infringements under the doctrine of equivalents. According to Section 14 of the Patents Act in connection with Article 69 of the European Patent Convention and the protocol for the interpretation of that article, the protection conferred by the patent is determined by the terms of the claims, whereby the description and the drawings shall be used to interpret the claims. In order to determine the

9 Federal Court of Justice of 20 January 2009, X ZB 22/07 (*Equipment for selecting medical examination methods*).
10 Federal Court of Justice of 22 April 2010, Xa ZB 20/08 (*Dynamic document generation*).
11 Federal Court of Justice of 19 October 2004, X ZB 33/03 (*Offer of interactive help*).
12 Federal Court of Justice of 24 February 2011, X ZR 121/09 (*Website display*).
13 Federal Court of Justice of 20 January 2009, X ZB 22/07 (*Equipment for selecting medical examination methods*).
14 Federal Court of Justice of 26 October 2010, X ZR 47/07 (*Display of topographic information*); Federal Court of Justice of 18 December 2012, X ZR 3/12 (*Route planning*).

literal scope of protection, the German courts focus on the technical meaning of the claims and the technical function of the claim features within the teaching of the patent.

3.2 Doctrine of equivalents

Even if the claim is not literally used, a device or a process may still be considered an infringement under the doctrine of equivalents (see Article 2 of the protocol on the interpretation of Article 69 of the European Patent Convention). According to established case law, the scope of protection of a patent may be extended to alternatives beyond the literal meaning of the claim if:

- the alternative fulfils the same function as the claimed feature;
- the alternative could have been discovered by persons skilled in the art without independent inventive activity; and
- the alternative appears as an equivalent solution against the background of the entire teaching of the patent.

According to the first prerequisite, the infringing embodiment must have the same effect as the patented invention. Determining 'same effect', it is not sufficient for it to generally achieve the patented invention's overall effect; rather, the specific technical effect of the feature that has not literally been used must be achieved.[15] The second prerequisite rules out an alternative solution that constitutes an invention in itself.[16] Finally, the last prerequisite is a filter that takes into account the language used in the patent to define the claims. If a patentee, by using certain language, limits the claimed solution as against what could have been claimed, this limitation cannot be overcome by the doctrine of equivalents.[17]

Pursuant to German case law, a defendant may use the so-called 'Formstein' defence, which involves contending that an infringing device is identical to a prior art embodiment and thus cannot be covered by the scope of protection of a later patent.[18]

3.3 Acts of infringement

Section 9 of the Patents Act prohibits third parties from using the invention without the consent of the patentee. In the case of product inventions, infringement may consist of making, offering, putting on the market or using a product, or importing or stocking the claimed product for these purposes.[19] In

---

15 Federal Court of Justice, decision of 14 December 2010 (*Crimp tool*).
16 Federal Court of Justice, decision of 17 March 1994, X ZR 16/93 (*Means for cutting trees*).
17 Federal Court of Justice, decisions of 13 September 2011, X ZR 69/10 (*Diglycid composition*) and of 10 May 2011, X ZR 16/09 (*Occluder*).
18 Federal Court of Justice, decision of 29 April 1986, X ZR 28/85 (*Formstein*).
19 Section 9(1) of the Patents Act.

the case of process inventions, an act of infringement may consist either of using the claimed process or, under certain circumstances, of offering the claimed process for use in Germany.[20] In addition, offering, putting on the market, using or importing or stocking for these purposes a product that is directly obtained by a claimed process will also constitute infringement.[21]

'Offering' a product is construed broadly and will cover any act that involves the possibility of providing an infringing device to interested parties. Thus, for example, even without a formal offer of sale, any indication of a device in a brochure or on the internet, or exposure of a device at a trade fair, may constitute an act of infringement.

By contrast, offering a claimed electronic process for use in Germany is more narrowly construed and requires not only an offer of the sale or provision of a device on which a patented process is running, but additionally requires proof that the purchaser is specifically enabled to use the protected process and the seller knows that the use of the process without consent of the patentee is prohibited.

The requirement in Section 9 of the Patent Act that infringement can only occur where there is no consent on the part of the patentee, excludes from infringement any products that have previously been put onto the market by the patentee or by a third party with the patentee's consent, for example a licensee (principle of exhaustion of rights). The relevant market is, however, limited to the European Union. Therefore, there will be no exhaustion if the patentee sells products outside the European Union and the products are subsequently resold in the European Union by a third party without the patentee's consent. In addition, exhaustion only occurs if the product as claimed in the patent is put onto the market by the patentee (or a third party with the patentee's consent). Thus the sale of only part of a claimed device by the patentee is not sufficient to allow third parties the use of the entire patented device. Similarly, in the case of process claims exhaustion does not automatically occur with sales of a device which performs a patented process (but such a sale may constitute an implied licence even if no exhaustion of rights has occurred).[22]

3.4 Contributory infringement

Contributory infringement according to Section 10 of the Patent Act covers certain acts that may precede the use of the invention. This provision is of great significance in the area of electronics and software, in particular where the patent covers a process.

---

20 Section 9(2) of the Patents Act.
21 Section 9(3) of the Patents Act.
22 BGH GRUR 1980, 38 (*Fullplastverfahren*).

The conditions for contributory infringement are that:

- a means relating to an essential element of the invention – other than a staple commercial product –
  - is supplied or offered in Germany for use in Germany;
  - to a party not entitled to use the invention; and
- the seller/offeror knows, or it is obvious in the circumstances, that the means is suitable and intended to put the invention into effect.

What is considered to be an essential element is construed quite broadly in the case law. Any claim feature that is expressly mentioned in the claim may be considered an essential element of the invention and thus this condition regularly is fulfilled if the means constitutes itself an element of the claim. For example, this could be the case for a hardware component mentioned as a component of a complex patented device or a patented system. Another example would be a device on which a claimed process can be used, for example a chipset on which a claimed software process is to be run.

Contributory infringement requires proof of positive knowledge of the intended patent-infringing use. Due to the difficulties of establishing that kind of proof, Section 10 allows a patentee to establish that it is obvious from the circumstances that the means is suitable and intended for the patent-infringing use. The German courts regularly conclude that such knowledge exists if sellers/offerors expressly refer to an infringing use in their marketing material.

Although Section 10 of the Patent Act can provide protection against acts prior to infringement taking place, available remedies will depend on the degree of risk that an infringement will take place. If the only possible use of the means offered or supplied will infringe a patent then the patentee will be entitled to an unlimited injunction against the sale and offer of the means. If, however, a means supplied may have many potential uses, only one of which would infringe a patent, the patentee may be granted only a limited injunction that prevents the sale or offer for sale of the means without a disclaimer regarding the patent-infringing use.

3.5 Network claims

System or network claims are typically directed to a combination of hardware (and software) components forming a patented system. Infringement of such claims is not limited to the use of the system as claimed, since contributory infringement of such claims can occur through the offer and sale of the components of such systems. For instance, a patent directed to a network comprising a base station (BS) and a mobile station (MS) defined by certain operations taking place between BS and MS may be infringed by operating the entire network and also, under contributory infringement, by offering a BS or MS that is intended to be used within such a network.

In principle, acts that constitute an infringement must take place in Germany. However, if acts take place both in Germany and abroad, for example parts of a network claim are used in Germany and parts outside Germany, the combination of domestic and non-domestic acts may establish an infringement of a German patent. To avoid unreasonably broad jurisdiction, the Higher Regional Court of Dusseldorf has, however, ruled that only non-domestic acts that aim directly at an infringing effect in Germany can result in an infringer being liable in Germany.[23]

4. Copyright protection

4.1 Copyright works and related rights

German copyright protects the authors of works in the literary, scientific and artistic domain. According to Sections 2(1), 3 and 4 of the Copyright Act,[24] protected works include, in particular:

- literary works, such as written works, speeches and computer programs;
- musical works;
- dramatic works, including works of dance;
- artistic works, including works of architecture and of applied art, and drafts of such works;
- photographic works, including works produced by processes similar to photography;
- cinematographic works, including works produced by processes similar to cinematography;
- illustrations of a scientific or technical nature, such as drawings, plans, maps, sketches, tables and 3D representations;
- adaptations; and
- collections and database works.

The German Copyright Act also provides protection for rights related to the above works. Performers enjoy protection for their interpretation of works (Section 73 and following), and the following related rights are granted for technical, financial and organisational effort: scientific editions (Section 70), posthumous works (Section 71), photographs (Section 72), producers of audio recordings (Sections 85, 86), broadcasting organisations (Section 87), makers of a database (Section 87 and following) and producers of films and moving pictures (Sections 94, 95).

Hybrid products (eg, video games) may comprise multiple types of copyright works as well as being protected by related rights. This raises the question of

---

23 Higher Regional Court of Dusseldorf, Decision of 10 December 2009, I-2 U 51/08 (*Prepaid telephone card*).
24 English version at www.gesetze-im-internet.de/englisch_urhg.

applicable law as the level of protection varies depending on the applicable provisions. In recent decisions concerning hybrid products, the German Federal Court of Justice set a high level of protection for rights holders for rights in hybrid works.[25] They enjoy protection against the distribution of devices designed to circumvent technological protection measures for hybrid works (Section 95(a)(3)), even though the software aspects of such works alone would be granted a lower level of protection (Section 69(f)).[26] Software analysis – ie, the user's right to observe, study or test the functioning of a computer program – is prohibited, if it involves loading, displaying etc of the audio-visual parts of a hybrid product, even though analysis of the software part alone would ordinarily be permitted according to Section 69(d)(3) of the Copyright Act.[27]

4.2 Copyright protection for software programs

Computer programs are, by legal definition in Section 2(1) of the Copyright Act, literary works. They are, however, subject to special provisions in Section 69(a)–(g) of the Copyright Act.[28] The basic requirement for copyright protection is the presence of a certain level of originality. However, the threshold for originality is not high (referred to in Germany as the 'small coin' requirement) and this test equally applies to computer programs. Where complex software is involved, however, there is a rebuttable presumption of originality.[29] German courts generally evaluate originality by requiring the author to have employed some degree of creativity in collecting, selecting and arranging the subject matter, that is, the information, and in writing the programs. As computer-aided design is not excluded, developers are free to use tools in the programming process. Section 69(a)(3) of the Copyright Act stipulates that other criteria, such as quality or aesthetics, are irrelevant. Therefore, expressions of computer programs in any form (eg, source and object code) are protected by copyright if they have 'small coin' originality. According to the official statement of reasons for the act implementing Council Directive 91/250 EEC (cited in fn 28), protectability of computer programs is the norm, non-protectability the exception.[30]

---

25 Federal Court of Justice, decision of 2 March 2017, I ZR 124/11 (*Videospiel-Konsolen III*); decision of 27 November 2014, I ZR 124/11 (*Videospiel-Konsolen II*); decision of 6 October 2016, I ZR 25/15 (*World of Warcraft I*). The BGH decisions refer to a preliminary ruling of the European Court of Justice (ECJ), decision of 23 January 2014, C-355/12 (*Nintendo/PC Box and 9 Net*).

26 Federal Court of Justice, decision of 27 November 2014, I ZR 124/11 (*Videospiel-Konsolen II*).

27 Federal Court of Justice, decision of 6 October 2016, I ZR 25/15 (*World of Warcraft I*).

28 Implementation of European software law, Council Directive 91/250 EEC of 14 May 1991 on the legal protection of computer programs, amended by Council Directive 2009/24 of 23 April 2009 on the legal protection of computer programs, https://eur-lex.europa.eu/legal-content/en/TXT/?uri=CELEX:32009L0024. Unless provided otherwise in these rules, the general provisions applying to literary works apply to computer programs as well.

29 Federal Court of Justice, decision of 3 March 2005, I ZR 111/02 (*Fash 2000*).

30 BT-Drucks. 12 /4022.

4.3 Limitations on copyright protection

In general, copyright protection is excluded for features which are technically required.[31]

Additionally, Section 69(2) of the Copyright Act explicitly states that the ideas and principles which underlie programs and their interfaces are not protectable. Abstract algorithms *per se* are therefore excluded from protection, as they are universally applicable methods rather than concrete implementations of code. Similarly, a website display via HTML is not protected by software copyright, as it is not an expression of code itself and its compilation does not constitute an individual programming effort.[32]

Copyright protection is also typically limited with regard to a single right of exploitation – eg, the right of distribution, Section 17 of the Copyright Act. The criteria of this so-called exhaustion principle were set by the Federal Court of Justice (BGH) upon a preliminary ruling of the European Court of Justice in a case about used software, where the BGH ruled that the right of distribution would be exhausted, if:

- the rights holder consents to the use of a program against payment of a reasonable remuneration;
- a perpetual licence was granted;
- the first buyer has entered into a service level agreement with the rights holder which covers updates and fixes to the copy of a subsequent buyer; and
- the first buyer has destroyed its copy.[33]

The BGH has subsequently expanded this jurisdiction to time-limited licences, where a licence was granted for the whole time the program would be functionable. The principle of exhaustion also applies in case of volume licences provided an initial purchaser destroys all copies.[34]

4.4 Acts of infringement – what constitutes copying?

Copying of software is constituted by permanent or temporary reproduction, in whole or in part, of a computer program by any means and in any form. Where the loading, displaying, running, transmission or storage of the computer program necessitates such reproduction, those actions too need the authorisation of the rights holder.[35] Accordingly, copying on data carriers as well as storing in RAM while uploading and downloading need the rights holder's consent. Browsing web pages and web caching, however, do not constitute acts of infringement, as they

31 Federal Court of Justice, decision of 12 May 2011, I ZR 53/10 (*Seilzirkus*).

32 Higher Regional Court of Frankfurt, decision of 22 March 2005, 11 U 64/04 (*Online job market*).

33 Federal Court of Justice, decision of 17 July 2013, I ZR 129/08 (*UsedSoft II*). Preliminary ruling of the European Court of Justice in decision of 3 July 2012, C-128/11.

34 Federal Court of Justice, decision of 19 March 2015, I ZR 4/14 (*Green-IT*).

35 Section 69(c)(1) of the Copyright Act.

do not have a separate economic value of their own.[36] In contrast to mere copies, adaptations[37] are protected works in themselves and can only be exploited with a rights holder's consent.[38] Since adaptations involve copying, the line between a mere copy and an adaption is defined by originality. Therefore, for example, making 3D copies of 2D plans can constitute an adaptation if it involves some originality. All changes that are due to the technical process of transferring a work into another dimension are ignored when assessing originality.

4.5 Permitted acts

Apart from the excluded subject matter, Section 69(d) and (e) of the Copyright Act contain special limitations on software protection (or permitted acts from a user's standpoint).[39] Basically, legitimate users are entitled to:

- reproduce software permanently or temporarily by any means and in any form, if it is necessary for the use of the computer program in accordance with its intended purpose (subject to contractual provisions);[40]
- make back-up copies;[41]
- observe, study or test the functioning of a program in order to determine the underlying ideas and principles (eg, black-box techniques and debugging);[42]
- obtain information on interfaces if necessary to achieve interoperability with other programs (decompilation); with regard to the guiding principles of 'Web 2.0', this provision has little effect, since most developers make their interface information available anyway to allow their program to be more widely taken up.[43]

5. Registered design rights

German registered design right is granted for the appearance of the whole or a part of a product resulting from the features of, in particular, the lines, contours, colours, shape, texture and/or materials of the product itself and/or its ornamentation.[44] 'Product' means any industrial or handicraft item, including among other things, parts intended to be assembled into a complex product,

---

36 Recital 33 of Directive 2001/29/EC of 22 May 2001 on the harmonisation of certain aspects of copyright and related rights in the information society: https://eur-lex.europa.eu/legalcontent/EN/ALL/?uri=CELEX%3A32001L0029.

37 Sections 3 and 69(c)(2) of the Copyright Act.

38 Section 23 of the Copyright Act.

39 Apart from these special provisions, only copying, distribution, exhibition in and communication to the public for use in proceedings before a court, an arbitration tribunal or authority are allowed (Section 45 of the Copyright Act).

40 Section 69(d)(1) of the Copyright Act.

41 Section 69(d)(2) of the Copyright Act.

42 Section 69(d)(3) of the Copyright Act. See fn 24.

43 Section 69(e) of the Copyright Act.

44 Section 1(1) of the Design Act. English version at www.gesetze-im-internet.de/englisch_geschmmg/index.html.

packaging, 'get-up', graphic symbols and typographic typefaces, but excluding computer programs.[45]

A design must be new and have individual character to enjoy protection.

According to Section 2(3) of the Design Act, a design will be treated as having individual character if the overall impression it produces on the informed user differs from the overall impression produced on such a user by any design which has been made available to the public before the date of filing of the application for registration of that design. In assessing individual character, the degree of creative freedom the designer enjoyed in developing the design is taken into consideration.

The registration procedure is set out in Section 11 (and the following sections) of the Design Act. An application can be filed at the German Patent and Trademark Office in Munich, Jena or Berlin, or at any of the 21 patent information centres across Germany.[46] No substantive examination takes place: that is to say, examination is limited to formal matters, in particular looking for excluded subject matter[47] – for example compliance with public policy and morality, checking for improper use of state emblems or other signs.[48]

Industrial designs can also be protected under the Copyright Act if the design meets the general criteria for copyright protection. Recently, the Federal Court of Justice confirmed that only 'small coin' creativity is required to enjoy such protection, abandoning its previous jurisprudence that required that industrial designs would only be a work of applied art, if it is 'highly' original.[49]

6. Confidential information

The misuse of trade secrets and industrial secrets may constitute an act of unfair competition and a crime according to Section 17 of the Unfair Competition Act. The misuse of confidential information may be either committed by an employee or by a third party if carried out for the purposes of competition, personal gain or with intent to cause damage to the owner of the business.[50] This provision can be of particular importance in cases of misappropriation of confidential software by employees of the rights holder.

'Confidential information' according to Section 17 would cover any commercial or technical fact linked to a business, which is not generally in the public domain, but instead known only to a limited group of people, and which the owner wishes to be kept secret.[51]

---

45 Section 1(2) of the Design Act.
46 See www.piznet.de.
47 Section 3(1)(3) of the Design Act.
48 Section 3(1)(4) of the Design Act.
49 Federal Court of Justice, decision of 13 November 2013, I ZR 143/12 (*Geburtstagszug*).
50 Federal Court of Justice, decision of 16 November 2017, I ZR 161/16 (*Knochenzement I*) concerning misuse of trade secrets by a former distributor for the benefit of his or her production of goods.
51 Federal Court of Justice, decision of 26 February 2009, I ZR 28/06 (*Insurance agent*).

7. Other forms of protection

7.1 Utility models

Utility models basically differ from patents as regards registration requirements, duration and scope. In contrast to patents, utility models do not undergo an examination process, but are simply registered. The term of protection of a utility model is only 10 years.

It is possible to separate out utility models from pending patent applications in order to obtain an enforceable right while a patent application is still pending (Section 5 of the Utility Model Act).

Utility models are only available for products and cannot be granted for methods and processes. That said, the term 'process' has been given a very narrow interpretation, with the German Federal Supreme Court allowing claims in respect of a signal sequence for a computer program flow on the grounds that such a signal does not constitute a process claim.[52] Further, the Court ruled that 'means plus function' claims do not violate the no process claim rule.[53]

The prior art for the purposes of German utility model law is slightly narrower than is the case for patents. This is because the law provides for a period of grace where use or publication in the six months prior to the priority date is permitted, provided it is based on the applicant's own work or that of his or her predecessor in title.

Utility model applications can be filed at the PTO's offices in Munich, Jena and Berlin (including e-filing), as well as the 21 patent information centres (Section 4(1) and (2) of the Utility Model Act). Examination is limited to checking whether an application fulfils the formal requirements for a utility model application: that is, whether the application includes a technical description of the application. As there is no substantive examination, a utility model rapidly proceeds to grant.

The current application fee is €40 and includes a renewal fee for the first three years. Renewal fees for subsequent periods are currently:

- €210 for years 4 to 6;
- €350 for years 7 to 8; and
- €530 for years 9 to 10.

Registration of a utility model gives the owner the exclusive right to make, sell or market the subject matter of the utility model. As in the case of a patent, the scope of protection afforded by a utility model is determined by the content of its claims which are interpreted in the same way as patent claims.

52 Federal Court of Justice, decision of 17 February 2004, X ZB 9/03 (*Signal sequence*).
53 Federal Court of Justice, decision of 29 July 2008, X ZB 23/07 (*Telecommunications systems*).

7.2 Trademarks and passing off/unfair competition law

Software can be subject to trademark as well as service brand protection.[54] In addition, software may be protected under the rules against unfair competition. A breach of the terms and conditions of a software agreement may also constitute a breach of Section 4(4) of the Act Against Unfair Competition,[55] which prohibits deliberate obstruction of competitors.[56] Unfair competition law might also come into play where a competitor exploits a brand's or product's reputation, such as in unfair comparative advertising or false advertising of market leadership.[57]

In addition to protection against unfair statements the rules of unfair competition law can also provide protection against slavish copying. In order to assess a violation of unfair competition law by slavish copying, the courts take into consideration the competitive individuality of the original, the type and intensity of any copying, and any special anti-competitive circumstances. These are all closely related and have reciprocal effects on each other. Therefore, the closer the copy is to the original, the lower the hurdle is to be cleared to affirm competitive individuality of the original and special circumstances.[58]

Finally, unfair competition law may come into play as a defence against unjustified IP law infringement threats. According to Federal Court of Justice jurisprudence, claims for injunction and damages may be based on tort[59] or on unfair competition law.[60]

7.3 Database rights

Database rights have been introduced into German copyright law as a result of European Community law.[61] Databases may be protected by copyright or by related rights. For the purposes of protection by related rights, a database is a collection of elements arranged systematically or methodically, which requires a 'substantial' investment. 'Substantial', however, excludes only 'commonplace' expenses. The producer of the database, who has made that investment, enjoys protection. The producer is protected against the use of substantial parts of the database as well as unsubstantial parts if the uses involve the repeated and systematic reproduction (eg, reproduction in alphabetical order), distribution or communication of those parts.[62] Private use is generally allowed, as are uses for

54 Federal Court of Justice, decision of 2 March 2017, I ZR 273/14 (*Videospiel-Konsolen III*).
55 English version at www.gesetze-im-internet.de/englisch_uwg/index.html.
56 Federal Court of Justice, decision of 12 January 2017, I ZR 253/14 (*World of Warcraft II*). Terms and conditions prohibited 'bots' and 'cheats'.
57 Federal Court of Justice, decision of 6 December 2007, I ZR 169/04 (*Imitation advertisement*); decision of 16 November 2017, I ZR 160/16 (*Knochenzement II*).
58 Federal Court of Justice, decision of 24 May 2007, I ZR 104/04 (*Garden chair*).
59 Federal Court of Justice, decision of 19 January 2006, I ZR 217/03.
60 Federal Court of Justice, decision of 19 January 2009, I ZR 123/06 (*Milling machine*).
61 Directive 96/9/EC of 11 March 1996 on the legal protection of databases was crafted to fight protection deficits, http://eur-lex.europa.eu/LexUriServ/LexUriServ.do?uri=CELEX:31996L0009:en:HTML. Also see Section 87(a)–(e) of the Copyright Act.
62 Higher Regional Court of Cologne, decision of 28 March 2014, 6 U 140/13 (*Photovoltaik-Datenbanken*).

personal scientific and non-commercial educational purposes. A special feature of the related right for database creators is the relatively short term of the right which only lasts 15 years from publication.

7.4 Semiconductor topographies

By definition in Section 1(1) of the Semiconductor Protection Act, protected topographies are 3D structures of micro-electronic semiconductors. They have to be the product of intellectual work rather than a replica of other topographies. Commonplace topographies are also excluded subject matter. The creator enjoys the exclusive IP right, but that is transferred automatically to employers in working relationships (subject to contractual provision to the contrary). Topographies have to be registered at the PTO. Their term is 10 years.

8. Ownership, inventors and compensation

8.1 Patents

According to Section 6 of the Patents Act, an inventor enjoys the right to file for a patent and is entitled to the grant of a patent after application. If several inventors have created an invention jointly, they can only transfer ownership of it jointly. However, each patentee can also dispose of his or her own share individually.

Inventors' remuneration is subject to a specific code of law, the Employee's Invention Act. Its principle rule is that intangible assets like know-how and other work created by the employee automatically vest with the employer, but inventions protected by patents or utility models are an exception to this rule, belonging initially to the employee as the inventor (Section 6 of the Patent Act), albeit they can be transferred to the employer.

Transfer of rights to an employer triggers a right for the employee to adequate compensation, as fleshed out in special remuneration guidelines. As an alternative to claiming ownership, it is also possible for employers to declare that the invention is in the free disposition of the employee (Section 8 of the act). Employees have a duty to inform their employers about an employment-related invention and an obligation to support their employer if the employer chooses to file a patent application – employers are entitled to file patent applications for inventions for which ownership is claimed: Section 13(1) – but under Section 13(3) the employee may file an application in the name of, and on behalf of, the employer if the employer does not file an application. The employer may be liable for damages in case of breach of the obligation to file a patent.

Litigation between the employer and the employee on matters of employee invention law may – and in some cases must – be filed with a special arbitral tribunal at the PTO before it can be filed with a civil court.

8.2 Copyright and other IP rights

The same general rules of ownership apply to computer programs as to other categories of protected works. German copyright is based on the principle that the creator owns all rights and can only grant exploitation rights to third parties, not transfer the rights. The creator of a computer program enjoys protection of both moral rights and exploitation rights. Moral rights comprise the right of publication, attribution and remedy against distortion. Exploitation rights are listed in Section 69(c) of the Copyright Act: reproduction, transformation, distribution and communication to the public, including making it available (eg, for download).

If several software engineers have jointly written a computer program and their contributions cannot be exploited separately, they are joint authors. Consequently, publication and exploitation accrue jointly to all of them.

As computer programs are mostly developed in employment relationships, Section 69(b) of the act leads to an automatic entitlement of the employer as regards the author's exploitation rights – subject to contractual provisions to the contrary. As a result, the employer will normally hold all exploitation rights. Limitations on joint authorship do not apply, since employees waive the right of publication in favour of their employer.

The person designated as the author in the usual manner on the copies of a released work or on the original of an artistic work is deemed to be the author unless the contrary is proven. The same applies to any designation which is known to be a pseudonym (Section 10(1) of the Copyright Act). This presumption of authorship does not apply if a designation fails to reveal the author to the public, for example using a corporate name as copyright notice.[63]

Related rights (Section 70 and following of the Copyright Act) accrue to performers and producers.

Trademark (joint) ownership arises from:

- entry of a sign as a trademark in the trademark register kept by the Patent Office; or
- the use of a sign in the course of trade insofar as the sign has acquired public recognition as a trademark within affected trade circles; or
- notoriety (Section 4 of the Trademark Act).[64]

Rights in a registered design belong to the designer or successor in title. Where several persons have jointly created a design, the rights in the registered design belong to them jointly (Section 7 of the Design Act).

---

63 Federal Court of Justice, decision of 18 September 2014, I ZR 76/13 (*CT-Paradies*).
64 English version at www.gesetze-im-internet.de/englisch_markeng/englisch_markeng.html#p0032.

9. Counterfeiting

A rights holder may use any means of enforcement against IPR infringement, regardless of whether the infringement may be characterised as counterfeiting of 'normal' IPR infringement. There is, however, an additional procedure which is typically used against counterfeiting (although it is not necessarily limited to counterfeiting), namely the possibility of requesting border seizures of infringing goods (eg, under Section 142 of the Patent Act). The seizure of infringing goods may be requested from the German customs authorities and is a purely administrative request.[65]

10. Enforcement

10.1 Obtaining information on the infringer and the infringement

According to Section 140b(1) of the Patents Act, a patentee is entitled to request information from the infringer about the origin and the distribution channels of the infringing products. This provision enables a patentee to identify and bring actions against infringers higher up the distribution chain. In order to be able to calculate damages for infringement, the patentee is further entitled to request detailed information on the number of infringing products manufactured or distributed as well as information about prices and relevant cost factors (Section 140b(3) of the act). Orders for this kind of information are common in patent infringement decisions.

If a patentee has insufficient information to be able to establish infringement, Section 140c provides the possibility of accessing the required information either by asking the alleged infringer to provide certain documents or by requesting an inspection order against that party. This provision may be useful, for instance, in patents cases which involve software and where proof of infringement requires access to the source code.

An inspection order may be requested in *ex parte* proceedings before infringement proceedings on the merits are initiated. Before issuing such an order the courts would appoint a technical expert. The inspection may include any technical measure necessary to obtain information about the potential infringement. The expert may thus inspect machines located in the premises of an alleged infringer, request access to any kind of data required to run a machine, and make copies of software. In addition, a patentee may request commercial information such as evidence as to the commercial value of the software, including financial information like invoices, purchase orders and bank statements. Based on the information obtained during inspection, the court-appointed expert will draft a detailed report. However, before that report

65 See www.zoll.de/EN/Businesses/Movement-of-goods/Import/Restrictions/Goods/Protection-of-industrial-property/protection-of-industrial-property_node.html.

can be made freely available to the patentee, the defendant has the opportunity to address confidentiality concerns. After clearance of such concerns the expert report will then be issued to the patentee, who may then use information in it to initiate infringement proceedings.

The right to information about copyright infringers is provided for in Section 101 of the Copyright Act. The alleged infringement must be on a commercial scale (ie, with a view to economic or commercial benefit). If a copyrighted computer program is distributed via P2P, the rights holder can seek disclosure of the identity of the distributor in the Regional Court of Cologne, which has jurisdiction for all of Germany for these purposes.

10.2 Interim relief

In certain cases, the court may grant a preliminary injunction based on the general provision of Section 935 of the Code of Civil Procedure. The court may decide on the request for a preliminary injunction *ex parte* or following a joint hearing. While in trademark and unfair competition cases *ex parte* preliminary injunctions are quite usual, such measures are rather exceptional in patent infringement cases. The prerequisites for obtaining a preliminary injunction are:

- a clear infringement situation;
- confirmed validity of the patent; and
- urgency.

The validity of the patent should be sufficiently clear; if the patent has already survived an opposition or a nullity action this may be sufficient to meet this condition. In order to establish the necessary urgency of the interim measure, patentees must establish that they have acted without undue delay on becoming aware of the infringement. Usually the courts consider a timeline of one month as sufficient to initiate an action on obtaining that knowledge. However, depending on the specific circumstances of the case, the timeline could be longer.

The preliminary injunction has immediate effect against any further infringing acts. Violations of the preliminary injunction can be sanctioned by fines of up to €250,000, or by imprisonment.

The defendant can appeal the preliminary injunction. Any damages caused by the enforcement of the preliminary injunction may give rise to damage claims against the claimant where the preliminary injunction is subsequently lifted.

In order to minimise the risk of a preliminary injunction being granted *ex parte*, alleged infringers have the option of filing a protective letter. Its purpose is to make the court aware of the reasons why an *ex parte* injunction would be inappropriate, so the letter must be filed before any preliminary injunction request is filed by the patentee.

10.3 Warning letters and pre-trial correspondence

There is no requirement for a warning letter or any other communication to be addressed to the defendant before initiating infringement proceedings. If in some cases a warning letter is sent to the other side this would be in order to avoid certain cost disadvantages in the event that the alleged infringer immediately accepts the claim and complies with all remedies requested.

A warning letter equally provides a sufficient basis for the alleged infringer to initiate a negative declaratory action requesting the confirmation of non-infringement in court. In addition, the warning letter may trigger defensive actions such as a protective letter or a torpedo action. And any interested party also has the option of filing a nullity action in the Federal Patent Court, irrespective of the existence of a warning letter and without needing to provide a specific reason for such a nullity action.

10.4 Remedies

The legal consequences of an infringement of IP rights can include orders imposing:

- injunctions;
- damages;
- requests for information on infringing activities;
- destruction of infringing products;
- recall of infringing products on the market; and
- reimbursement of costs of proceedings.

Normally an injunction is granted against an infringer automatically following a finding of infringement.[66] However, in the case of patent infringement a defendant may raise the so-called 'FRAND[67] defence' against the issuance of a requested injunction – which can be of particular importance in electronics cases. This defence applies if a patentee is seeking to enforce patents that are essential for compliance with a standard. A FRAND defence entails submitting to the claimant a licence offer that complies with FRAND terms. Because the patentee may have an obligation to grant licences under FRAND conditions for patents that are essential to a standard either because of express agreement with the standard-setting body or because of general antitrust rules, the refusal to grant such a licence can result in loss of the right to request an injunction.

According to the leading case law of the Federal Court of Justice 'Orange Book', a standard essential patent owner must not turn down an unconditional offer by an alleged infringer if complying with FRAND conditions.[68] The

66 See, eg, Section 139(1) of the Patents Act for patents.
67 Fair, Reasonable and Non-Discriminatory.
68 Federal Court of Justice, decision of 6 May 2009, KZR 39/06 (*Orange Book standard*).

requirements for such an unconditional FRAND offer have been developed in more detail in a number of decisions by the lower instance courts. As a rough guideline the alleged infringer is under an obligation to act like a good faith licensee in order to comply with the Orange Book standard.

In order to obtain damages for infringement, a rights holder must follow a two-step process:

- First liability for damages has to be established.
- Then the rights holder has to request payment of a specific amount of damages, based on the information provided by the alleged infringer.

A rights holder can choose between three options in order to calculate an appropriate amount of damages:

- reasonable royalty;
- infringer's profit; or
- lost profit.

The computer program provisions of the Copyright Act also set out special remedies which are wider in scope than the general provisions. According to Section 69(f) of the Copyright Act, a rights holder may require the owner or proprietor to destroy all unlawfully produced or distributed copies or all copies intended for unlawful distribution. In addition, according to Section 98(3) of the act, an injured party may require that infringing copies are handed over to them against payment of an equitable remuneration (which may not exceed production costs).

11. Hot topics

The enforcement of standard essential patents and the obligations flowing from Article 102 of the TFEU has been a much-debated issue in Germany since the decision of the CJEU in *Huawei v ZTE*. This relates to the rights and obligations of patentees that participate in standard-setting bodies and limitations on the rights of such patentees requesting injunctions based on such patents. According to a decision of the Appeal Court Düsseldorf,[69] a patentee who has declared its readiness to offer licences of a standard essential patent under FRAND terms must fulfil certain requirements, before being able to enforce an injunction request for infringement of the patent. Such a patentee must inform the defendant, identifying the patent and the alleged infringing act. If the defendant responds indicating a readiness to take a licence of the patent under FRAND terms, the patentee must provide the defendant with an offer for a licence under FRAND terms, specifying in particular the basis for the determination of the FRAND character of the offered royalty. If the patentee

69 Appeal Court Düsseldorf of 30 March 2017, I 15 U 66/15 (*Mobiles Kommunikationssystem*).

does not fulfil these requirements, a defendant can prevent an injunction request from being made as being in breach of Article 102 of the TFEU.

Other hot topics relevant to patents discussed in recent litigation include the scope of protection of a patent under the doctrine of equivalents and the scope of contributory infringement. Recent Federal Court of Justice decisions relate to the methodology for determining the scope of patents if the description refers to several technical solutions, but the claim only refers to one of the solutions. If an interpretation of a patent leads to the conclusion that certain elements have been disclosed, but do not fall within the patent's claimed scope, then a use of those elements will not establish infringement of the patent under the doctrine of equivalents.[70]

The definition of 'means leading to contributory infringement' is usually interpreted broadly. Means relating to crucial elements of an invention fall within the scope of a patent, if they are suitable to interact with one or more elements of a patented claim. According to established case law, this is regularly the case when a means is mentioned in the patented claim itself.[71] Recently, however, the Federal Court of Justice has declined contributory infringement in the case of a method directed at decoding encoded video files with regard to the distribution of DVDs containing encoded data.[72] Thus, if a means does not contribute to an invention's claimed solution, its use cannot establish contributory infringement.[73] On the other hand, if a means is sufficient to lead to contributory infringement in Germany, supplies of such means outside Germany may also lead to a liability for subsequent contributory patent infringement in Germany.[74]

70 Federal Court of Justice of 10 May 2011, X ZR 16/09 (*Occlusion apparatus*); Federal Court of Justice of 13 September 2011 X ZR 69/10 (*Diglycidverbindung*); Federal Court of Justice of 14 June 2016, X ZR 29/15 (*Pemetrexed*).

71 Federal Court of Justice of 27 February 2007, X ZR 38/06 (*Pipette system*).

72 Federal Court of Justice of 21 August 2012, X ZR 33/10 (*MPEG-2-video signal encoding*).

73 Federal Court of Justice of 3 February 2015, X ZR 69/13 (*Audiosignalcodierung*).

74 Federal Court of Justice of 3 February 2015, X ZR 69/13 (*Audiosignalcodierung*).

Italy

Elena Martini
Martini Manna Avvocati

Italy is one of the world's major economic powers (third largest economy in the European Union and eighth in the world[1]) and represents a strategic place to invest in. In fact, Italy offers excellence in research and innovation, thanks to a highly qualified and professional workforce, and it is strongly open to foreign investments: companies investing in hi-tech electronics and software can take advantage of tax credits, concessional taxation for intangible property, concessional financing for innovative start-ups and other benefits. Finally, under a jurisdictional point of view, Italian legislation offers effective means to protect intellectual property and the Italian courts are able to intervene promptly and effectively against IP infringements.

1. National patent procedure

1.1 Filing

In Italy, a patent can be obtained by submitting the relevant application before the Italian Patents and Trademarks Office (UIBM) or by submitting an application for a European patent and validating the latter in Italy.

1.2 Novelty-based examination system

Before 2008, the UIBM only used to check that the application fulfilled the necessary formal requirements. As of 1 July 2008, patent applications not claiming priority also undergo a novelty examination. Specifically, the European Patent Office (EPO) carries out a prior art search on behalf of UIBM, the costs of which are borne by UIBM. The EPO search report is made available to the applicant who is entitled to submit observations.

In the event that EPO's search reports show a lack of substantial requirements of the application (such as lack of novelty), applicants can choose between making amendments to the description or the claims, withdrawing the application or requesting UIBM to proceed with the examination.

In addition, the Italian authorities check whether the invention is of interest

1 https://investorvisa.mise.gov.it/index.php/it/home-it/10-buoni-motivi-per-investire-in-italia.

to the national defence. For this reason, Italian residents cannot file patent applications abroad before 90 days after having submitted the relevant application in Italy.

1.3 Timescales

Patent applications are published 18 months after filing. The applicant can request prior publication, but in any case no earlier than 90 days after filing.

The EPO search report is generally provided to the UIBM within nine months of the filing day.

Patents are typically granted within 30–36 months of the filing day. In urgent cases, applicants can request the UIBM to accelerate prosecution, but the UIBM has the right to refuse. In any case, the patent will not be granted before the application has been published.

The UIBM decision to refuse an application can be appealed within 60 days. In addition, the validity of a granted patent may be challenged before the Italian courts after grant.

1.4 Outline of the costs

The main official fees for filing and prosecuting an application are as follows:

- application fee: €50 for e-filing; from €120 to €600 in case of filing of hard copy, depending on the number of pages. Additional €45 shall be added for each claim exceeding the tenth;
- search fee, if an English translation of the claims is not provided: €200; and
- renewal fees are due as from the fifth year from the filing date, and range from €60 for the same fifth year, to €650 for the fifteenth and following years.

2. Examination of software patents

Under Italian law, in line with the European Patent Convention (EPC), 'software as such' is not patentable. Hence, software-related inventions are patentable only when the execution of the computer program "has a technical effect that goes beyond the usual effects resulting from running a computer program".

When assessing whether the patent application meets all the necessary formal requirements, the UIBM also checks that the claimed invention is not a software as such and, if this is not the case, it refuses to grant the application. The EPO, in turn, when requested to do a novelty search by the UIBM, can refuse to make the search if it believes that the patent is directed to software as such.

Patents in Italy are regulated by Articles 45 and following of the IP Code (Legislative Decree 30/2005). Article 45(1) sets forth the requirements for an invention to be patented, namely:

- novelty;
- inventive step; and
- capability to have an industrial application.

Article 45(2) states that the following cannot be considered as inventions:
- discoveries, scientific theories and mathematical methods;
- schemes, rules and methods of performing a mental act, playing a game or doing business, and computer programmes; and
- the presentation of information.

Article 45(3), however, points out that the preceding provision shall prevent the things listed therein from being patented only to the extent that a patent or patent application relates to that thing as such.

2.1 Italian courts' approach to software inventions

With Decision 3196 of 1981, the Italian Supreme Court found that a software could be patented in the framework of a combination invention patent, but it did not elaborate on the specific issue of when a software can be patented. The same happened in a decision of 10 February 1993 by the Court of Turin which simply recalled the previous Supreme Court decision and stated that the content (as opposite to the form of expression) of a software "cannot be protected under copyright law, but might be patented when the requirements posed by the law are met, and when it is part of a combination invention".

After that, decisions have been issued acknowledging the validity of software patents, but still they generally do not elaborate on the issue. This is due to the fact that Italian judges do not have a technical background, hence they ground their decision on the opinion of an expert witness appointed to assess validity and infringement of the patent. The actual examination of the patent is therefore made by the expert witness and is included in his or her opinion which is not public. As the court decision will generally only recall the conclusions of the expert witness' opinion, the part of the opinion elaborating on whether that specific software is patentable or not is seldom available. On the other hand, it can be safely stated that expert witnesses are chosen among the Italian and European Patent Attorneys, and that – also as a consequence of this – Italian courts apply the principles set forth by the EPO and are in line with its case law.

In one of the few decisions elaborating on patentability of software inventions (Decision 1623/2014), the IP Court of Turin found for the validity of a software patent relating to an electronic program guide (EPG) which was found to have a technical effect going beyond the usual effects resulting from running a computer program. In doing so, the appointed expert witness in fact referred to the EPO Guidelines, Part G, Chapter II, Section 3.7.2 ("The technical effect achieved might be a more efficient man-machine interface") and found that a technical effect was given by the following features of the claimed invention, that:

> *... the claimed subject matter causes a change in the way of working of specific devices that show EPG information on the screen and respond to commands given by the user in order to display more information and subsequently tune television programs. It requires specific hardware to be implemented (for example, a programming control device, a memory, a tuner and a monitor). Hence the patent is technical and does not refer to a computer program as such.*

This decision was later confirmed by the Turin Court of Appeal in its Decision 30/2017.

2.2 Methods of doing business

Business methods are generally considered not patentable in Italy.

In its Decision 7597 of 2004, the Italian Supreme Court confirmed the previous case law ruling out the possibility to patent business methods because "they realise the right to private initiative, and therefore the aim of competition, hence they cannot be monopolised". The business method in issue in this case related to the use of areas above the railway venues for car parks, offices and commercial premises.

The Bologna IP Court in its Decision 2921 of 2010 invalidated a patent which concerned selecting "which tools can be chosen in order to display, on a screen, advertising images and audio messages 'delivered' by a central server and disseminated through the mobile cellular network to an unlimited number of users". According to the court, in fact, the invention merely consisted of a method of exploiting communications technologies (GSM, UMTS) and generic hardware devices belonging to the state of the art, indicating operations and steps that alone or altogether were not inventive, while ultimately failing to describe the technical applications of the same method.

2.3 Presentation of information

Article 45 of the IP Code forbids patent presentations from consisting of the mere display of information. Nevertheless, systems for presentations might be patented when meeting the patentability requirements.

In its Decision 1623/2014 mentioned above, the IP Court of Turin found for the validity of a patent relating to an electronic program guide (EPG) stating the following:

> *As for the presentation of information, the patent concerns the particular (grid or list) configuration of EPG information and how this information can be made accessible to and used by the user, and not just what information is displayed. It concerns how EPG information, arranged in a certain way, can be displayed on the screen in different forms depending on the commands given by the user.*

2.4 Mathematical methods

Under Italian law, mathematical methods can be patented only when they are susceptible of an industrial application. This was stated by the Italian Court of Cassation in its Decision 7083 of 1988, which highlighted that, by law:

> *only mathematical methods as such are not patentable, since they are a common heritage of humanity, which means (otherwise the whole provision would be useless), that the solution to an industrial technical problem described in the patent application, which is an immediate application of a mathematical method, is patentable.*

This decision recalled the same court's previous Decisions 2632 of 1959, 821 of 1968, 3855 of 68 and 208 of 1972 and reaffirmed the principles already set forth by the latter, as follows:

> *The new solution to a technical problem satisfying a typically industrial interest is patentable, although the solution itself can be implemented directly and immediately with the sole application to the industrial technique of a scientific principle (that is, discoveries, scientific theories and mathematical methods) without the use of a specific material entity as a physical support. In this case it is possible to patent not the scientific principle as such, but the solution to the specific industrial technical problem that inventor indicated as the immediate technical application of the scientific principle discovered or elaborated by him.*

More recently, this was confirmed by the Milan IP Court in its Decision 9805 of 2013, which stated as follows: "The element which distinguishes a patentable mathematical method is the fact that the description of the latter is accompanied by claims determining the industrial application of those methods".

3. Infringement

3.1 Acts of infringement

Acts of infringement are regulated by Article 66 of the IP Code, based on which the patent holder is granted with the following exclusive rights:

- where the subject matter of the patent is a product, the right to forbid any third party to manufacture, use, market, sell or import to those aims the product; and
- where the subject matter of the patent is a process, the right to forbid any third party to use the process or use, market, sell or import the product obtained through that process.

As a consequence, persons doing the above in absence of the patent holder's authorisation infringe the patent.

In addition, the patent is infringed by a person who supplies to third parties, or offers to supply, the means relating to an essential element of the invention and necessary to implement the invention in a state where the invention is patented, when he or she knows, or could know with ordinary diligence, that those means are capable of and directed to implement the invention.

3.2 Claim interpretation

Under Article 52 of the IP Code, the scope of a patent is that specified in the patent claims. The provision reads as follows:

1. *In the claims it is specifically indicated what is meant to be the object of the patent.*
2. *The extent of protection shall be determined by the claims. Nevertheless the description and drawings shall be used to interpret the claims ...*

In applying this provision, the Turin IP Court in a decision of 28 September 2010 stated that: "Patent claims should not be interpreted separately, but must be integrated into a single interpretative process that must take into account, in establishing the limits of patent protection, all claims and the description of the invention".

The Milan IP Court, in a decision of 4 November 2010, found that "The finding of the Court expert witness which, in order to rule out the infringement, based his decision on a re-interpretation of the claims (initially considered 'ambiguous') supported by the description and by the drawings, complies with the principles set out in the Interpretation Protocol of Art. 69 of the Munich Convention, reproduced in Art. 52 IP Code".

Most recently, the Supreme Court confirmed this interpretation with Decision 6373/2019, stating that: "The scope of a patent, which can be reconstrued according to its claims under Article 52(2) IP Code, can be also interpreted through the description and the drawings, when the sole claims are not sufficient to clearly define the scope".

3.3 Contributory infringement

Under Article 66(2)*bis* of the IP Code, a patent is infringed by a person who supplies to third parties, or offers to supply, the means relating to an essential element of the invention and necessary to implement the invention in a state where the invention is patented, when he or she knows, or could know with ordinary diligence, that those means are capable of and directed to implement the invention. This provision was only included in the Italian IP Code by Law 214 of 2016, but the relevant principles had been set and were applied by the Italian courts before that date by way of interpretation of the existing provisions.

The Bologna IP Court, in its decision of 3 March 2006, stated as follows:

Contributory infringement occurs when individual parts of the patented machine are produced, which parts are not patented and therefore are in the public domain. This is not expressly sanctioned by the Italian law, but it is sanctioned by the

Italian case-law: see the Supreme Court decision of 24 October 1956, the Turin Court decision of 12 October 1981, the Milan Court decision of 25 October 1984, the Milan Court of Appeal decision of 23 June 1992, the Supreme Court decision no. 9140 of 1994.

The Venice IP Court, in its decision of 29 October 2008, stated that: "Contributory infringement definitely occurs when the alleged infringer does not directly produce the patented good, but produces a different good the specific functionality of which is to realize the patented one".

On this matter, the Court of Turin recently specified that:

although the contribution to the infringement must focus on "an indispensable element of the patented invention" and concern "means necessary for its implementation", it does not require, however, that the destination of the 'means' be univocal or exclusive, but only requires that the third supplier of the means knows, or may know with ordinary diligence, that these means are directed to implement the invention and are suitable for the purpose (order of 20 December 2018).

4. Copyright protection

4.1 Requirements for protection

Software copyright protection was introduced in the Italian Copyright Law by Decree 518 of 1992, which implemented EC Directive 91/250/EEC.

The Italian courts had previously been granting copyright protection to software for a decade prior to when these provisions entered into force.

By way of example, the Turin Court, in a decision of 15 July 1983, found that videogames were to be considered as cinematographic works and thus enjoined copyright protection. Similarly, the Pisa Court, in a decision of 11 April 1984, stated that not only software, but also the relevant instruction and installation manuals enjoined copyright protection when original. Other decisions in those years stated that original software was protected by copyright being intellectual works belonging to the science field and falling within the scope of the copyright law. The Court of Cassation finally confirmed that software enjoys copyright protection in a decision of 6 February 1987.

The provision indicating the requirements for protection is Article 2(8) of the Copyright Law, according to which copyright exists on:

computer programs expressed in any form, provided that they are original, being the author's intellectual creation. The ideas and principles which underlie any element of a computer program, including those which underlie its interfaces, are excluded from the protection afforded by this law. The term 'computer program' also includes its preparatory design material.

As expressly stated by this provision, in order for a software to qualify as original it is sufficient that it is the author's intellectual creation, ie, it was not copied from another work. According to the Italian case law:

creativity and originality exist even if the work consists of simple ideas and notions, included in the intellectual heritage of people who have experience in the subject matter of the work, provided that they are formulated and organized in a personal way, autonomous from the previous ones (see, among others, Supreme Court Decisions 581/2007 and 13524/2014).

Graphical aspects of user interfaces can also enjoy copyright protection, as was confirmed among others by the IP Court of Milan in its Decision 1744/2012. Finally, copyright protection can also be awarded to the software's user manuals, as confirmed by the above-mentioned Pisa Court decision of 11 April 1984 and by a decision of the Milan Court of 2 February 1990.

Original software enjoys copyright protection as from its creation, with no need to register it. Nevertheless, registration at the Italian collecting society (SIAE) is possible, with the aim of ensuring evidence that the software was created by its author before the registration date. This goal is currently achievable also by means of a combined use of electronic signatures and timestamps.

Electronic devices might in turn theoretically enjoy copyright protection as industrial design works under Article 2(10) of the copyright law, provided that they have "creative character and artistic value". While the threshold for creativity is rather low, in line with the above case law, the 'artistic value' requirement significantly limits the number of goods that can enjoy this kind of protection. In fact, Italian courts generally find that this additional requirement is met when evidence is submitted showing that the product was exhibited in museums, mentioned in design books, awarded with prizes precisely for its design. Nevertheless, these criteria typically apply only in respect of goods which have been on the market for a while. For goods which have just been launched on the market, or goods in respect of which the above evidence is not available, there is large room for discretion on the part of the judge. To date, apparently no court decision has granted copyright protection to the design of an electronic device.

4.2 Limitations on copyright protection

As mentioned above, Article 2(8) of the Copyright Law provides that the ideas and principles which underlie any element of a computer program, including those which underlie its interfaces, are excluded from protection.

In applying this provision, the Bologna IP Court in an order of 17 January 2006 stated that:

The purpose of the software, that is, the aim of the software as a whole and of its so-called 'modules', is not protected, nor are protected the mathematical algorithms that implement the functions that the program must perform, or the so-called flowchart, which describes in detail how the different parts interact with each other.

The Milan IP Court in an order of 6 July 2012 stated:

The copyright protection of computer programs only concerns their form of expression – their so-called formal aspect – not their content (that is, ideas, principles, methods and techniques or formulas), nor their functions.

4.3 Acts of infringement

Pursuant to Article 64bis of the copyright law, which basically reproduces Article 4 of Directive 91/250/EEC, copyright in software includes:

the right to do or to authorize:

(a) the permanent or temporary reproduction of a computer program by any means and in any form, in part or in whole. Insofar as loading, displaying, running, transmission or storage of the computer program necessitate such reproduction, such acts shall be subject to authorization by the right holder;

(b) the translation, adaptation, arrangement and any other alteration of a computer program and the reproduction of the results thereof, without prejudice to the rights of the person who alters the program;

(c) any form of distribution to the public, including the rental, of the original computer program or of copies thereof. The first sale in the European Community of a copy of a program by the right holder or with his consent exhausts the distribution right within the European Community of that copy, with the exception of the right to control further rental of the program or a copy thereof.

Typically, the most important software cases relate to the partial copying of a computer program and the commercialisation of the software thus obtained. In order to ascertain whether the two computer programs in issue overlap, typically the relevant source code is compared, unless either the original or the alleged infringing one is not available (typically, due to the fact that the party owning it did not submit it in court). In this respect, the Bologna IP Court, in an interim order of 17 January 2006, stated that "graphical representations on monitors or on paper are merely visual transpositions of the source code, which is the only element to be taken into consideration, given that only the source code is protected by the copyright law". The court dismissed the petitioner's motion for preliminary injunction based on the fact that it had not submitted its software's source code.

However, the Milan IP Court, in an interim order of 28 September 2005 (*SIA v Innovazione srl*), found that copyright infringement had occurred based on the fact that the contested software partially reproduced the "structure, organisation, field names, types and sizes, pages and formulas" of the previous one. In the absence of the contested software's source code, which had not been provided by its holder, the court in fact stated that, "given the strong similarity between the executable files, it can be assumed that, if a certain result is reached at the end of a complex logical process, the achievement of the same results derives from the use of the same procedure".

The Venice IP Court, in an interim order of 14 April 2004 (*MBM Italia srl v SGM srl and Magnetek spa*), dealt with the alteration of a computer program by its licensee, finding that this amounted to copyright infringement.

Again, in the matter of licences, the Supreme Court in its Decision 16041/2011 stated that the buyer of a company's business unit to which a software was licensed does not automatically acquire the relevant licence if the licence agreement forbids any transfer by the licensee, hence, use of the software by that buyer amounts to infringement.

4.4 Permitted acts

Article 64*ter* of the Copyright Law in line with Article 5 of Directive 91/250/EEC, provides for a number of exceptions to copyright infringement. This provision reads as follows:

1. *In the absence of specific contractual provisions, the acts referred to in Article 64bis (a) and (b) shall not require authorization by the right holder where they are necessary for the use of the computer program by the lawful acquirer in accordance with its intended purpose, including for error correction.*
2. *The making of a back-up copy by a person having a right to use the computer program may not be prevented by contract insofar as it is necessary for that use.*
3. *The person having a right to use a copy of a computer program shall be entitled, without the authorization of the right holder, to observe, study or test the functioning of the program in order to determine the ideas and principles which underlie any element of the program if he does so while performing any of the acts of loading, displaying, running, transmitting or storing the program which he is entitled to do.*

The provision adds that "any contractual clause breaching numbers 2 and 3 above are null".

Article 64*quater* of the Copyright Law regulates decompilation in line with Article 6 of Directive 91/250/EEC and reads as follows:

1. *The authorization of the right holder shall not be required where reproduction of the code of the computer program and translation of its form within the meaning of Article 64bis (a) and (b), made to the aim of altering the code form, are indispensable to obtain the information necessary to achieve the interoperability of an independently created computer program with other programs, provided that the following conditions are met:*
   - *(a) these acts are performed by the licensee or by another person having a right to use a copy of a program, or on their behalf by a person authorized to do so;*
   - *(b) the information necessary to achieve interoperability has not previously been readily available to the persons referred to in subparagraph (a);*
   - *(c) these acts are confined to the parts of the original program which are necessary to achieve interoperability.*

2. The provisions of paragraph 1 shall not permit the information obtained through its application:

(a) to be used for goals other than to achieve the interoperability of the independently created computer program;

(b) to be given to others, except when necessary for the interoperability of the independently created computer program;

(c) to be used for the development, production or marketing of a computer program substantially similar in its expression, or for any other act which infringes copyright.

3. Any contractual clause breaching numbers 1 and 2 above are null.

4. In accordance with the provisions of the Berne Convention for the protection of Literary and Artistic Works, the provisions of this Article may not be interpreted in such a way as to allow its application to be used in a manner which unreasonably prejudices the right holder's legitimate interests or conflicts with a normal exploitation of the computer program.

5. Design right protection

5.1 Registered design right

The appearance of electronic devices can be protected by design rights – in addition to any protection by copyright, as mentioned above – when meeting the relevant requirements.

The definition of 'design' under the IP Code follows that of EU Regulation 6/2002, ie, "the appearance of the whole or a part of a product resulting from the features of, in particular, the lines, contours, colours, shape, texture and/or materials of the product itself and/or its ornamentation".

A registered design can be obtained by submitting the relevant application before the Italian Patents and Trademarks Office (UIBM) or before the EU Intellectual Property Office (EUIPO): a European registered design will be valid in all EU countries, including Italy. The procedures for obtaining the two registrations are similar: an application shall be submitted together with a set of drawings and the payment of a fee the amount of which will depend on the number of designs included. The application will then undergo a formality check, while no check will be made regarding whether the design meets the novelty and individual character requirements posed by the law. The registration lasts five years from the date of the application, but it can be renewed up to four times (thus extending the protection to up to 25 years) upon payment of renewal fees.

A registered (Italian or EU) design confers on its holder the exclusive right to use it and any other design which does not produce a different overall impression on the informed user, and to prevent any third party not having his or her consent from using it. The scope of the protection conferred by a registered design in fact includes any design which does not produce on the informed user a different overall impression. In assessing the scope of protection, the degree of freedom of the designer in developing his or her design shall be taken into consideration.

Under certain circumstances, infringement of registered design rights might also amount to unfair competition pursuant to Article 2598 of the Civil Code.

5.2 Unregistered design right

Italy does not provide for a national unregistered design right, but the EU unregistered design right established by the EC Regulation 6/2002 is available. Based on this, an unregistered design is protected for a period of three years as from the date on which it was first made available to the public within the European Union. Under certain circumstances, infringement of unregistered design rights might also amount to unfair competition pursuant to Article 2598 of the Civil Code.

6. Confidential information

Technical and commercial confidential information is protected by Articles 98 and 99 of the IP Code, based on which, in order to enjoy protection, the information shall:

- be secret, in the sense that they are not, altogether or in the precise configuration and combination of their elements, generally known or readily accessible to experts and operators in the field;
- have economic value as long as they are secret;
- be subject to measures to be considered as reasonably appropriate for keeping them secret, put in place by the persons to whose control the information is subject.

Where these requirements are met, the legitimate holder of the information is entitled to prohibit third parties from acquiring, disclosing to third parties or abusively using such information, unless it has been obtained by the third party independently.

In addition, under Article 2598 of the Civil Code, the unlawful acquisition, disclosure or use of confidential information amounts to unfair competition even if the information was not subject to appropriate security measures, where it was in any case not to be disclosed outside the company. The Civil Code also provides for a general obligation of confidentiality on the part of the employee (Article 2105).

Finally, the disclosure of confidential information is punished by Articles 621, 622 and 623 of the Criminal Code.

A combination of the above provisions allows action to be taken, particularly in those cases in which rights to confidential information are infringed, that is when former employees start their own company or join a competitor (including as an external contractor).

7. Other forms of protection

7.1 Utility models

Utility models are available in Italy. As with patent protection, utility model protection can be obtained by submitting an application to the UIBM. Utility models can be granted for "new models designed to confer particular effectiveness or convenience of application or use to machines, or parts of them, instruments, tools or objects in general, such as new models consisting of particular conformations, arrangements, configurations or combinations of parts".

The registration procedure only entails a formal examination of the application. No prior art search is carried out.

In addition, under Article 58 of the Italian IP Code, revoked European patents, as well as rejected or withdrawn applications for European Patents, can be converted into national utility model applications, provided that they meet the patentability requirements set by the Italian law for utility models. This conversion can be obtained by filing with the UIBM an application for registration of the utility model, together with an Italian translation of the European patent. Applicants must pay registration fees within two months of the filing.

7.2 Database rights

Directive 96/9/EC on the Legal Protection of Databases was implemented in Italy by Legislative Decree 169/99, which added new provisions to the Copyright Law.

Under Articles 1 and 2 of the Copyright Law, copyright protection is granted to creative databases which, by reason of the selection or arrangement of their contents, constitute the author's own intellectual creation. The protection does not extend to their contents and shall be without prejudice to any rights subsisting in those contents themselves.

In addition, Article 102*bis* of the Copyright Law provides for a *sui generis* right granted to the maker of a database which, although not possessing the requirements to benefit from copyright protection, implied a substantial investment on the part of their creator in obtaining, verifying or presenting the contents of the database. This right expires 15 years after the first of January of the year following the date of completion of the database.

In practice, copyright protection is rarely granted to databases as they are typically considered to lack creativity. However, the Milan Court of Appeal, upholding a first instance decision by the Milan IP Court, granted copyright protection to a database consisting of an app which collected users' data from Facebook and organised them in an original way (decision of 16 April 2018). Nevertheless, far more frequently, databases are instead covered by this *sui generis* protection.

Thus by example, in its Decision 7825/13, the Milan IP Court refused to grant copyright protection to a flight information database stating that, to receive copyright protection, "the choice of the relevant data and their disposition inside the database must be considered". In this framework, "the choice and organization of data regarding availability and prices of flights are dictated exclusively and necessarily by technical and functional considerations, which leave no room for the expression of a creative freedom, so that copyright protection under articles 1 and 2 of the copyright law cannot be granted". Nevertheless, the court found that the database in issue could be granted the *sui generis* protection under Article 102*bis*, "the purpose of which is to safeguard, beyond copyright, the creators of databases from undue appropriation of the financial and professional investment made to obtain and collect the content", which investment "is given by the whole of the means used to search for existing independent elements and to include them in the database".

Once it is ascertained that the database creator enjoys the *sui generis* protection, the creator is entitled to prevent extraction and/or reutilisation of the whole or of a substantial part of the contents of that database, where:

- 'extraction' is "the permanent or temporary transfer of all or a substantial part of the contents of a database to another medium by any means or in any form"; and
- "the definition of 'substantial part' requires the consideration of quantitative as well qualitative criteria. Under the first criteria, reference is made to the volume of data extracted and/or reused from the database and must be evaluated in relation to the volume of its total content … Under the second criteria, reference is made to the relevance of the investment connected with the achievement, verification or presentation of the contents which were extracted and/or re-used, irrespective of whether they represent a quantitatively substantial part of the content of the protected database".

A number of preliminary court decisions in Italy granted the *sui generis* protection to a database of judicial auctions created by a private company by collecting and organising judicial documents and information. The relevant data were chosen, extracted, verified and processed (eg, formatted, anonymised and indexed) by the company before including them in the database, which therefore required significant financial and professional investments. Hence, extraction and reuse of large parts of that database were found to constitute infringement of the *sui generis* right under the Copyright Law and unfair competition under Article 2598 of the Civil Code (Florence IP Court, preliminary decisions of 5 October 2015 and – on appeal – 21 March 2016; Rome IP Court, preliminary decisions of 18 September 2013 and – on appeal – 19 December 2013; Milan IP Court, preliminary decision of 20 November 2012 based on unfair competition only). In one case, a criminal investigation spun off the civil lawsuit and led to the criminal seizure of the infringer's website (Lucca Criminal Court, decisions of 7 September 2017 and – on appeal – 10 October 2017). However, this decision was later overturned by the Supreme Court (Decision 6734/2018), which stated that the criminal provisions concerning copyright protection under criminal law (namely, Articles 171 and following of Copyright Law) will not apply to the *sui generis* database right.

7.3 Semiconductor topographies

Legal protection of topographies of semiconductor products was introduced in Italy by Law 70/1989, which implemented Directive 87/54/EC.

According to Article 87 of the IP Code, a semiconductor topography consists of a series of related images, however fixed or encoded:

- representing the three-dimensional pattern of the layers of which a semiconductor product is composed; and
- in which series, each image totally or partially reproduces a surface of the semiconductor product at any stage of its manufacture.

According to Articles 88 and 90, this shall be protected against reproduction or commercial exploitation by third parties insofar as:

- it is the result of its creator's own intellectual effort and is not commonplace in the semiconductor industry; or
- it combines elements that are commonplace in the semiconductor industry, if that combination, taken as a whole, fulfils the above conditions.

7.4 Trademarks

Brands used on software and electronic goods can be registered as trademarks. Under some circumstances, a trademark can also cover the very shape of the good, ie, basically when this shape has distinctive character, is not functional and does not give substantial value to the good.

The application for registration in Italy can be submitted before the Italian Patents and Trademarks Office (UIBM), before the World Intellectual Property Organisation (WIPO) in the framework of an application for international registration, or before the EU Intellectual Property Office (EUIPO): the European registered trademark will be valid in all EU countries.

The registration lasts 10 years from the date of the application, but it can be renewed for further periods of 10 years upon payment of the relevant fees, with no time limits.

The rules governing national registered trademarks almost perfectly correspond to those set forth by the EU Regulation 1001/2017 on the European trademark.

Italian law also provides for the protection of unregistered trademarks. These can be enforced against third parties when, besides meeting the usual requirements posed for registered trademarks (ie, mainly novelty and distinctive character), they are known in a large part of the national territory thanks to their use. When this latest requirement is not met, ie, the unregistered trademark is known only in a small part of the territory, this cannot be enforced against any third party but it allows the proprietor to continue using it, within the limits of their prior use.

Under certain circumstances, infringement of (registered or unregistered) trademark rights might also amount to unfair competition pursuant to Article 2598 of the Civil Code.

7.5 Counterfeiting

The Italian law provides two extremely efficient tools aimed at detecting and seizing counterfeits. First, Regulation (EU) 608/2013, which allows the IP right holder to apply for a customs' monitoring programme: the (Italian and/or EU) customs, upon the IP right holder's application and receipt of proper information, will halt goods suspected of infringing IP rights which are detected by them.

The other tool is the anti-counterfeiting portal of the Fiscal Police (*Guardia di Finanza*). This allows exchange of information between the IP right holder and the Fiscal Police, eg, in order to provide details of the IP right holder's rights and of counterfeits found on the market. The Fiscal Police might in turn seize infringing products detected on the national territory.

Both Italian Customs and the Fiscal Police have proven to be extremely efficient and proactive. Once the counterfeits are seized by them, the IP right holder will be entitled to decide whether they want to commence proceedings against the infringer or not. The Customs/Police will generally in any case commence criminal proceedings on their own motion.

8. Ownership, inventorship and compensation

Under Article 63 of the IP Code, the right to apply for a patent belongs to the inventor and to his or her successors in title. If a patent is registered by someone who is not the person entitled to file it, the latter can take legal action to have the counterparty's patent declared null or transferred to him or her pursuant to Article 118 of the IP Code.

Article 64 of the IP Code regulates employee inventions, providing that:

- The invention belongs to the employer if it was made in the course of the employee's normal duties, the inventive activity is provided by the employment contract and the employee is paid to this aim. Nevertheless, the inventor always has the right to be recognised as the author of the invention, which right cannot be transferred to anyone (Article 63 of the IP Code); and
- If the invention was made in the course of the employee's normal duties, but the inventive activity is not specifically paid under the employment contract, the invention belongs to the employer but, if the latter obtains a patent or uses the invention as secret know-how, the employee is entitled to compensation (further than recognition as the author of invention).

Article 65 of the IP Code provides an exception to Article 64: where the employer is a university or a public administration body having research among its statutory purposes, a researcher is the exclusive holder of the rights on the invention that he or she made. In addition, a researcher is entitled to no less than 50% of the incomes deriving from the exploitation of the invention. These provisions, however, do not apply when the research is sponsored by a private company or a public company different from the one where the researcher works.

Copyright generally belongs to the author of the work. Where a work has been created by two or more authors jointly, with no possibility to distinguish their single contributions, copyright belongs to all the authors jointly.

However, pursuant to Article 12*ter* of the Copyright Law (reproducing Article 2(3) of Directive 2009/24/EC), where a work is created by an employee in the execution of his or her duties or following the instructions given by his or her employer, the employer shall be exclusively entitled to exercise all economic rights in the program so created, unless otherwise provided by contract.

This provision applies only to employees, and not to external contractors. Hence, in the case of external contractors, copyright will remain vested in the author unless the parties expressly agree that it belongs to the commissioner or at least provided for a creative activity to be carried out by the contractor in order for the commissioner to exploit the relevant results.

9. Enforcement

9.1 Obtaining information on the infringer and the infringement

Under Italian law, each party is obliged to provide evidence of the rights enforced. However, parties are not obliged to disclose documents which negatively affect their case.

An IP right holder is, however, entitled to request the court to order the counterparty to disclose information on the infringement.

In particular, pursuant to Article 121 of the IP Code (in line with Article 210 of the Civil Procedure Code), a party who provided significant circumstantial evidence of the infringement and identified documents or information confirming that evidence held by the counterparty, can ask the court to order the counterparty to:

- submit the documents (including accounting records) or provide the information; and
- identify the persons involved in the production or commercialisation of the goods/services constituting infringement of IP rights.

Pursuant to Article 121*bis* of the IP Code, an IP right holder, both in preliminary and in full-scale proceedings, can ask the court to order any person using an infringing product/service on a commercial level to provide information on the origin and the distribution network of the infringing goods/services, including the details of the persons involved, the amounts of products sold and their price.

9.2 Interim relief

Interim relief is available in Italy. Such relief includes:

- search orders: giving an IP right holder the authority to inspect (even *ex parte*, ie, without giving prior notice) the premises of a suspected infringer under supervision of court officials, so as to collect evidence of the infringement;
- seizure orders: giving an IP right holder a right to seize the allegedly infringing goods. Such orders may be limited to seizing samples to secure evidence of the alleged infringement (often granted *ex parte*) or may be granted to provide a right to seize all infringing products in stock and the material/tools used to manufacture the products. Orders of the latter type are rarely granted *ex parte*; and
- preliminary injunction orders: these are rarely granted *ex parte*; they can be assisted by penalty provisions in case of breach of the injunction and/or delay in complying with the same. The IP right holder is also entitled to request that the infringer is ordered to publish the injunction order at his or her costs on one or more newspapers and/or magazines. However, most Italian courts will not grant such publication orders in preliminary proceedings but only grant such orders at the end of proceedings on the merits.

Search orders and seizure orders require the subsequent commencement of proceedings on the merits within around one month. A failure to do so causes such orders to expire. In contrast, preliminary injunction orders will become final if they are not appealed and overturned, with no need for the IP right holder to subsequently commence proceedings on the merits.

9.3 Threats

Based on Italian case law, under certain circumstances, threatening to initiate legal proceedings might amount to a crime either under Article 393 (arbitrary enforcement of one's rights) or under Article 629 (extortion) of the Criminal Code. However, to date it would appear that the Italian courts have never had to deal with this issue in the context of an IP rights enforcement case.

At the civil level, unjustified threats of legal proceedings can amount to unfair competition under Article 2598 of the Civil Code, if they interfere with someone's lawful business and are likely to affect it. In a decision of 29 March 2007, the IP Court of Milan expressly stated that "the threat of actions made in bad faith or with negligent ignorance of the evident invalidity of the patent inadmissibly disturbs the market and might therefore amount to unfair competition under article 2598 (3) of the Italian Civil Code".[2]

9.4 Remedies

Typical remedies awarded following a successful infringement action include:

- a final injunction from any further infringement;
- an order of withdrawal from the market of the infringing goods;
- an order of destruction of the infringing goods and production tools; and
- damages.

Article 125 of the IP Code sets the main criteria to be applied by Italian Courts in assessing damages:

- the right holder's lost profit and other negative consequences of the infringement, including moral damages;
- the reasonable royalty that the infringer would have paid if they had obtained a regular licence to commercialise the product (which criterion constitutes an alternative way to determine the right holder's lost profit); and
- the infringer's profit, alternatively to the right holder's lost profit or insofar as they exceed it.

Typically, a damages claim would include a request for compensation for lost profit plus a request for restitution of the infringer's profits in the amount exceeding the lost profit. The latter would amount at least to the reasonable

2 Giurisprudenza Annotata di Diritto Industriale, 2009, p305.

royalty, which in turn, according to a large part of the Italian case law, should be calculated by raising the standard royalty applied in the specific market sector by a given amount, in order to take account of the fact that the infringer had not obtained the IPR holder's authorisation to exploit the IP right in issue (even if Italian law excludes punitive damages).

10. Hot topics

10.1 Unified Patent Court – the Milan seat and the impact of Brexit

The Unitary Patent system should have been up and running in 2017, but delays were caused by Brexit and by a constitutional issue which has recently risen in Germany. Italy ratified the Unified Patent Court agreement with Law 214 of 3 November 2016, and requested to have only one local division in Milan. However, Milan might ultimately host one of the seats of the Central Division, ie, the one which was originally intended to be based in London, as a consequence of Brexit were the Unified Patent Court to proceed without the United Kingdom as Italy was the EU state with the highest number of validated European patents after the United Kingdom, and this was the criteria on which the countries hosting the seats of the local divisions were chosen.

United Kingdom

Nicholas Fox

One of the most important jurisdictions in Europe – the United Kingdom – is distinguished by having a common law system rather than the civil law system used throughout continental Europe. This provides a legal system which has powerful tools enabling owners to challenge or establish infringement of intellectual property rights but one which can sometimes sit apart from mainstream developments in the rest of Europe.

1. Outline of UK national patent procedure

1.1 Filing

In the United Kingdom, a patent can be obtained via a national UK patent application or via a European patent application designating the United Kingdom. A UK national patent application can be derived from a patent application filed directly at the UK Intellectual Property Office (IPO) or, alternatively, via an international patent application filed under the Patent Cooperation Treaty (PCT).

1.2 An examination system

The UK IPO like the European Patent Office (EPO) is a full examination system. When a patent application is filed, provided the application fee is paid, the UK IPO reviews the paperwork to confirm that the application fulfils the formal requirements for a patent application.[1] A search request must be filed within 12 months from the filing date or the priority date if priority is claimed.[2] If a search request is included in an initial filing for a new patent application not claiming priority, normally the UK IPO will issue a search report within about three months.

All patent applications are published 18 months from the filing date or priority date if priority is claimed.[3] The publication of the application sets a six-

1 Patents Act 1977, Section 15A.
2 Patents Act 1977, Section 17 and Rule 27(3) Patents Rules 2007.
3 Patents Act 1977, Section 16.

month deadline for filing an examination request.[4] Alternatively, if the search and examination requests are filed and the related fees are paid simultaneously, the UK IPO will issue a combined search and examination report. Such a combined search and examination report resembles an extended European search report issued by the EPO.

1.3 Time scales

Backlogs at the UK IPO vary depending upon the technical subject matter of a patent application. If no action is taken to accelerate proceedings typically the first examination report will issue approximately three years from the filing or priority date. All UK patent applications are required to be in order for grant within four-and-a-half years from the filing date or priority date if claimed. This deadline is extended to 12 months from the issuance of the first examination report, if the issuance of that report is delayed.[5] Due to the existence of this deadline, the UK IPO sets progressively shorter terms for responding to the examination report as the deadline approaches.

It is possible to request accelerated prosecution and the UK IPO will accept any such request if an applicant can provide a good reason for wanting to achieve faster grant. No additional fees need to be paid. If prosecution is accelerated, grant is normally possible within a few months. In any event, however, the UK IPO will only grant an application once three months have passed following publication of the application. An applicant can request early publication of an application to avoid this delay in grant.

1.4 Outline of costs[6]

The main official fees for filing and prosecuting an application online[7] are as follows:

Application fee	£60
Search fee	£150 plus £20 for each claim over 25
Examination fee	£100 plus £10 for each page of description over 35

Renewal fees are due after grant and range from £70 for the fifth year from the filing date to £610 for the twentieth year.

2. Examination of software patents

Issues of patentability of computer software can be raised at any stage in the patent process. If the UK IPO is of the opinion that the patent application is

4 Rule 28(2) Patents Rules 2007.
5 Patents Act 1977, Sections 18(4) and 20(1) and Rule 30 Patents Rules 2007.
6 Fees as of April 2018.
7 The Application Fee, Search Fee and Examination Fee are increased to £90, £180 and £130 plus excess claims fees and excess pages fees in the case of paper filings.

directed towards excluded subject matter, they may refuse to search the patent application until the claims have been amended so that they are directed to allowable subject matter.[8] A decision of an examiner to refuse an application can be contested at a hearing and the decisions of the UK IPO can be appealed to the English High Court.[9] Additionally, the validity of a granted patent may be challenged in the UK courts after grant.

3. Relationship with EPO law

The United Kingdom is a common law jurisdiction where the doctrine of precedent applies. This means that the earlier decisions of higher courts are binding authorities which must be subsequently followed. In principal, the Court of Appeal is bound by its earlier decisions unless a previous decision is inconsistent with a subsequent decision of the Supreme Court or was reached without reference to a relevant statute or other authority.[10]

As EPO decisions are not decisions of an English court, they are only persuasive and do not constitute binding authority. The effect of this is that in order to be binding on UK courts, the approach of the EPO must be expressly endorsed by a UK court decision.[11] This has resulted in the United Kingdom's approach to software inventions diverging from the current position of the EPO. In particular, in the United Kingdom inventions lacking technical character are rejected as relating to unpatentable subject matter whereas equivalent inventions in the EPO are considered to lack inventive step as they fail to provide a technical solution to a technical problem.

Although there are differences in approach, in practice it is difficult to identify circumstances where this has any practical consequences as it does not matter to an applicant whether his or her application is rejected on the grounds of unpatentable subject matter or lack of inventive step; either way he or she does not get a patent.

As UK law is in principle at least harmonised with that of the EPC, the hurdles an application needs to overcome in the United Kingdom are very similar to those in the EPO. A few years ago the UK IPO issued practice notices[12] which took a stricter line against the patentability of computer-readable media claims and the interpretation of the mental acts exclusion than is the case in the EPO. However, both aspects were subsequently overturned by the UK courts.[13]

---

8 Patents Act 1977, Section 17(5).

9 Patents Act 1977, Section 97.

10 *Young v Bristol Aeroplane Co Ltd* [1944] KB 718.

11 This remains the case even following *Actavis UK Limited v Merck and Co Inc* [2008] EWCA Civ 444 where the Court of Appeal acknowledged that UK law must now be construed in a European context and that the Court of Appeal could depart from its previous decisions where it was satisfied that the EPO Boards of Appeal had formed a settled view on a point of law which differs from earlier UK decisions.

12 *Patents Act 1977: Patentable subject matter* (2 November 2006) and *Patents Act 1977: Patentability of Computer Programs* (8 December 2008).

13 *Astron Clinica and Others' Applications* [2008] EWHC 85 (Pat) and *Halliburton's Application* [2011] EWHC 2508 (Pat).

This does suggest that the UK IPO has previously taken a harder line on these matters than the UK courts.

4. National patent law

The basic requirements for a patentable invention in the United Kingdom are set out in the Patents Act 1977, Section 1(1):

A patent may be granted only for the invention in respect of which the following conditions are satisfied, that is to say:

- the invention is new;
- it involves an inventive step;
- it is capable of industrial application; and
- the grant of a patent is not excluded by sub-sections (2) and (3) or section 4(A) below; and references in this act to a patentable invention shall be construed accordingly.

Inventions which are excluded from patentability are then listed in the Patents Act 1977, Section 1(2):

It is hereby declared that the following (among other things) are not inventions for the purposes of this Act, that is to say anything which consists of:

- *a discovery, scientific theory or mathematical method;*
- *a literary, dramatic, musical or artistic work or any other aesthetic creation whatsoever;*
- *a scheme, rule or method of performing a mental act, playing a game or doing business, or a programme for a computer; and*
- *the presentation of information;*

but the foregoing provision shall prevent anything from being treated as an invention for the purposes of this Act, only to the extent that a patent or application for a patent relates to that thing as such.

In order to ensure conformity with the European Patent Convention, Sections 1(1) and 1(2) are declared by Section 130(7) of the Patents Act 1977 to be framed so as to have, as nearly as practical, the same effects in the United Kingdom as the corresponding provisions of the European Patent Convention.

4.1 UK court's approach to software inventions

Throughout the 1980s and 1990s the question of the patentability of software has been considered by the English Court of Appeal on four occasions, in *Genentech's Patent* [1989] RPC 147, *Merrill Lynch's Application* [1989] RPC 561, *Gale's Application* [1991] RPC 305 and *Fujitsu's Application* [1997] RPC 608.

In *Merrill Lynch*, the Court of Appeal endorsed the then current approach of the EPO which was established in T208/84 *Vicom/computer-related invention* which required, as a matter of patentable subject matter, the presence of a 'technical effect'.

Previously the EPO Board of Appeal in *Vicom* had stated that:

12. The Board is of the opinion that a claim directed to a technical process in which the process is carried out under the control of a programme (be this implemented in hardware or in software), cannot be regarded as relating to a computer programme as such within the meaning of Article 54(3) EPC as it is the application of the programme for determining the sequence of steps in the process for which in effect protection is sought. Consequently, such a claim is allowable under Article 52(2) EPC and (3) EPC.

The board continued:

16. In arriving at this conclusion the Board has additionally considered that making a distinction between embodiments of the same invention carried out in hardware or in software is inappropriate as it can be fairly said that the choice between these two possibilities is not of an essential nature but is based on technical and economic considerations which bear no relationship to the inventive concept as such. Generally speaking, an invention which would be patentable in accordance with conventional patentability criteria should not be excluded from protection by the mere fact that for its implementation modern technical means in the form of a computer program are used. Decisive is what technical contribution the invention as defined in the claim when considered as a whole makes to the known art. Finally it would seem illogical to grant protection for a technical process itself controlled by suitably programmed computer but not the computer itself when set up to execute the control.

The requirement that a software invention gives rise to a 'technical effect' was then crystallised into English law by Fox LJ in *Merrill Lynch* who said:[14]

... it seems to me to be clear ... that it cannot be permissible to patent an item excluded by section 1(2) under the guise of an item which contains that item – that is to say in the case of a computer programme, the patenting of a conventional computer containing that programme. Something further is necessary. The nature of the addition is, I think to be found in the Vicom *case where it is stated: 'Decisive is what technical contribution the invention makes to the known art'. There must I think be some technical advance over the prior art in the form of a new result (eg, a substantial increase in processing speed as in* Vicom*).*

The requirement for patentable inventions to have a 'technical effect' was restated in subsequent cases with, for example, the Court of Appeal in *Fujitsu* stating:[15]

... it is and always has been a principle of patent law that mere discoveries or ideas are not patentable, but those ideas and discoveries which have a technical aspect

14 *Merrill Lynch*, at page 569, lines 2–10.
15 *Fujitsu*, at page 614, lines 40–46.

or make a technical contribution are. Thus the concept of what is needed to make an excluded thing patentable is a technical contribution is not surprising. That was the basis of the decision in the Vicom. *It has been accepted by this court and by the EPO and has been applied since 1987. It is a concept at the heart of patent law.*

The Court of Appeal revisited the question of the patentability of computer software in 2006 in the conjoined appeals *Aerotel Limited v Telco Holdings Limited; Macrossans' Patent Application* [2006] EWCA Civ 1371.

By that time the approach of the EPO had modified the test in *Vicom* through the decisions of *T931/95 PBS partnership*, *T258/03 Hitachi* and *T641/00 COMVIK* and the requirement that an invention provide a technical solution to a technical problem no longer formed part of the assessment of patentable subject matter in the EPO. Rather, the matter was assessed solely under inventive step.

Due to the doctrine of precedent in the United Kingdom, however, the Court of Appeal considered itself bound by its own earlier pronouncements on patentable subject matter and hence it declined to follow the EPO's lead. Instead, it restated a test for patentable subject matter which retains the requirement for a 'technical effect' to be present.

This test[16] (now known as the *Aerotel* test) is as follows:

- properly construe the claim;
- identify the actual contribution;
- ask whether it falls solely within the excluded subject matter; and
- check whether the actual or the alleged contribution is actually technical in nature.

The Court of Appeal then went on to explain[17] each of these steps as follows:

No one would quarrel with first step, construction. You have to decide what the monopoly is before going on the question of whether it is excluded. Any test must involve this first step.

The second step – identify the contribution – is set to be more problematical. How do you assess the contribution? Mr Birss submits the test is workable – it is an exercise in judgement, probably involving the problem said to be solved, how the invention works, what its advantages are. What has the inventor really added that perhaps best sums up the exercise. The formulation involved looking at substance not form – which is surely what the legislator intended.

Mr Birss added the words 'or alleged contribution' in his formulation of the second step. That will do at the application stage – where the office must generally accept what the inventor says is his contribution. It cannot actually be conclusive, however. If an inventor claims a computer when programmed with his new

16 *Aerotel*, at paragraph 40.
17 *Id*, at paragraphs 42–46.

programme, it will not assist him if he alleges wrongly that he has invented the computer itself, even if he specifies all the detailed elements of a computer in his claim. In the end the test must be what contribution has actually been made, not what the inventor says he has made.

The third step – is the contribution solely of excluded matter? – is merely an expression of the 'as such' qualification of Article 52(3). During the course of argument Mr Birss accepted a reformulation of the third step – ask whether the contribution thus identified consists of excluded matter as such? We think either formulation will do – they mean the same thing.

The fourth step – check whether the contribution is 'technical' – may not be necessary because the third step should have covered that. It is a necessary check, however, if one is to follow Merrill Lynch *as we must.*

In 2009 the Court of Appeal had an opportunity to realign UK law with that of the EPO in *Symbian's Application* [2009] RPC 1, as by that date the court had in *Actavis UK Limited v Merck and Co Inc* [2008] EWCA Civ 444 ruled that the Court of Appeal could depart from its previous decisions where it was satisfied that the EPO Boards of Appeal had formed a settled view on a point of law which differs from earlier UK decisions.

The court, however, declined to do so and repeated its position that software inventions would continue to be allowable based on the existence of a 'technical contribution' such as was the case in *Vicom* and which was reflected in the early EPO case law such as *T6/83 IBM/data processor network* and T115/85 *IBM/computer-related invention* as well as in the UK case law.[18]

Subsequently, in *AT&T Knowledge Ventures and CVON Innovations Limited Applications* [2009] EWHC 343 (Pat), the English High Court provided further guidance on the correct approach to determining technical contribution under the *Aerotel* test. Noting that it was impossible to define the meaning of 'technical' in the context of the *Aerotel* test, the court identified that there were a number of signposts as to what might be a relevant technical effect, namely[19]:

- whether the claimed technical effect has a technical effect on a process which is carried out outside the computer;
- whether the claimed technical effect operates at the level of the architecture of the computer, that is to say whether the effect is produced irrespective of the data being processed or the applications being run;
- whether the claimed technical effect results in the computer being made to operate in the new way;
- whether there is an increase in the speed or reliability of the computer; and

18 Such a decision was taken prior to the ruling of the Enlarged Board of Appeal in *G3/08* which upheld the current EPO practice and arguably gives this area the status of 'settled law' which would permit the Court of Appeal to act.

19 *AT&T*, at paragraph 40.

- whether the perceived problem is overcome by the claimed invention as opposed to merely being circumvented.

Lewison LJ reviewed these signposts in the Court of Appeal in *HTC Europe Co Ltd v Apple*[20] and approved the comments of Mann J in the judgment in *Gemstar-TV Guide International Inc v Virgin Media Ltd*[21] that the fourth signpost set out above should be the less restrictive question whether a program makes a computer a better computer in the sense of running more efficiently and effectively as a computer.

Subsequently, Kitchin LJ in *Lantana's Application*,[22] after stating that there is no comprehensive test for determining technical contribution, repeated and commented favourably on Lewison LJ's summary in *HTC Europe Co Ltd v Apple* as to the following being a useful starting point:

- *First, it is not possible to define a clear rule to determine whether or not a program is excluded, and each case must be determined on its own facts …*
- *Second, the fact that improvements are made to the software programmed into the computer rather than hardware forming part of the computer does not make a difference … the analysis must be carried out as a matter of substance not form.*
- *Third, the exclusions operate cumulatively. So, for example, the invention in* Gale *related to a new way of calculating a square root of a number with the aid of a computer and Mr Gale sought to claim it as a ROM in which his program was stored. This was not permissible. The incorporation of the program in a ROM did not alter its nature: it was still a computer program (excluded matter) incorporating a mathematical method (also excluded matter) …*
- *Fourth, it follows that it is helpful to ask: what does the invention contribute to the art as a matter of practical reality over and above the fact that it relates to a program for a computer? If the only contribution lies in excluded matter then it is not patentable.*
- *Fifth, and conversely, it is also helpful to consider whether the invention may be regarded as solving a problem which is essentially technical, and that is so whether that problem lies inside or outside the computer. An invention which solves a technical problem within the computer will have a relevant technical effect in that it will make the computer, as a computer, an improved device, for example by increasing its speed. An invention which solves a technical problem outside the computer will also have a relevant technical effect, for example by controlling an improved technical process. In either case it will not be excluded by Art 52 as relating to a computer program as such.*[23]

---

20 [2013] EWCA Civ 451.
21 [2009] EWHC 3068 (Ch).
22 [2014] EWCA Civ 1463.
23 *HTC v Apple*, at paragraphs 45–49.

4.2 Inventive step

The UK courts' approach to inventive step was developed by the Court of Appeal in *Windsurfing v Tabur Marine* [1985] RPC 59 (CA) and restated in a slightly re-ordered form in *Pozzoli SpA v BDMO SA* [2007] EWCA Civ 588. This approach differs significantly from the problem-and-solution approach adopted by the EPO.[24]

When assessing inventive step, having identified the art or field of endeavour in which an invention arises the UK courts then proceed to:

- Identify:
  - the notional 'person skilled in the art'; and
  - the relevant common general knowledge of that person;
- Identify the inventive concept of the claim in question or if that cannot readily be done construe the claim;
- Identify what, if any, differences exist between the matter cited as forming part of the 'state of the art' and the inventive concept of the claim or the claim as construed; and
- Decide whether those differences, viewed without any knowledge of the alleged invention as claimed, constitute steps which would have been obvious to the person skilled in the art or whether they require any degree of invention.[25]

Notably, unlike the EPO where inventive step is assessed based on a single identified closest piece of prior art, any suitable item of prior art can form the starting point for assessment of inventive step. The assessment of inventive step is then a factual enquiry based on the evidence before the court as to differences between the claimed invention and a particular item of prior art and the common general knowledge of the relevant person skilled in the art. Such an assessment is purely factual and is not dependent upon the formulation of a technical problem which can be said to have been solved by an invention, as the presence of a 'technical contribution' for the basis of a patent is assessed separately as part of the test for patentable subject matter.[26]

4.3 Methods for doing business

There is little case law in the United Kingdom on the patentability of business methods. What limited jurisprudence exists suggests that it is very difficult to persuade the courts that a patent application should be granted where a claimed invention is a computer-implemented business method.

The leading case is *Merrill Lynch's Application* [1989] RPC 561 where the

---

24 The EPO problem-and-solution approach is discussed in detail in the "Patents in Europe" chapter, section 2.3.

25 *Pozzoli*, at paragraph 23.

26 See 4.1 above.

Court of Appeal rejected an application directed to a computerised trading system which accommodated real-time order flow. In rejecting the case, the court considered that the only possible technical advance arising from the invention was the production of a novel trading system – that is, a data-processing system for implementing a specific business. The court went on to reject the application stating that although normally a data-processing system operating to produce a novel technical result would be patentable, this would not be the case when the end result was a method of doing business and hence itself a prohibited item under Section 1(2) of the Patents Act 1977.

A similar approach was taken in *Oneida Indian Nation's Application* [2007] EWHC 954 (Pat) which related to a computer system for online gambling. In that case, the court accepted that the system gave rise to a number of technical advantages in that the number of processing steps per bet were reduced, the number or data transmission steps were reduced, the system dispensed with the need to make an account check for every bet, and hence the system was more secure and robust. However, the court still rejected the case despite these advantages because all of the advantages were the direct result of implementing a new business method.

4.4 Mental acts

The mental acts exclusion is construed narrowly in the United Kingdom. Here, the exclusion only applies to acts carried out mentally. This was recently confirmed by the High Court in *Halliburton's Application* [2011] EWHC 2508 (Pat).

Previously in *Fujitsu's Application*[27] the Court of Appeal proposed taking a broad view of the exclusion which would exclude from patentability any methods which were capable of being performed mentally regardless of whether the methods were in fact performed in that way. This broad view was, however, doubted by the Court of Appeal in *Aerotel*[28] and was finally rejected in the *Halliburton* case. As a result, the mental acts exclusion is of very limited relevance to the patentability of computer programs as the exclusion can easily be avoided by identifying that a method is undertaken by a computer.

4.5 Presentation of information

The application of the *Aerotel* test to presentations of information was considered in *Gemstar TV Guide International v Virgin Media* [2009] EWHC 3069 (Ch) in a case concerning electronic program guides. The approach taken confirmed that the UK courts, like the EPO, will make a distinction between content and delivery of information and any technical aspects of that delivery. In *Gemstar* the court accepted that the actual contribution of the claimed

27 *Fujitsu*, at pages 619 line 10 to 621 line 30.
28 *Aerotel*, at page 98.

invention was a better, more intuitive electronic program guide which meant that the invention had an independent effect which went beyond the mere presentation of information.

The same approach was followed in *HTC v Apple* [2012] EWHC 1789 (Pat) in a case involving four patents including Apple's 'slide to unlock' patent. Here, although ultimately the claimed invention was found to be obvious in the light of the prior art, the court was happy to accept that the provision of a novel and intuitive user interface for activating a mobile phone was clearly suitable subject matter for a patent.

4.5 Mathematical methods

Purely abstract mathematical methods are not patentable under UK law. However, mathematical methods having practical application are not excluded.

This follows from the Court of Appeal decision in *Gale's Application* [1991] RPC 305 where the court stated that:

> *In the present case Mr Gale claims to have discovered an algorithm. Clearly that, as such is not patentable. It is an intellectual discovery which, for good measure, falls squarely within one of the items, mathematical method, listed in s.1(2). But the nature of this discovery is such that it has a practical application, in that it enables instruction to be written for conventional computers in a way which will, so it is claimed, expedite one of the calculations frequently made with the aid of a computer. In my view the application of Mr Gale's mathematical formulae for the purpose of writing computer instruction is sufficient to dispose of the contention that he is claiming a mathematical method as such.*

Following *Gale's Application*, the UK IPO will only reject patent applications relating to mathematical modelling or the implementations of algorithms if no benefit arising from the algorithm can be identified or the algorithm is entirely unconnected with real world phenomena.[29]

4.6 Scheme, rule or method for playing a game

Where an invention relates solely to the rules of a game, a patent application for the invention will be rejected as relating to unpatentable subject matter. *IGT's Application* [2007] EWHC 1341 is an example of such a case. There the claimed invention related to a gaming apparatus where success on a main game led to a chance to play a bonus game. The invention therefore related solely to the rules of the game being played and hence was rejected as being unpatentable for relating to excluded subject matter as such.

---

29 Compare *Schlumberger Holdings* BL O/016/09 – provision of quality control indicators based on processing sonic well logging data – allowed; *Western Gecco* BL O/135/07 – determining one or more parameters relating to physical properties of the earth's interior from seismic data – allowed. But note also *Institut du Français Petrol & Elf* BL O/201/03 – case rejected due to lack of any link to physical benefit or measurement.

4.7 Other matters

The UK IPO follows the EPO practice of allowing claims for computer-readable media storing computer programs which enable an otherwise patentable invention to be implemented. Although acceptance of such claims was briefly doubted following the issuance of a practice note by the UK IPO in 2006, that practice note was revoked following the appeal in *Astron Clinica's Application* [2008] EWHC 85 (Pat). The current practice is therefore that such claims are permissible if a claimed invention is otherwise allowable.

5. Infringement

5.1 Acts of Infringement

Acts of infringement are defined by Section 60(1) of the Patents Act 1977. The wording of Section 60 is modelled on Article 25 of the Community Patent Convention and provides that where the patented invention is a product, a patent is infringed by a person who makes, disposes of, offers to dispose of, uses, imports or keeps the product. Where the patented invention is a process, the patent is infringed by a person who uses or offers the process for use in the United Kingdom when he or she knows, or it would be obvious to a reasonable person, that the use of the process would amount to an infringement of the patent. A process patent is also infringed by a person who disposes of, offers to dispose of, uses, imports or keeps any product obtained directly by means of the patented process.

In order to infringe a patent, it is therefore necessary to undertake one of these acts in relation to a product or process which falls within the scope of the claims of a patent.

5.2 Claim interpretation

The scope of an invention for the purpose of UK law is defined by Section 125 of the Patents Act 1977 as being that specified in the claims of a patent. Section 125(3) of the Patents Act 1977 expressly states that the extent of protection is that set out in Article 69 of the European Patent Convention. This provides that:

> *The extent of protection conferred by a European patent or a European patent application shall be determined by the claims. Nevertheless the description and drawings shall be used to interpret the claims.*

The Protocol on the interpretation of Article 69 then goes on to explain that:

> *Article 69 should not be interpreted as meaning that the extent of the protection conferred by a European patent is to be understood as that defined by the strict, literal meaning of the wording used in the claims, the description and drawings being employed only for the purpose of resolving an ambiguity found in the claims. Nor should it be taken to mean that the claims serve only as a guideline and that the actual protection conferred may extend to what, from a consideration of the*

description and drawings by a person skilled in the art, the patent proprietor has contemplated. On the contrary, it is to be interpreted as defining a position between these extremes which combines a fair protection for the patent proprietor with a reasonable degree of legal certainty for third parties.

It goes on further:

For the purpose of determining the extent of protection conferred by a European patent, due account shall be taken of any element which is equivalent to an element specified in the claims.

Prior to 2017 the leading UK case on the interpretation of patent claims was *Kirin-Amgen Inc v Hoechst Marion Rousell* [2005] RPC 9. In that case, Lord Hoffmann in the House of Lords reviewed the earlier UK case law on the interpretation of claims and confirmed the primacy of Article 69, stating:

The determination of the extent of protection conferred by a European patent is an examination in which there is only one compulsory question, namely that set by article 69 and its Protocol: what would a person skilled in the art have understood the patentee to have used the language of the claim to mean?

However, in 2017, the Supreme Court in *Actavis UK Limited & Ors v Eli Lilly and Company*[30] effectively overruled *Kirin-Amgen*, replacing the assessment in *Kirin-Amgen* with a two-step test stating that:

... a problem of infringement is best approached by addressing two issues, each of which is to be considered through the eyes of the notional addressee of the patent in suit, i.e. the person skilled in the relevant art. Those issues are: (i) does the variant infringe any of the claims as a matter of normal interpretation and, if not, (ii) does the variant nonetheless infringe because it varies from the invention in a way or ways which is or are immaterial? If the answer to either issue is 'yes', there is an infringement; otherwise, there is not.[31]

The Supreme Court provided the following guidance[32] as to the scope of the additional protection for determining whether a variant infringes a patent:

(i) Notwithstanding that it is not within the literal meaning of the relevant claim(s) of the patent, does the variant achieve substantially the same result in substantially the same way as the invention, ie, the inventive concept revealed by the patent?

(ii) Would it be obvious to the person skilled in the art, reading the patent at the priority date, but knowing that the variant achieves substantially the same result as the invention, that it does so in substantially the same way as the invention?

(iii) Would such a reader of the patent have concluded that the patentee nonetheless intended that strict compliance with the literal meaning of the relevant claim(s) of the patent was an essential requirement of the invention?

---

30 [2017] UKSC 48.

31 *Actavis UK Limited v Eli Lilly*, at paragraph 54.

32 *Actavis v Eli Lilly*, at paragraph 66. The questions being a reformulation of the *'Improver'* questions previously adopted by the House of Lords, prior to the decision in *Kirin-Amgen*.

In order to establish infringement in a case where there is no literal infringement, a patentee would have to establish that the answer to the first two questions was 'yes' and that the answer to the third question was 'no'.

5.3 Network claims

The UK courts have taken a robust approach to direct infringement of patent claims where the benefits of an invention occur in the United Kingdom but only part of a computer network is present in the jurisdiction. This issue arose in *Menashe Business Mercantile Ltd v William Hill Organization Ltd* [2002] EHWC 397 (Pat). The case concerned a gaming system and the claims required the presence of a host computer, a terminal computer and a communications system between them. In a decision on a preliminary point, the defendant claimed that their activities could not infringe the patent because their server was passed in the Netherlands Antilles. The High Court vigorously rejected the defendant's arguments, a view which was later confirmed on Appeal.[33]

5.4 Contributory infringement

A patent is also infringed by a person who supplies or offers to supply in the United Kingdom the means to make the patented product or use the patented process, if that person knows, or it is obvious to a reasonable person, that those means are suitable for putting, and are intended to put, the invention into effect in the United Kingdom. The means supplied must relate to an "essential element of the invention".[34]

It should be noted that in order to infringe both the offer or supply and the putting of the invention into effect must occur in the United Kingdom. This can have the effect of narrowing the scope of infringement. Thus, for example, importing a patented apparatus into the United Kingdom in a dismantled form could constitute infringement, whereas supplying the same apparatus for export would, however, not infringe.

This double territoriality requirement can be of particular importance when considering contributory infringement of inventions involving computer networks where only part of a computer network is based in the United Kingdom. Thus, for example, in *Research In Motion v Motorola* [2010] EWHC 118 (Pat), the court found there to be no contributory infringement of a patent to a computer network where only a messaging gateway system was operated in the United Kingdom. The situation in *Research In Motion v Motorola*, contrasts with that previously discussed in *Menashe v William Hill*[35] where the location of a server was considered immaterial for the purposes of direct infringement. The

---

33 *Menashe Business Mercantile Ltd v William Hill Organization Ltd* [2002] EWCA Civ 1702.
34 Patents Act 1977, Section 60(2).
35 See 5.3 above.

key distinction between the two cases is that in *Menashe v William Hill* the benefit of the invention was derived in the United Kingdom, whereas in *Research In Motion v Motorola* the gateway system was only an entry point for initiating the invention and the benefit of the invention was ultimately derived elsewhere.

Cases on contributory infringement often turn on the meaning of the 'essential element of the invention'. In *Cranway v Playtech* [2009] EWHC 1588 (Pat), the High Court explained that an essential element was not necessarily the same thing as is used to work an invention. Rather the test is an objective one, whether on not an element is suitable for putting a particular invention into effect. This contrasts with a subjective assessment of the knowledge or intention of the supplier of the essential elements where the court held in *Grimme v Scott* [2010] EWCA Civ 1110, that the knowledge and intention requirements of Section 60(2) were satisfied if, at the time of supply or offer of supply, the supplier knows, or it is obvious in the circumstances, that ultimate users (not necessarily the person directly supplied) will intend to put the invention into effect in the United Kingdom.

6. Ownership, inventors and compensation

Any person may apply for a patent application. However, a patent may only be granted to:

- the inventor or joint inventors;
- other persons entitled to property in the invention; or
- successors in title.[36]

In practice this means that whereas anyone may file for a patent application, title to the invention should be secured before an application grants as a patent may be revoked if it is granted to a person who was not entitled to that grant.[37]

Sections 39–43 of the Patents Act 1977 deal with employee inventions. These provide that ordinarily an invention made an employee in the United Kingdom belongs to the inventor's employer if either:

- it was made in the course of the normal duties of the employee or in the course of duties falling outside his or her normal duties, but specifically assigned to him or her; or
- the invention was made in the course of the duties of the employee and the employee had a special obligation to further the interest of the employer.[38]

36 Patents Act 1977, Section 7(2).
37 Patents Act 1977, Section 72(1)(b).
38 Patents Act 1977, Section 39(1).

The first limb generally applies to anyone who is an employed inventor in the sense of someone who is employed to invent. The second limb is normally applicable to directors and higher managers in companies who although they may not be employed to invent are by virtue of the nature of their relationship with a company under an obligation to forward the interest of the company.

In the event that an employer is entitled to an invention, an employee can claim compensation for the invention where the invention of the patent is of outstanding benefit to the employer.[39] The test for outstanding benefit is a high one and is assessed in the context of the size and nature of the employer's undertaking and an award is only made if in all the circumstance it is just to do so. As a result, successful applications for compensation are very rare.[40]

7. Copyright protection

7.1 Requirements for protection

Frequently, copyright is the most significant intellectual property right in software or electronics design. Section 1(1)(a) of the Copyright, Designs and Patents Act 1988 provides that copyright exists in original literary, dramatic, musical or artistic works. Literary works are defined[41] as including computer programs and the preparatory design materials for a computer program and databases. Artistic works are defined to include graphic works and sculpture irrespective of artistic quality and works of artistic craftsmanship.[42] Separate copyrights therefore potentially exist in both a computer program and individual graphic images generated by a program as well as potentially in the design of a product and in design details applied to the surface of a product.

In order for a work to qualify as original it is not necessary for the work to have artistic or literary merit. It is only necessary that the author creates the work through their own skill and judgement rather than merely copying from another work. In the case of computer programs, the Software Directive indicates that a program is protected if it is original in the sense that it is the author's own intellectual creation.[43]

A database, being a collection of pre-existing data, is protected by copyright[44] if the data is collected and arranged in a systematic or methodical way and is accessible by electronic or other means.[45]

---

39 Patents Act 1977, Section 40.

40 *Kelly and another v GE Healthcare Ltd* [2009] EWHC 181 (Pat) being a rare example of a successful claim for compensation under the pre-amended wording of Section 40 which specifically required that outstanding benefit was derived directly from the patent rather than the patent or the invention as is now the case.

41 Copyright, Designs and Patents Act 1988, Section 3(1).

42 Copyright, Designs and Patents Act 1988, Section 4.

43 Software Directive, Article 1(3).

44 As distinct from 'database rights'.

45 Copyright, Designs and Patents Act 1988, Section 3A.

The scope of copyright protection for databases was considered by the ECJ in *Football Dataco v Yahoo* (Case C-604/10). In that case the ECJ held that national courts are required to interpret the EU Database Directive (96/9/EC) as precluding the grant of national copyright protection which goes beyond that established by the directive. As a result, as protection under the Database Directive does not extend to devising the data in a database, the effort involved in doing so is not taken into account when determining whether or not a database is protected by copyright and any such copyright protection is limited to the structure of a database and not its contents.

A work must be recorded in order for copyright in a literary work to come into existence.[46] On occasions this can cause problems in the computer field where, for example, elements in user interfaces are transitory and not recorded as such in the underlying source code. This was the case in *Navitaire Inc v Easyjet Airline Company* [2004] EWHC 1725 (Ch), where the judge considered that keyboard command codes entered by a user and interpreted by a computer would not be protectable because they were not recorded.

7.2 Limitations on copyright protection

It has always been the case that copyright protection is limited to the expression of ideas and not protection of the underlying ideas themselves. This is often an important limitation on copyright protection for computer programs. In *SAS Institute Inc v World Programming Ltd* [2010] EWHC 1829 (Ch), the court stated that copyright in computer programs could not protect, for example, programming languages or interfaces or the functionality of a program itself. However, graphical aspects of such user interfaces might still themselves constitute independently protectable graphic works.

A separate consideration needs to be taken into account when considering copyright in the design of electronic goods. In relation to an artistic work, copying includes making a copy in three dimensions of a two-dimensional work.[47] However, in practice the scope to which copyright in engineering drawings can be used to protect against the copying of a physical product is severely limited by Section 51 of the Copyright, Designs and Patents Act 1988 which limits the protection of such design documents to infringement of design right which is discussed in 9.1 below. For this reason, in the electronic design field in the United Kingdom, copyright protection is generally only of relevance to reproduction of graphic works themselves, for example the reproduction of surface decoration, graphical displays or designed graphical elements on a user interface.

---

46 Copyright, Designs and Patents Act 1988, Section 3(2).
47 Copyright, Designs and Patents Act 1988, Section 17(3).

7.3 Acts of infringement

Acts of infringement are defined by Section 16(1) of the Copyright, Designs and Patents Act 1988. For the purposes of electronics and software the most important of these acts are copying a work and issuing copies of a work to the public.

In order to infringe copyright a substantial part of a work must be copied.[48] Assessment of whether copying has occurred is a two-step test.[49] Initially, the courts identify whether there has been any copying and, if so, which features have been copied. Then, the courts ask whether what has been copied represents a substantial part of the original.

In *Cantor Fitzgerald International and another v Tradition* (UK) Limited and another [2000] RPC 95, the court emphasised that in assessing what is a substantial part the court must consider the skill and labour involved in creating the aspects of the original work which have been copied. In doing so, the court indicated that in cases involving the copying of computer code, it is the importance of the code which is copied rather than the amount of copying which is determinative. Similarly, in *Cantor Gaming Ltd v Gameaccount Global Ltd* [2007] EWHC 1914, the High Court found that the copying of a number of tables from a database constituted substantial copying. This was because what was taken was functionally useful and hence substantial.

In the case of graphic works, it is the visual significance of the elements which have been taken which need to be assessed to determine whether or not a substantial part of an artistic work has been copied. The assessment is made on a feature-by-feature basis rather than based on asking whether the alleged infringement looks on the whole similar to the original work. The fact that an artist may be using commonplace techniques is not relevant. What is of importance is the extent to which an original work is used to provide guidance to achieve an end result.[50]

The most straightforward cases concern literal copying of a work. In the case of computer software such literal copying can occur merely by loading or running a program on a computer without authorisation.

More complex issues arise where the abstract features or architecture of a computer program are copied. This issue was considered in *Ibcos Computers Ltd & another v Poole* [1994] FSR 275. Although accepting that copyright could only protect the expression of an idea and not the idea itself, the court nevertheless went on to state that even so copyright protection extends to the non-literal reproduction of a sufficiently detailed idea. In doing so the court emphasised the importance of the benefits derived from the unauthorised copying and in

48 Copyright, Designs and Patents Act 1988, Section 16(3).
49 *Designer Guild Ltd v Russell Williams (Textiles) Ltd* [2000] UKHL 58.
50 See, eg, *Temple Island Collections Ltd v New English Teas Ltd* [2012] EWPCC 1.

particular equated the taking of a substantial part of a copyright work with the notion of over-borrowing of skill, labour and judgement.

Probably the most controversial issues relating to the copying of software revolve around non-literal copying where an alleged infringer has copied the structure or manner of operation of an earlier program but there is no direct copying of the program itself.

An early example of this type of case was *John Richardson Computers v Flanders* [1993] FSR 497 where it was alleged that the copying of general design features and the look and feel of an earlier program could constitute copyright infringement. Relying largely on earlier US case law, the judge in that case held that copyright could be infringed by copying a computer program's menu design and hence its look and feel. The court therefore considered that substantial copying had occurred despite there being no literal similarities between the original program code and the alleged copy.

However, in subsequent decisions the courts have rejected the contention that non-textual copying can amount to infringement. So, for example, in *Navitaire Inc v Easyjet* [2004] EWHC 1725 (Ch) it was held that a program that processed the same or similar inputs and achieved similar results did not infringe copyright. Similarly in the computer gaming field in *Nova Productions Ltd v Mazooma Games Ltd* [2006] EWCA Civ 1044 the copying of high-level concepts, in that case a gambling game based on the game of pool, did not constitute infringement. In that case, emphasis was placed on the abstract nature of the concepts said to be copied and the fact that they had little connection with the amount of skill and effort expended by the programmer.

A similar result occurred in *SAS Institute Inc v World Programming Ltd* [2010] EWHC 1829 (Ch), which involved an allegation that a program for executing applications written in a particular programming language infringed the copyright in an earlier program for undertaking the same task. In that case the court held that there was a distinction between on the one hand the structure, sequence and organisation of a program which might be protected against copying and on the other the functionality of a program which would be protected, a view which was confirmed in a reference to the ECJ.[51]

7.4 Permitted acts

Although in general the unauthorised copying of a computer program will constitute an act of infringement, certain acts are expressly permitted. The permitted acts include making back-up copies of computer programs,[52] decompilation of computer programs in order to create an interoperable

---

51 *SAS Institute Inc v World Programming Ltd, Case* C-406/10, 2 May 2012.
52 Copyright, Designs and Patents Act 1988, Section 50A.

program,[53] copying a program in order to test its functioning,[54] and copying or adapting a program in so far as that is necessary for its lawful use.[55] These expressly permitted acts are, however, narrowly drawn and are frequently only permitted when undertaken for specified purposes.

7.5 Ownership and authorship

The general rule is that the initial owner of copyright in a work is the author,[56] who is generally the person who creates the work or, in the case of a computer-generated work, the person who makes the arrangements necessary for creating the work. The main exception to this rule is that if a work is created in the course of an employee's employment, copyright will be owned by the employer unless there is an agreement to the contrary.[57]

It should be noted that the automatic vesting of copyright in an employer only applies to works created by employees and not by external contractors. Hence where someone commissions a work copyright will remain vested in the author unless expressly assigned to the commissioner. In the absence of an express assignment no implied transfer of copyright is normally imposed by the courts.[58] An exception to this rule occurs in the case of someone who has a fiduciary relationship with the commissioner.[59]

Joint ownership of a copyright work occurs where it is produced in collaboration with two or more authors and the individual contributions of each author cannot be distinguished.[60] As the writing of a program and the preparatory documentation for a program are separate copyright works, the writing of one will not normally cause someone to be considered to be the author of the other.[61]

8. Confidential information

The elements required by the English courts to found an action for breach of confidence or protect trade secrets where summarised in *Coco v AN Clark (Engineers) Ltd* [1969] RPC 41 as:

> *First, the information itself … must 'have the necessary quality of confidence about it'. Secondly, that information must have been communicated in circumstances*

---

53 Copyright, Designs and Patents Act 1988, Section 50B.
54 Copyright, Designs and Patents Act 1988, Section 50BA.
55 Copyright, Designs and Patents Act 1988, Section 50C.
56 Copyright, Designs and Patents Act 1988, Section 9(1).
57 Copyright, Designs and Patents Act 1988, Section 11.
58 See, eg, *Clearsprings Management Ltd v Businesslink* Ltd [2005] EWHC 1487, *Meridian International Services Limited v Ian Richardson* [2007] EWHC 2539 (Ch) and *Infection Control Enterprises Ltd v Virrage Industries Ltd* [2009] EWHC 2602 (QB), where no implied transfer of copyright was found.
59 *Vitof Limited v Antony John Altoft* [2006] EWHC 1678 (Ch) where the founder and subsequent director of a company established after code had been written was found to hold the copyright of the code on trust for the company.
60 Copyright, Designs and Patents Act 1988, Section 10(1).
61 *Flyde Microsystems Ltd v Key Radio Systems Ltd* [1998] FSR 449.

importing an obligation of confidence. Thirdly, there must have been an unauthorised use of the information to the detriment of the party communicating it.

The key to information having the necessary quality of confidence has long been established by defining what confidential information is not, that is "it must not be something which is public property and public knowledge".[62]

Whether or not information is communicated in circumstances importing an obligation is a matter of fact specific to an individual case, However, the High Court in *Vestergaard v Bestnet* [2009] EWHC 657 (Ch), helpfully listed the following as factors to be taken into account:

- the nature of the work being undertaken and whether the nature of the duties were such that confidential information might reasonably be expected to result;
- the nature of the information itself;
- the attitude of an employer towards the alleged confidential information;
- the steps taken at the time to protect the information;
- the separability of the information;
- the commercial value of the information; and
- uses and practices in the trade.

Frequently issues of breach of confidence arise when an employee leaves a firm and sets up or joins a competitor. The courts consider that obligations of confidence and an obligation not to misuse trade secrets are continuing obligations.[63] In such cases it is possible that the earlier employer could use the common law of confidence against the ex-employees should they attempt to use business confidential information or trade secrets. Additionally, in recent years patent entitlement actions have been used as a way to claim rights in inventions filed by ex-employees following a move from an employer if the earlier employer considers that inventions made whilst the ex-employees were in their previous employment have been taken and exploited elsewhere[64] including where such ex-employees act as contractors.[65]

9. Other forms of protection

9.1 Unregistered Design Right

A separate UK unregistered design right exists alongside the EU unregistered

62 *Saltman Engineering Co Ltd v Campbell Engineering Co Ltd* [1948] 65 RPC 203 at 215.

63 See *Faccenda Chicken v Fowler* [1985] FSR 105.

64 *Markem Corporation and another v Zipher Limited and Markem Technologies Limited and others v Buckby and others* [2005] EWCA Civ 267.

65 As was the case in *Vestergaard v Bestnet* [2009] EWHC 657 (Ch).

design right established by the Community Design Regulation.[66] A UK unregistered design right is a right similar to copyright and comes into existence automatically when a new design is created. However, it is of much shorter duration. An unregistered design right lasts for 10 years from the date when a design is first marketed,[67] up to an overall limit of 15 years from the date when the design was created.[68]

In addition, in order for an unregistered design right to come into existence, the designer, commissioner or the person first marketing the design or their employer must be a citizen or a resident or a company incorporated in a qualifying country.[69] Qualifying countries include the United Kingdom, the rest of the EEA and certain British overseas territories and other countries providing similar rights.[70] At present the only commercially significant country recognised as providing such rights outside of the EEA is New Zealand.

An unregistered design right only applies to aspects of shape and configuration of three-dimensional designs.[71] It does not subsist in surface decoration.[72] A design right also does not subsist in features of designs which are necessary for a design to interact with or match other articles, the so-called 'must fit' and 'must match' exceptions.[73] In order for any unregistered design right to come into existence, it is necessary for a design to be original which means that it must not be commonplace in the design field in question at the time of its creation.[74]

When a design right comes into existence it initially provides the owner with the exclusive right to reproduce a design for commercial purposes.[75] However, in the final five years of its existence an unregistered design right only provides the right to charge a royalty for others wishing to copy the design and in the absence of an agreement between the parties licensing terms will be imposed.[76]

9.2 Registered design right

A national design registration system also exists in the United Kingdom.[77] The procedures for obtaining a design registration in the United Kingdom are similar to those for obtaining an EU registered design in that a set of design drawings needs to be submitted to the UK Design Registry together with an

---

66 Council Regulation (EC) No 6/2002.
67 Copyright, Designs and Patents Act 1988, Sections 216(1)(b) and 216(2).
68 Copyright, Designs and Patents Act 1988, Section 216(1)(a).
69 Copyright, Designs and Patents Act 1988, Section 217(1).
70 Copyright, Designs and Patents Act 1988, Section 217(3).
71 Copyright, Designs and Patents Act 1988, Section 213(2).
72 Copyright, Designs and Patents Act 1988, Section 213(3)(c).
73 Copyright, Designs and Patents Act 1988, Section 213(3)(b).
74 Copyright, Designs and Patents Act 1988, Section 213(1) and (4).
75 Copyright, Designs and Patents Act 1988, Section 226.
76 Copyright, Designs and Patents Act 1988, Section 237.
77 Registered Designs Act 1949 as amended.

application form and an application fee. Designs which can be registered include the lines, colours, texture or shape of the whole or part of a product or its ornamentation and includes graphic symbols and logos.[78]

After checking that the application fulfils the formal requirements the application is registered and a design registration certificate is issued. Registration is automatic and no check is made before registration that a design is valid and novel in view of the prior art. A design registration lasts for up to 25 years from the date of application subject to payment of renewal fees every five years.[79]

A design registration gives the proprietor an exclusive right to use the design and any other design which does not produce a different overall impression on the informed user.[80] A registered design therefore provides much stronger protection compared with copyright or an unregistered design right because it is not necessary to establish copying in order to enforce a registered design.

The scope of protection given by a design will depend on whether or not an alleged infringement gives a different overall impression to the registered design. The approach taken in the UK courts for UK registered designs and EU registered designs is identical in that the design is assessed from the viewpoint of the informed user based on an assessment of the design, the alleged infringement and knowledge of the existing design corpus and the degree of design freedom available in the field.[81]

9.3 Trademarks and passing off

The branding of software and electronics goods like any other commercial products can be protected via the registered trademark system. UK registered trademark law[82] is largely harmonised with trademark law across the European Union by virtue of a series of harmonisation directives.[83] Although normally concerned with brand names and logos, occasionally trademark law can provide protection for the copying of the look or shape of goods when these are sufficiently distinctive to act as a badge of origin.

In addition to registered trademark law, English law also provides for protection of unregistered trademarks through the tort of passing off. The classic requirements for passing off were set out in the *Jif Lemon* case[84] as being:

- the presence of goodwill or reputation attached to goods or services;
- a misrepresentation by the defendant leading or likely to lead the public

---

78 Registered Designs Act 1949, Section 1.
79 Registered Designs Act 1949, Section 8.
80 Registered Designs Act 1949, Section 7(1).
81 Compare *Dyson Limited v Vax Limited* [2010] EWHC 1923 (Pat) and *Samsung Electronics (UK) Ltd v Apple Inc* [2012] EWHC 1882 (Pat).
82 Trademarks Act 1994
83 EC Directives 89/104/EEC, 2008/95/EC.
84 *Reckitt & Colman Products Ltd v Borden Inc* [1990] RPC 341.

to believe that the goods or services offered by him or her are the goods or services of the claimant; and

- damage or likely damage to the claimant by virtue of that erroneous belief.

9.4 Database rights

The provisions of the EC Directive on the Legal Protection of Databases[85] have been implemented in the United Kingdom by virtue of the Copyright and Rights in Databases Regulations 1997. These regulations bring into effect a *sui generis* right which arises when there has been substantial investment in obtaining, verifying or presenting the contents of the database.[86] Such a right lasts for 15 years from the end of the calendar year in which the making of the database was completed.[87]

The scope of rights under the Copyright and Rights in Databases Regulations 1997 was considered in *British Horseracing Board v William Hill* [2001] RPC 612. In that case the court held that in order to infringe database rights it is only necessary to copy the content and not the form of data contained in a database.

Again, as with copyright, the determination of what amounts to substantive copying depends upon whether the taking of data involves a substantial taking of the investment involved in creating a database. In *Beechwood v Guardian* [2010] EWPCC 12, the Patents County Court considered that taking 6,000 records from a database of 43,000 records (approximately 14%) was significant particularly because of the work involved in the original compilation of data.

9.5 Semiconductor topographies

Semiconductor topographies enjoy protection under the Design Right (Semiconductor Topographies) Regulations 1989. These regulations provide a special right of protection for the chip design of integrated electronic circuits. The rights under the Design Right (Semiconductor Topographies) Regulations 1989 are similar to those enjoyed by UK unregistered designs.

10. Counterfeiting

The EU Anti-Counterfeiting Regulation[88] provides customs authorities across the European Community with the power to detain goods suspected of infringing intellectual property rights which are defined to include infringements of registered trademarks, copyright or design rights or patents. Customs can detain these goods either of their own volition or (more usually) pursuant to a notification by the rights holder to 'watch' for potentially counterfeit goods.

---

85 EC Directive 96/9 EC 11 March 1996.
86 Copyright and Rights in Databases Regulations 1997, Regulation 13(1).
87 Copyright and Rights in Databases Regulations 1997, Regulation 17.
88 EC Council Regulation 1383/2003.

Following the detention of suspected counterfeit goods, the rights holder has an opportunity to inspect the goods. In certain circumstances the rights holder is also provided with samples for analysis if this is required to demonstrate infringement.

The rights holder then has a 10-day period (extendable to 20 days) in which to commence national infringement proceedings. If no action is brought the seized goods are then released.

However, the seizure of goods travelling through EU customs by the external transit procedure cannot automatically be considered counterfeit and can only be seized if sufficient evidence is provided to establish that such goods will ultimately be sold in the European Union rather than merely transiting through the European Union. This restriction follows from the decisions in *Koninklijke Philips Electronics NV v Lucheung Meijing Industrial Company Ltd* (C-446/09) and *Nokia Corporation v Her Majesty's Commissioners of Revenue and Customs* (C-495/09).

11. Enforcement

11.1 Obtaining information on the infringer and the infringement

The general rule in English litigation is that parties to litigation are obliged to identify and provide copies of documents relevant to the litigation.[89] More specifically, a party is obliged to disclose documents which adversely affect his or her own case or adversely affect or support another party's case.[90] The definition of document is a broad one and includes anything in which information of any description is recorded.[91] The duty of disclosure is a continuing one until the proceedings are concluded.[92]

The obligation of disclosure imposes on parties to litigation an obligation to undertake a reasonable search for relevant documents and the extent of what is or is not reasonable will depend upon the number of documents involved, the nature and complexity of the proceedings, the ease and expense or retrieval of documents and the likely significance of the documents which are identified.[93] Issues of proportionality are particularly important in the Patents County Court where ordinarily disclosure, if permitted at all, is limited to specified documents identified by the parties.

It is possible for parties to request disclosure of specific documents even before proceedings have been initiated.[94] A request for pre-action disclosure has to be supported by evidence that proceedings are likely to occur and must

89 Civil Procedure Rules, Part 31.
90 Civil Procedure Rules, Part 31.6.
91 Civil Procedure Rules, Part 31.4.
92 Civil Procedure Rules, Part 31.11.
93 Civil Procedure Rules, Part 31.7.
94 Civil Procedure Rules, Part 31.16

identify the documents or classes of documents which are relevant to the issue of infringement and additionally the court must be persuaded that such disclosure is desirable to dispose fairly of the anticipated proceedings; to assist the dispute to be resolved without proceedings; or to save costs. An order for pre-action disclosure will not be made if the application is purely speculative.[95]

It is also possible to apply to the courts to oblige a non-infringing third party, to identify infringers or infringing goods.[96] Such orders are useful to claimants aware of infringing goods entering the United Kingdom but unable to ascertain their identity. In these cases it is likely that certain governmental agencies or other third parties could possess information that would assist the patentee. The UK courts would be sympathetic to the granting of an order placing the third parties under a duty to assist the patentee by giving full information and disclosing the identity of the infringer.

Disclosed documents normally may only be used for the purposes of the proceedings in which it is disclosed.[97] Documents may only be used for other purposes if they have been read to or by the court or referred to at a public hearing or if the court gives permission for the document to be used for other purposes or in other proceedings. It is, however, possible for third parties to obtain documents held on the court file; normally restricted to pleadings.[98]

Although potentially wide-reaching the obligations of disclosure are subject to a number of limitations. There is no obligation to provide copies of privileged documents created by lawyers for the purpose of giving legal advice on a matter or documents created specifically for the purposes of ongoing litigation. Importantly, the obligation of disclosure is limited by the issues raised in a pleaded case. That is to say whether or not a patent is novel and inventive in view of specific identified documents rather than valid or invalid in an abstract sense.[99] This means that disclosure is limited to documents which are relevant to the questions of novelty and obvious in view of those documents and would not, for example, extend to questions about other arguments which have not been pleaded. Additionally, in patent cases disclosure on commercial success is only of relevance if it is put in issue.[100] Further, in patent cases disclosure on invalidity is limited to documents created in a four-year window centred on the priority date of the patent.[101] The court rules also provide that in patent cases the obligation to provide disclosure can be discharged by an alleged infringer providing a full written description of the product or process alleged to infringe.[102]

---

95 *BSW Ltd v Balltec Ltd* [2007] FSR 1.
96 *Norwich Pharmacal Co v Commissioners of Customs & Excise* [1974] RPC 101.
97 Civil Procedure Rules, Part 31.22.
98 Civil Procedure Rules, Part 5.4(c).
99 *Avery v Ashworth* (1915) 32 RPC 569 (CA).
100 Civil Procedure Rules, Part 63 Practice Direction 6.3.
101 Civil Procedure Rules, Part 63 Practice Direction 6.1(2).
102 Civil Procedure Rules, Part 63 Practice Direction 6.1(1).

11.2 Interim relief

In the English courts, interim injunctions are granted on the basis of the test set out by the House of Lords in *American Cyanamid Co v Ethicon Ltd* [1975] RPC 513. This provides that when deciding whether or not to grant an interim injunction, initially a court needs to determine whether damages would be an adequate remedy and whether the defendant is in a position to pay them. If this is the case then no injunction will be granted. If damages would not be an adequate remedy, the court then goes on to consider whether, if the defendant were to succeed at trial, he or she would be adequately compensated by a claimant's undertaking as to damages. If this is the case then an injunction will be granted. Finally, if there is doubt as to the adequacy of damages, the court then considers the 'balance of convenience' between the parties and if other factors appear to be evenly balanced then the court will take measures to preserve the status quo.

11.3 Threats

The mere act of threatening to initiate legal proceedings based on an allegation of intellectual property rights can result in serious commercial damage. For this reason a number of provisions in the intellectual property statutes provide parties aggrieved by unjustified threats of intellectual property infringement with the right to obtain a declaration that the threats were unjustified, an injunction against the continuance of the threats and damages for any losses suffered. The UK legislation has, however, developed in a piecemeal fashion and although there are threats provisions in respect of patents,[103] trademarks[104] and designs,[105] no such provisions exist in respect of copyright or passing off.

The threats provisions in respect of all types of intellectual property rights are broadly similar. Although what is considered to be a potentially actionable threat has to be interpreted broadly, the law provides that it is never a threat merely to bring the existence of rights to someone's attention. Further, certain activities are exempted from the threats provisions where they constitute primary acts of infringement, for example making or importing a product, using a process or applying a mark to a product or supplying services under a mark. The rationale behind these exemptions is that a rights holder should be able to place the main offenders on notice that he or she intends to sue but that retailers and other traders further down the supply chain should be protected against such threats.

It is a defence to an allegation of unjustified threats to demonstrate that the

---

103 Patents Act 1977, Section 70.

104 UK registered trademarks – Trademarks Act 1994, Section 21, Community trademarks – Community Trademark Regulations 2006, paragraph 6.

105 UK registered designs – Registered Designs Act 1949, Section 26, UK unregistered design right – Copyright, Designs and Patents Act 1988, Section 253, Community registered designs – Community Designs Regulation 2005, Regulation 2

acts complained of constitute acts of infringement. In the event that a patent is subsequently found to be invalid, it is also a defence under the threats provisions if a patentee is able to show that at the time of making the threats he or she did not know and had no reason to know that his or her patent was invalid. The assessment of whether a patentee has reason to suspect that their patent is invalid is an objective one based on what the patentee knows at the time the threat was made.[106]

11.4 Remedies

Typical remedies awarded following a successful infringement action include:

- an injunction restraining further infringement;
- an order for delivery up or destruction of the infringing products; and
- damages; or
- an account of profits.

Damages in intellectual property proceedings are assessed based on the usual tortious scale. That is to say, in the quantum of any damages award the court will seek to restore the patentee to the position it would have been in but for the wrongful acts of the infringer. Where a rights holder licenses their rights, damages can be assessed on the basis of an assessment of a reasonable royalty rate which the defendant would have had to have paid had they sought a licence. In other cases, an assessment of a rights holder's overall losses will be made taking all relevant factors into account such as any price reduction necessitated by the infringer's competing product being on the market and losses in respect of sales of goods normally sold together with the patented goods.

As an alternative to requesting damages for losses incurred, a rights holder may instead ask for an assessment of the profits made by the infringer. The purpose of an account of profits is to deprive an infringer of the benefits they received from their infringing acts. An assessment of profits is made by determining the net profit derived from the infringement after deducing allowable costs such as manufacturing and distribution costs from the gross profit.

12. Hot topics

12.1 FRAND

Since 2014, a significant proportion of UK Patent Court time has been taken up by the ongoing dispute between Unwired Planet, Samsung and Huawei with Unwired Planet looking to enforce its portfolio of standard essential patents against Samsung and Huawei.

---

106 *FNM Corporation Ltd v Drammock International Ltd* [2009] EWHC 1294 (Pat).

Unwired Planet originally acquired most of its patent portfolio from Ericsson and then proceeded to assert six of its European patents in Germany and the United Kingdom saying that this was necessary to "expedite ongoing licensing negotiations".

The first judgment[107] in a technical trial was handed down in November 2015 with Birss J, finding EP (UK) 2 229 744 to be valid and infringed by wireless telecommunications networks operating in accordance with the LTE standard.[108] A second technical trial in 2015 then found two further patents invalid.[109] Whereas a third technical trial held in February 2016 found EP (UK) 1 230 818 to be valid and infringed.[110]

After this second success, the technical trials in relation to the other patents were put on hold and the court was asked to assess whether and to what extent the various terms on offer by the parties seeking to settle the case constituted FRAND offers (Fair Reasonable And Non-Discriminatory) as required by participation in the LTE standard. In a lengthy judgment[111] in April 2017, Birss J held that none of the offers made by Unwired Planet or Huawei were FRAND and the court set out a series of licence rates for a worldwide licence of the Unwired Planet portfolio on terms that the court considered were FRAND with the court then issuing a further judgment[112] in June 2017 detailing how an injunction (a FRAND injunction) would be issued which would be discharged upon the defendant entering into a FRAND licence. These decisions are now under appeal and the outcome is keenly awaited.

12.2 Brexit

On 23 June 2016, 51.9% of the participating UK electorate voted for the United Kingdom to leave the European Union. A letter invoking Article 50 of the Treaty on the European Union was delivered to the European Council President on 29 March 2017. Originally the United Kingdom was scheduled to cease to be a member of the European Union on 29 March 2019. However, an extension to 31 October 2019 was agreed shortly before that date. Whether or not the deadline will be further extended is a matter of considerable political controversy.

Britain's departure from the European Union will have a significant impact on many aspects of UK intellectual property law. Following Britain leaving the European Union, European trademarks, registered and unregistered Community designs and Community plant variety rights will cease to have

---

107 *Unwired Planet International v Huawei Technologies & Others* [2015] EWHC 3366 (Pat).

108 A decision subsequently upheld by the Court of Appeal in *Unwired Planet International v Huawei Technologies & Others* [2017] EWCA Civ 266.

109 *Unwired Planet International v Huawei Technologies & Others* [2016] EWHC 94 (Pat).

110 *Unwired Planet International v Huawei Technologies & Others* [2016] EWHC 576 (Pat).

111 *Unwired Planet International v Huawei Technologies & Others* [2017] EWHC 711 (Pat).

112 *Unwired Planet International v Huawei Technologies & Others* [2017] EWHC 1304 (Pat).

effect in the United Kingdom, which will no longer be part of the European Union and the Regulations under which they were granted and enforceable will no longer have force in the United Kingdom. The UK government will need to legislate for substitute rights and appropriate transitional provisions. The same will apply to EU designations of origin and geographical indications.

The UK IPO will need to take over the administration of the registrations which substitute in the United Kingdom for the former EU-wide rights. It is currently expected that this will be on an opt-in basis, almost certainly on payment of a fee. The UK IPO will need to recruit the examiners and other staff to administer the greatly expanded workload.

Brexit will also affect the geographical scope of trademark use. Owners of EU trademarks who rely wholly or substantially on use in the United Kingdom to establish genuine use in the European Union in the event of a revocation claim, may find their EU-wide registrations more vulnerable to revocation for non-use after Brexit. Conversely, when the UK portion of an EU trademark, which is used only outside the United Kingdom, is converted to a national registration, the national registration will similarly be vulnerable. It is not expected that the UK statute law which implements harmonised EU law in this area will be changed, although over time UK jurisprudence is likely to diverge from the EU jurisprudence, since the Court of Justice and the General Court decisions will no longer bind the UK courts.

The impact of Brexit on patent protection will be very limited as the United Kingdom will continue to be a member of the European Patent Convention. The UK government has stated its intent to participate as a member of the Unified Patent Court if and when that is established, although the details as to how that will be achieved are as yet unclear.

Customs detention of (allegedly) infringing goods at the point of importation is an important weapon in the hands of rights owners, not only against counterfeiters, but also in standards-based industries such as electronics and telecoms. Following the United Kingdom's departure from the European Union, there will be a border between the United Kingdom and the European Union where detention of goods can take place, certainly of goods passing from the United Kingdom to the European Union, but also, if successor legislation brings customs detention into UK law, of all goods entering the United Kingdom. If the United Kingdom adopts an EU-style scheme, it will be potentially very significant, as it will be almost impossible to circumvent.

This added border may affect how the market for parallel imports will develop, which is particularly affected by the rules on the exhaustion of rights. The doctrine of exhaustion of rights applies to the whole EEA, so if the United Kingdom leaves the European Union but re-joins first EFTA, and then the EEA, the position with regard to exhaustion and parallel imports will not change. However, if the United Kingdom does not re-join the EEA, then the existing rules on exhaustion of rights will cease to apply, in which case no goods covered by UK intellectual property rights which entered the market overseas can be parallel imported into the United Kingdom unless the rights owner consents. This could effectively close the UK market to parallel imports. Since parallel imports compete chiefly on price with products first placed on the market in the United Kingdom, one factor restraining price inflation would be removed. This may be an area for legislation, depending on how significant the government perceives this factor to be.

Australia

Peter Leonard
Data Synergies and UNSW Business School
Chris Williams
Gilbert + Tobin

Australia has a well-regarded intellectual property (IP) regime that consistently ranks in the top tiers of international comparisons.[1] Australia is a member of the key international IP treaties, such as the Madrid Protocol, Paris Convention and Patent Cooperation Treaty, which provide a streamlined option for applicants seeking international protection of their IP rights. Multinational corporations can therefore transfer technology to Australia and invest in R&D and develop innovative products in Australia, with confidence as to stability and predictability of IP laws and jurisprudence.

Australia has small but active domestic electronics and software industries, principally focused on industry verticals and in particular medical technologies, mining, gaming and agriculture. The Australian IP legal framework generally reflects international best practice and affords similar protection for foreign corporations doing business in Australia and Australian corporations. There are no mandatory technology transfer regimes.

In 2016 the Australian government tasked the Productivity Commission with undertaking a comprehensive inquiry into the Australian IP system.[2] The

---

1 Australia was given a regional rank of 4 and global rank of 10 in the International Property Rights Index report IPRI 2017: www.internationalpropertyrightsindex.org/countries; and ranked 17th for IP protection in the World Economic Forum's *The Global Competitiveness Report 2017–2018*, p 51.

2 The Australian government directed the Productivity Commission to investigate Australia's IP arrangements and produce its report following a strong recommendation to do so by the Harper Review. The Harper Review was a major review of Australia's competition laws. The Productivity Commission's final Report was published in December 2016 and is available at www.pc.gov.au/inquiries/completed/intellectual-property%23report#report. The government subsequently announced its intention to implement many of the Commission's recommendations: see www.industry.gov.au/data-and-publications/government-response-productivity-commission-inquiry-into-intellectual-property-arrangements.

3 The first stage response was the Intellectual Property Laws Amendment (Productivity Commission Response Part 1 and Other Measures) Act 2018, made on 24 August 2018, and the Intellectual Property Laws Amendment (Productivity Commission Response Part 1 and Other Measures) Regulations 2018 which commenced on 16 October 2018. Part 1 of the act repeals a provision in the Copyright Act and amends the Trademarks Act 1995 (Trademarks Act) to clarify the circumstances in which the parallel importation of trademarked goods does not infringe a registered trademark. Part 2 amends the Plant Breeder's Rights Act 1994 (PBR Act) to allow an application for essentially derived variety (EDV) declaration to be made in instances where the plant variety subject of the EDV application is not a registered Plant Breeder's Right (PBR) or undergoing application for PBR. Part 3 amends the Trademarks Act to change the period that must elapse before certain actions of non-use actions of trademark can be taken. Part 4 amends the Patents Act to remove a requirement for patentees to provide the Secretary of the Department of Health with certain data relating to pharmaceutical patents with an extended term.

government's response to the Productivity Commission's report is reflected in staged legislative changes which commenced in 2018[3] and are expected to continue in 2019.[4] Some key changes that the government has already signalled it intends to implement in relation to patents that may impact on the electronics and software industries include:

- the introduction of an objects clause to the Patents Act 1990 (Cth) (Patents Act) to provide a clear statement of legislative intent to guide the courts in interpreting the Patents Act;[5]
- raising the bar for an inventive step under Section 7(2) and (3) of the Patents Act "to put beyond doubt that the assessment of inventive step in Australia is consistent with the [standards adopted by the] European Patent Office";[6] and
- the innovation patent system will be brought to end.[7]

1. Outline of national patent procedure

1.1 A two-tier system – for now

Australia presently has a two-tier patent system, providing for the grant of

4 In 2018, IP Australia sought public comment on the Exposure Draft of the Intellectual Property Laws Amendment Bill (Productivity Commission Response Part 2 and Other Measures) Bill 2018. IP Australia then considered submissions and published a response: www.ipaustralia.gov.au/about-us/public-consultations/consultation-intellectual-property-laws-amendment-bill-2018.

5 Australian Government Response to the Productivity Commission Inquiry into Intellectual Property Arrangements, August 2017, response to Recommendation 7.1. The exposure draft Intellectual Property Laws Amendment Bill (Productivity Commission Response Part 2 and Other Measures) Bill 2018 proposed a new objects clause be inserted into the Patents Act as Section 2A, stating: "The object of this Act is to provide a patent system in Australia that promotes economic wellbeing through technological innovation and the transfer and dissemination of technology. In doing so, the patent system balances over time the interests of producers, owners and users of technology and the public." The exposure draft also proposes to provide that "the term 'technological innovation' … is not intended to narrow or change the subject matter eligibility threshold for grant of a patent" and "is intended to be interpreted broadly to mean the applications of scientific knowledge for practical purposes".

6 *Ibid*, response to Recommendation 7.2. The government had accepted the Productivity Commission's recommendation to raise the threshold for inventive step, in line with the approach taken by the European Patent Office (EPO). The government noted the EPO's filter against low-value patents by application of the so-called 'problem-solution approach'. However, the government later noted that the EPO's 'Raising the Bar' initiative of 2010 included other measures to achieve higher patent quality. Although the exposure draft of the Intellectual Property Laws Amendment (Productivity Commission Response Part 2 and Other Measures) Bill 2019 included provisions raising the inventive step requirements, following the consultations conducted by IP Australia on the exposure draft the government announced the government's decision to remove this measure from the Part 2 Bill, pending further analysis and consultation with stakeholders. This may be a temporary halt, while the government reviews the EPO's November 2018 revision to Section 3.3.1 (Artificial intelligence and machine learning) of its Guidelines for Examination. As at May 2019 it was not clear whether the Part 2 Bill when introduced into the Australian Parliament will include provisions raising the inventive step requirements.

7 *Ibid*, response to Recommendation 8.1; see below at 1.1. Innovation patents were introduced in 2001 and intended to stimulate innovation of small and medium-sized businesses in Australia by providing a faster and more affordable means for SMEs to protect their innovations. The system still required novelty and an innovative step, but with a lower threshold than the inventive step of a standard patent, and a shorter period of protection (eight years). Since inception in 2001, less than 5,000 innovation patents were granted. The system was criticised by the Productivity Commission for granting "too easy" protection for inventions and resulting in a "proliferation of low-quality patents", which hinders innovation, stifles competition, and needlessly shifts the burden to consumers of technology.

standard patents having a term of 20 years (or 25 years for some pharmaceutical patents[8]) and 'innovation' patents having terms of eight years. Standard patents are required to undergo a substantive examination process before grant. Innovation patents, which are intended for low-level inventions or developments, do not automatically undergo a substantive examination before grant.

While the government accepted a recommendation of the Productivity Commission to abolish the second-tier innovation patent system, the proposed legislative amendments will include "appropriate arrangements to maintain existing rights".[9] The system will continue to operate for innovation patents that are filed before amendments giving effect to the proposed phasing out the innovation patent are introduced into the Parliament and enacted.[10] As new innovation patents filed in the 12 months following assent and prior to the commencement will benefit from an eight-year term, Australia will continue to recognise such innovation patents until 2028.

1.2 Standard patents

Standard patents in Australia have a term of 20 years from the effective date of filing.[11] Annual renewal fees are payable from the fifth to 19th anniversaries, inclusive. The process of obtaining a standard patent usually takes up to five years. Fees are payable at various stages during the patent prosecution process and also for annual renewals once granted. Fees incurred in the application process are set out on IP Australia's website[12] and can otherwise be found in Schedule 7 to the Regulations.[13]

Details of the unexamined standard patent application (including the invention title) are published in the Australian Official Journal of Patents approximately 18 months after the application's earliest priority date.[14] Examination must be requested within five years of the filing date and examination times vary, but typically a reply is received within about 12 months.

---

8 Pharmaceutical substances which have experienced a delay in market approval can receive patent extensions, granting up to 25 years' protection.

9 Australian Government Response to the Productivity Commission Inquiry into Intellectual Property Arrangements, August 2017, response to Recommendation 8.1.

10 On 25 July 2019 the Intellectual Property Laws Amendment (Productivity Commission Response Part 2 and Other Measures) Bill 2019 was introduced into parliament. Once the legislation comes into effect, a commencement date will be set 12 months and one day after the legislation receives royal assent from the Governor General and is signed into Australian law.

11 Section 67 of the Patents Act. The 'date of the patent' is the date of filing of a complete application. A provisional patent application (known as a 'basic' application in Europe) can act as a priority application, but it cannot mature into a patent in its own right and needs to be converted to a complete application in which the invention is fully claimed within 12 months. If the standard patent is derived from a PCT application, then the effective date of filing is the International Filing Date; Section 88(4) of the Patents Act.

12 IP Australia website "Time and Costs" available at: www.ipaustralia.gov.au/patents/understanding-patents/time-and-costs.

13 Patent Regulations 1991 (Cth).

14 Level AU-A publication. Published patent applications are available on AusPat.

In the first office action all thresholds of patentability are considered, including novelty and inventive step (non-obviousness). International Preliminary Examination Reports (IPERs) are in practice heavily relied upon by examiners for identification of relevant prior art. Following examination either an adverse report or a notice of acceptance is sent to the applicant. The applicant then has 12 months to satisfy the Office that the application is in order for acceptance. A response can be filed and a hearing may be sought to overcome objections. Once all objections are overcome and the application for a standard patent is accepted, the application is republished[15] and other parties have three months within which to commence opposition proceedings.

1.3 Innovation patents

Australia has enjoyed a relatively long history of second-tier patent protection. Australia's initial second-tier patent system, known as the petty patent, was first introduced in 1979 and this was subsequently replaced by the innovation patent in 2001.[16] Innovation patents were introduced to overcome the problems experienced with the petty patent system, which the Australian Council on Intellectual Property (ACIP) had found at the time was underutilised and not providing sufficient protection for incremental innovations.[17]

Innovation patents have a term of eight years from the effective date of filing.[18] Annual renewal fees are payable from the second to seventh anniversaries, inclusive. Innovation patents are limited to five claims.[19] The principal difference in requirements is that whereas standard patents require an invention to demonstrate an inventive step over the prior art, an innovation patent need only be directed to an invention having an 'innovative step' over the prior art.[20] An innovative step is a very low threshold in practice and exists if the variation over the prior art makes a "substantial contribution to the working of the invention".[21]

Innovation patents can be obtained relatively quickly, within three to six months, as an innovation patent application proceeds directly to acceptance and grant without any substantive examination.[22] Fees are approximately half of those associated with standard applications at each stage and for annual renewal after grant.[23] An innovation patent cannot be enforced until it has been 'certified'.[24]

---

15 Level AU-B publication. If amended the application becomes an AU-C level publication.
16 S Moritz and A Christie, "Second-Tier Patent Systems: The Australian Experience" [2006] 4 *European Intellectual Property Review* 230.
17 *ACIP Review of the Petty Patent System* (1995), p 15.
18 Section 68 of the Patents Act.
19 Section 40(2)(c) of the Patents Act.
20 Section 18(1A)(b)(ii) of the Patents Act.
21 Section 7(4) of the Patents Act.
22 Sections 52 and 62 of the Patents Act.
23 IP Australia website "Time and Costs" available at: https://www.ipaustralia.gov.au/patents/understanding-patents/time-and-costs.
24 Section 120(1A) of the Patents Act.

Once certified, however, an innovation patent provides the same remedies for breach as a standard patent. Innovation patents are therefore of considerable assistance to patentees particularly interested in early enforcement, their patent life-cycle management or simply obtaining shorter-term protection for inventions that represent non-inventive steps over the prior art.

As noted earlier, as at August 2019 there are current proposals that Australia abolish its innovation patent system, with legislation to that effect now introduced into the Australian Parliament.[25]

1.4 Outline of costs

Costs of patent prosecution will vary depending on the type of patent applied for and how the fee is paid: that is, whether it is a standard or innovation patent application and whether the application is made online or by other means. The main official fees for filing and prosecuting a patent application by approved means as at May 2019 are:[26]

Innovation patent	
Filing fee:	AU$280 (AU$180 if filed online)
Examination fee (optional):	AU$500

Standard patent	
Filing fee:	AU$470 (AU$370 if filed online)
Examination fee:	AU$490
Acceptance fee:	AU$250 plus AU$110 for each claim in excess of 20

Annual maintenance fees are payable on pending applications and on granted patents. For a standard patent, these start at AU$300 from the fourth anniversary, then AU$550 from the 10th anniversary and AU$1,250 from the 15th anniversary. An innovation patent, which has a maximum term of eight years, requires payment of AU$110 annually from the second-year anniversary through to AU$220 from the seventh year.

2. Examination of software patents

2.1 Patentable subject matter

The Patents Act provides that an invention is patentable if it:

- is a "manner of manufacture"[27] – that is, the invention is appropriate subject matter for patent protection;

---

25 See footnote 10 above.

26 A full list of fees is set out in Schedule 7 to the Patents Regulations 1991. Prices are in Australian dollars and payment by 'approved means' means using the Australian Patent Office's eServices facility at: www.ipaustralia.gov.au/eservices or by post.

- is novel;[28]
- involves an inventive or innovative step;[29]
- is useful;[30] and
- has not been used secretly within Australia before the priority date of the patent application.[31]

These fundamental requirements apply to all patent applications, but lead to particular issues in relation to certain areas of technology.

While software code will not on its own be regarded as a "manner of manufacture", the operative effect of that software may nevertheless be patentable. Whereas abstract information (eg, mere schemes, algorithms, ideas or intellectual information) is not generally patentable,[32] any new and inventive software application, which provides a commercially useful effect and results in an improvement in the functioning of the computer, may be patentable.[33] Where software results in an improvement to the functioning of the computer's processor, for example, the means of achieving that result may be patentable.

The *Australian Patent Office Manual of Practice and Procedure* (Examiners Manual) published by IP Australia is closely followed by examiners during patent prosecution. The Examiners Manual records at 2.9.2.7 "Computer Implemented Inventions – Schemes and Business Methods" that:

> *Under Australian law, there are no specific exclusions for software or methods that are implemented as computer software or a related product. However, they are only patentable if what is claimed "as a matter of substance" meets the requirements for a manner of manufacture and in particular is not a mere scheme, abstract idea or mere information.*[34]

The Examiners Manual lists the following examples of potentially patentable computer-related inventions:

---

27 Section 18(1)(a) of the Patents Act somewhat anachronistically still requires as a threshold for patentability that an invention be a "manner of manufacture within the meaning of Section 6 of the Statute of Monopolies". This threshold test for patentability keeps alive the common law exclusions of collocations, discoveries, fine arts, laws of nature, schemes with no practical implementation, methods of treating humans, working directions, mere printed matter and others. For a comprehensive list of historical categories, see Australian Council on Intellectual Property, "Patentable subject matter – Issues paper" July 2008 at pp 22–23, available at: https://www.ipaustralia.gov.au/sites/g/files/net856/f/acip_final_report_patentable_subject_matter_archived.pdf. This can be contrasted with the numerous jurisdictions that expressly recite in patent legislation any non-patentable subject matter categories.

28 Section 18(1)(b)(i) of the Patents Act for standard patents and Section 18(1A)(b)(i) for innovation patents.

29 Section 18(1)(b)(ii) of the Patents Act for inventive step/standard patent and Section 18(1A)(b)(ii) for innovative step/innovation patent. Also see Section 7 of the Patents Act regarding prior art base. See further footnote 6 above noting the Australian government's proposal to amend the definition of 'inventive step', in line with the approach taken by the EPO.

30 Section 18(1)(c) of the Patents Act for standard patents and Section 18(1A)(c) for innovation patents.

31 Section 18(1)(d) of the Patents Act for standard patents and Section 18(1A)(d) for innovation patents.

32 *Re Application by Honeywell Bull Inc* (1991) 22 IPR 463. See also guidelines for examination in *Australian Patent Office Manual of Practice and Procedure* at 2.9.2.7.

33 *CCOM v Jiejing* (1994) 28 IPR 481; *International Business Machines Corporation v Commissioner of Patents* (1991) 33 FCR 218; *Encompass Corporation Pty Ltd v InfoTrack Pty Ltd* [2018] FCA 421 at [193].

34 Available at: http://manuals.ipaustralia.gov.au/patents/Patent_Examiners_Manual.htm.

- a processing apparatus characterised by its method of operation;
- software or programs in a machine-readable form causing a computer to operate in a particular way;
- a computer, when programmed with code (source or executable) to operate in a particular way; and
- a computer-implemented method.

Despite the above guidance and as elaborated further below, the assessment and examination of patents and applications for computer-implemented inventions in Australia currently present considerable difficulties.

2.2 Inventiveness requirements

An invention is taken to involve an inventive step when compared with the prior art base unless the invention would have been obvious to a person skilled in the relevant art in the light of the common general knowledge before the relevant priority date, either considered separately or together with certain specified information that is taken to be part of the prior art base.[35] The Raising the Bar[36] amendments that came into effect in Australia on 15 April 2013[37] removed previous parochialisms that had required that:

- the relevant common general knowledge be limited to that which existed in Australia; and
- publicly available information could only be considered if the skilled person could be reasonably expected to have ascertained, understood and regarded it as relevant.

The unfortunate effect of the law change is that whether the previous restrictions upon the relevant common general knowledge and prior art information continue to apply depends upon whether examination of the patent application was requested before or after 15 April 2013. It is therefore necessary to check the examination request date, so as to apply the appropriate law accordingly. As noted above, the Productivity Commission has recommended to, yet again, amend Australia's inventive step test and IP Australia has proposed amending Section 7(2) and (3) of the Patents Act in line with Article 56 of the EPC.

The test for innovative step required for an innovation patent is quite different to that for inventive step required for a standard patent. An invention does not involve an innovative step when compared with the prior art if it only varies from the "prior art information" in ways that make no substantial

35 Section 7(2) and (3) of the Patents Act.
36 The Intellectual Property Laws Amendment (Raising the Bar) Act 2012 (Cth) took effect on 15 April 2013.
37 The Intellectual Property Laws Amendment Act 2012 (Raising the Bar Act).

contribution to the working of the invention.[38] The assessment of the variation is made from the point of view of the person skilled in the relevant art and in the light of the common general knowledge in Australia at the priority date of the claim. In practice, the Full Federal Court has confirmed that this threshold is not a high one.[39]

2.3 Court's approach to software patents

The Federal Court and state and territory Supreme Courts share original (ie, first instance) jurisdiction over matters relating to the exploitation and enforcement of patent rights, including challenges to patent rights, infringement proceedings and compulsory licences.[40] All cases are decided by judges (alone). Patent cases are almost invariably brought in the Federal Court of Australia due to that court's acknowledged expertise in IP matters and its dedicated National Practice Area having a sub-area of Patents and Associated Statutes. Appeals from a single judge of the Federal Court are made to the Full Court of the Federal Court, usually comprising three judges. The final court of appeal from all Australian courts is the High Court of Australia.[41]

In a landmark case regarding patentability of subject matter, *National Research Development Corporation v Commissioner of Patents (NRDC)*,[42] the High Court of Australia held that:

> *The point is that a process, to fall within the limits of patentability which the context of the Statute of Monopolies has supplied, must be one that offers some advantage which is material, in the sense that the process belongs to a useful art as distinct from a fine art ... that its value to the country is in the field of economic endeavour.*[43]

This liberal statement by the High Court opened the gates for patents in a range of new areas, including for those regarded as 'business methods'. More recently, the High Court, in *D'Arcy v Myriad*[44] clarified that while the test in NRDC remains relevant, it is not to be applied as a rigid rule for all cases. Broader policy factors, such as whether conferring patentability in the field could result in potentially negative effects on innovation, may also be relevant. Furthermore, in determining patentable subject matter, it is appropriate to have regard to the substance of the invention rather than the form in which it is claimed.

---

38 Section 7(4) of the Patents Act.

39 *Dura-Post (Australia) Pty Ltd v Delnorth Pty Ltd* (2009) 177 FCR 239; 81 IPR 480; [2009] FCAFC 81.

40 Section 154 of the Patents Act gives the Federal Court jurisdiction to hear appeals from the Patent Office. The Federal Court has recognised expertise in intellectual property matters and the vast majority of patent cases are brought in the Federal jurisdiction. There are Federal Courts in the capital cities of each state and territory in Australia. The New South Wales (Sydney) and Queensland (Brisbane) Federal Court registries presently have established panels of Judges to hear and determine patent matters.

41 Section 158 of the Patents Act; Section 87 of the Designs Act; Section 195 of the Trademarks Act.

42 (1959) 102 CLR 252.

43 At 275 (citation omitted).

44 [2015] HCA 35.

In Australia, there were two early court decisions handed down in relation to computer software related inventions that established their patentability; the first of which was *IBM v Commissioner of Patents*.[45] The *IBM* case concerned the patentability of a method and apparatus for producing curves on a computer graphics display. The invention developed by IBM utilised a particular mathematical algorithm which enabled the creation of smoother and more accurate computer-generated curves. It was held that IBM was entitled to the grant of a patent as the production of an improved curved image was a "commercially useful effect" in computer graphics.

The Full Federal Court, in *CCOM v Jiejing*,[46] subsequently considered application of the reasoning in NRDC to claims directed to a computer-implemented invention. In that case the claimed invention involved an apparatus for assembling text in Chinese language characters. The relevant field of economic endeavour was held to be:

> *... the use of word processing to assemble text in Chinese language characters. The end result achieved is the retrieval of graphic representations of desired characters, for assembly of text. The mode or manner of obtaining this, which provides particular utility in achieving the end result, is the storage of data as to Chinese characters analysed by stroke-type categories, for search including 'flagging' (and 'unflagging') and selection by reference thereto.*[47]

The Full Court restated the test applicable to software as whether there is "a mode or manner of achieving an end result which is an artificially created state of affairs of utility in the field of economic endeavour". This essentially is a two-step test for considering the steps or features provided in a patent application:

- is the result an artificially created state of affairs; [48] and
- is it useful in the field of economic endeavour?[49]

This test has allowed for the patenting of business methods in certain circumstances. In a further decision involving a patent for a method and a device using smart cards to operate a loyalty scheme, *Welcome Real-Time SA v Catuity Inc*[50] the Federal Court gave prominence to the physical aspects of the scheme in finding that:

---

45 (1991) 22 IPR 417.

46 (1994) 28 IPR 481.

47 Cooper J also cited the UK decisions in *International Business Machines Corporation's Application* (1980) FSR 564 and *Burroughs Corporation's Application* [1974] RPC 147, which are authority for the proposition that a method embodied in a computer program, that has the effect of controlling a computer to operate in a particular way, may be patentable subject matter.

48 *National Research Development Corporation v Commissioner of Patents* (1959) 102 CLR 252.

49 IP Australia on its webpage titled "Patents for computer related inventions" gives the following practical test on its website: "Software inventions must be industrially applied. Software that is merely a procedure for solving a given type of mathematical problem is not patentable." Available at: www.ipaustralia.gov.au/patents/understanding-patents/types-patents/what-can-be-patented/patents-computer-related.

50 [2001] FCA 445.

... what is involved here is not just an abstract idea or method of calculation. Moreover this result is beneficial in a field of economic endeavour – namely retail trading – because it enables many traders (including small traders) to use loyalty programs and thereby compete more effectively for business. Such competition is in turn beneficial to consumers, both in the general sense that competition is good and in the sense that they can obtain benefits in the form of discounts and free goods and services.

By contrast, in 2006 the Full Federal Court in *Grant v Commissioner of Patents*[51] considered a mere scheme for protection of an asset from the claims of creditors was not patentable, stating that: "A physical effect in the sense of a concrete effect or phenomenon or manifestation or transformation is required." In doing so, however, the Full Court expressly approved of earlier cases where the business method was implemented in a computer environment such that performance of the patented method resulted in a "change in state or information in a part of a machine".[52]

However, the law relating to patent-eligibility of computer-implemented inventions remains far from clear. In further and more recent decisions of the Full Court, *Research Affiliates LLC v Commissioner of Patents*[53] and *Commissioner of Patents v RPL Central Pty Ltd*,[54] the Full Court found that the presence of computing hardware or processing steps within a claimed method or system was insufficient to confer patentability. The effect of these decisions is that an invention which, in substance, is a mere scheme, algorithm, idea or intellectual information does not constitute patentable subject matter, notwithstanding any assertion of computer implementation in the claims. If, for example, the specification reveals that the computer implementation asserted in the claims involves use of standard hardware or software or is not necessary or intrinsic to the process claimed, then the invention may not be a manner of manufacture.[55] Conversely, an invention that involves an improvement in the operation of, or effect of the use of, the computer or computerised process and is integral and intrinsically linked to the method or product claimed, is more likely to constitute patentable subject matter.[56]

In *Encompass Corporation Pty Ltd v InfoTrack Pty Ltd*,[57] a single judge of the Federal Court applied the Full Court's decisions in *Research Affiliates LLC v*

---

51 (2006) 154 FCR 62; 69 IPR 221; [2006] FCAFC 120.

52 Including *Catuity* and *CCOM* (above) as well as US decisions: *State Street Bank & Trust Co v Signature Financial Group Inc* 149 F 3d 1368; 47 USPQ 2d 1596 (Fed Cir 1998); and *AT&T Corp v Excel Communications, Inc* 172 F 3d 1352; 50 USPQ 2d 1447 (Fed Cir 1999).

53 [2014] FCAFC 150.

54 [2015] FCAFC 177.

55 *Research Affiliates LLC v Commissioner of Patents* [2014] FCAFC 150 at [110]; *IBM v Commissioner of Patents* (1991) 22 IPR 417 at [16].

56 *Research Affiliates LLC v Commissioner of Patents* [2014] FCAFC 150 at [96]–[99]; *IBM v Commissioner of Patents* (1991) 22 IPR 417 at [16]; *CCOM v Jiejing* (1994) 28 IPR 481at 292–3.

57 [2018] FCA 421.

Commissioner of Patents[58] and *Commissioner of Patents v RPL Central Pty Ltd*,[59] and found in that the patents in suit did not involve a manner of manufacture because the combination of the method claimed did not result in an "improvement in the computer".[60] That decision was appealed, with the appeal heard in November 2018 by an enlarged bench (constituted by five judges) of the Full Federal Court. IP Australia intervened in the appeal hearing and submitted that a computer software invention was only patentable if programmed with 'some ingenuity'. To further muddy the waters, shortly after the *Encompass* appeal hearing, a judge of the Federal Court in *Rokt Pte Ltd v Commissioner of Patents*[61] reversed a Patent Office rejection of a patent for a computer-implemented invention and found that the patent translated a business problem into a technical problem of how to utilise computer technology to address the business problem.[62] The Commissioner of Patents has appealed the decision.[63]

2.4 Patent Office practice

These recent decisions of the Full Court and the *Encompass* decision have regrettably provided little clarity for patent applicants regarding patent-eligibility of computer-implemented inventions. To further confuse matters, the Patent Office (IP Australia) has been applying an adaptation of the England and Wales Court of Appeal's four-step test for patentable subject matter, as set out in *Aerotel Ltd v Telco Holdings Ltd*.[64] The 'Australianised' test set out in the Examiners Manual is:[65]

- construe the claim;
- identify the substance of the claim (ie, the alleged or actual contribution);
- ask whether the substance of the claim lies within established principles of what does not constitute a patentable invention (eg, a scheme, plan, rules of gameplay, intellectual or genetic information); and
- if not, consider whether the substance otherwise lies outside of existing concepts of manner of manufacture and is to be treated as a 'new class' of subject matter.

---

58 [2014] FCAFC 150.
59 [2015] FCAFC 177.
60 *Encompass Corporation Pty Ltd v InfoTrack Pty Ltd* [2018] FCA 421 at [192]–[197].
61 [2018] FCA 1988.
62 *Rokt Pte Ltd v Commissioner of Patents* [2018] FCA 1988 at [207].
63 As well as the *Encompass* and *Rokt* appeals, there are other appeals currently before the Federal Court in respect of IP Australia Patent Office decisions refusing applications for patents on computer-implemented inventions: *Aristocrat Technologies Australia Pty Ltd v Commissioner of Patents* (NSD1343/2018) is an appeal from *Aristocrat Technologies Australia Pty Limited* [2018] APO 45, and is scheduled to be heard on 2–4 September 2019. *Repipe Pty Ltd v Commissioner of Patents* (WAD323/2018) is an appeal from *Repipe Pty Ltd* [2018] APO 42, and was heard on 25–27 June 2019 (decision pending).
64 [2006] EWCA Civ 1371.
65 2.9.2.2 Principles for Examination – Examination Practice.

The correctness of this approach in the absence of clear judicial support remains questionable. The Examiners Manual further sets out a number of considerations that may be relevant to whether a computer-related invention is in substance a manner of manufacture, including:[66]

- whether the contribution of the claimed invention is technical in nature;
- whether the invention solves a technical problem within the computer or outside the computer or whether it results in an improvement in the functioning of the computer, irrespective of the data being processed;
- whether the claimed method merely requires generic computer implementation;
- whether the computer is merely an intermediary or tool for performing the method while adding nothing of substance to the idea;
- whether the ingenuity in the invention is in a physical phenomenon in which an artificial effect can be observed rather than in the scheme itself;
- whether the alleged invention lies in the way the method or scheme is carried out in a computer; and
- whether the alleged invention lies in more than the generation, presentation or arrangement of intellectual information.

In practice, the application of this approach has led to the rejection of many computer-implemented inventions by the Patent Office on subject matter grounds. In view of numerous recent decisions, applicants seeking patent protection in Australia should be cautious as to their prospects of success.

3. Infringement

3.1 Claim interpretation

Infringement is determined by reference to the claims of the patent.[67] The approach to claim interpretation in Australia has been influenced by UK authorities. The 'pith and marrow' doctrine and the need for a purposive construction propounded by the House of Lords in *Catnic Components Ltd v Hill & Smith Ltd*[68] have long been recognised by Australian courts. Even though a product or process may not fall literally within the words of a claim, a purposive interpretation may still lead to a finding of (non-literal) infringement under Australian law.[69]

---

66 Examiners Manual, 2.9.2.7 Computer Implemented Inventions – Schemes and Business Methods.
67 *Décor Corp Pty Ltd v Dart Industries Inc* (1988) 13 IPR 385 at 400.
68 [1982] RPC 183 at 242–3 *per* Lord Diplock; *Kirin-Amgen v Hoechst Marion Roussel* [2005] RPC 9.
69 See 'Improver' or 'Protocol' questions in *Improver Corp v Remington Consumer Products Ltd* [1990] FSR 181; *Populin v HB Nominees Pty Ltd* (1982) 41 ALR 471 at 475.

A non-literal infringement will result under a purposive interpretation when:

- a variant does not have a material effect upon the way an invention works;
- that variant would have been obvious at the date of publication to a person skilled in the art; and
- the person skilled in the art would not have understood from the language of the claim that strict compliance with its primary meaning was an essential requirement.

However, the Full Court of Appeal has confirmed that a purposive construction does not mean that one is extending or going beyond the definition of the technical matter for which the patentee seeks protection in the claims.[70]

3.2 Acts of infringement

Whilst 'infringement' is not defined in the Patents Act, Section 13 gives the patentee "the exclusive rights, during the term of the patent, to exploit the invention and to authorise another person to exploit the invention". For a patentee to enforce a patent for direct infringement therefore requires proving that an alleged infringer has 'exploited' the claimed invention or authorised another to do so.

The term 'exploit' is defined in the Dictionary[71] to the Patent Act as follows:

'exploit', in relation to an invention, includes:

(a) where the invention is a product – make, hire, sell or otherwise dispose of the product, offer to make, sell, hire or otherwise dispose of it, use or import it, or keep it for the purpose of doing any of those things; or

(b) where the invention is a method or process – use the method or process or do any act mentioned in paragraph (a) in respect of a product resulting from such use.

Infringement will typically require consideration of whether there has been a 'making' [72] or 'using'[73] of a patented device or method. The central question in any enforcement action will be whether the respondent, in exploiting the patented invention, has "taken all of the essential features or integers"[74] of one or more of the claims. Whereas early judicial decisions considered 'authorise' to have a narrow meaning of "to give legal or formal warrant to (a person) to do; to empower, permit

70 *Kimberly-Clark Australia Pty Ltd v Multigate Medical Products Pty Ltd* [2011] FCAFC 86 at [14]; and see also the similar finding of a differently comprised Full Court in *Australian Mud Company Pty Ltd v Coretell Pty Ltd* [2011] FCAFC 121 at [69]–[72].

71 Schedule 1 to the Patents Act.

72 *Schutz (UK) Ltd v Werit (UK) Ltd* (2013) 100 IPR 583.

73 *Bedford Industries Rehabilitation Association Inc v Pinefair Pty Ltd* (1998) 42 IPR 330; *Noakes v Meharey* (1915) 32 RPC 307.

74 *Populin v HB Nominees Pty Ltd* (1982) 41 ALR 471 and *Fresenius Medical Care v Gambro* [2005] FCAFC 220 at [49].

authoritatively",[75] later cases have approved a wider meaning comparable to expression found in Australia's copyright legislation of "sanction, approve or countenance".[76] Those factors include: the extent (if any) of the person's power to prevent the doing of the act concerned; the nature of any relationship between the person and the person who did the act concerned; and whether the person took any reasonable steps to prevent or avoid the doing of the act, including whether the person complied with any relevant industry codes of practice.

3.3 Contributory infringement

Under Australian law, a person may, in certain circumstances, be liable for contributory infringement if that person supplies a product to another, where the use of that product would infringe a third party's patent rights.[77]

Section 117 of the Patents Act 1990 (Cth) provides:

117 Infringement by supply of products

(1) If the use of a product by a person would infringe a patent, the supply of that product by one person to another is an infringement of the patent by the supplier unless the supplier is the patentee or licensee of the patent.

(2) A reference in subsection (1) to the use of a product by a person is a reference to:

(a) if the product is capable of only one reasonable use, having regard to its nature or design – that use; or

(b) if the product is not a staple commercial product – any use of the product, if the supplier had reason to believe that the person would put it to that use; or

(c) in any case – the use of the product in accordance with any instructions for the use of the product, or any inducement to use the product, given to the person by the supplier or contained in an advertisement published by or with the authority of the supplier.

Contributory infringement has been upheld in cases where a product has been supplied with instructions or inducements to a third party to infringe a patent.[78] Further guidance on the operation of the section was provided in the High Court's decision, *Collins v Northern Territory of Australia*:[79]

---

75 *Advanced Building Systems Pty Ltd v Ramset Fasteners (Aust) Pty Ltd* (1995) AIPC 91–129, 39,200; *Bristol-Myers Squibb Co v F H Faulding & Co Ltd* (1998) 41 IPR 467, 488.

76 Section 101(1A) factors listed in the Copyright Act 1968; see Bennett J in *Apotex Pty Ltd v Les Laboratories Servier (No 4)* [2010] FCA 1202 at [14]; and *Inverness Medical Switzerland GmbH v MDS Diagnostics Pty Ltd* (2010) 85 IPR 525, 568 at [194]. Authorisation under Section 13(1) may also extend to circumstances that satisfy the requirements for the common law doctrines of joint-tortfeasorship through procurement or in pursuance of a common design.

77 Contributory or indirect infringement has been a feature of Australian law since the introduction of the Patents Act in 1990. The stated legislative intention was to harmonise Australian contributory infringement law with the contributory infringement law of major trading partners, and particularly with that of the United States. *Dunlop Pneumatic Tyre Co Ltd v David Moseley & Sons Ltd* [1904] 1 Ch 612.

78 *Bristol-Myers Squibb Co v FH Faulding & Co Ltd* (2000) 46 IPR 553.

79 [2006] 70 IPR 614.

- 'supply' is to be interpreted in an intuitive and common-sense fashion;[80] and
- if a product is supplied commercially and has several uses, it will be considered a 'staple commercial product'.[81]

3.4 Cross-border issues

Although IP rights are territorial, Australian courts have demonstrated a readiness to find jurisdiction in cases that involve conduct in or by persons from Australia. The High Court of Australia held in a leading defamation case[82] that material that had been written in New York, uploaded to a server in New Jersey, United States and downloaded in Victoria, Australia, was nevertheless defamatory in Australia, where the reputation of the individual was ultimately damaged. In the case of rights that are exercised through the use of the internet, such as the right of communication to the public, Australian courts have found (infringing) use of such rights where at least some of the relevant activity takes place geographically within Australia or on Australian computer network assets. For example, in *Cooper v Universal Music Australia Pty Ltd*[83] it was found that the transmission of copyright content from a jurisdiction outside Australia into Australia constituted an infringement of the communication right in Australia and therefore an infringement of Australian copyright law. Although the full extent of the territorial reach of Australian courts in IP cases has not been fully tested, it has been suggested by some commentators that Australian courts may be both capable of and prepared to extend the type of jurisdiction over extra-territorial copyright infringements, as the UK Supreme Court acknowledged in *Lucasfilm Limited v Ainsworth*.[84]

Where a territorial nexus is established, Australian courts have been receptive to providing litigants with the means to effectively enforce internationally protected IP rights, which are being infringed in a number of locations including Australia.[85]

4. Ownership, inventorship and compensation

Section 15(1) of the Patents Act provides that a patent for an invention may only be granted to a person who is the inventor or a person who has taken an assignment or derives title to the invention from the inventor. That is, the

---

80 *Id* at [55] *per* Hayne J (with whom the majority of other judges effectively agreed), finding that a licence to cut and take timber was to be regarded as a 'supply' of the timber.

81 *Id* at [145] *per* Crennan J. As liability cannot arise under Section 117(2)(b) if the product supplied is a "staple commercial product", the contributory infringement claim failed and the Northern Territory Government's appeal was allowed.

82 *Dow Jones & Company Inc v Gutnick* [2002] HCA 56.

83 [2006] FCAFC 187; (2006) 156 FCR 380.

84 [2011] UKSC 39; the Supreme Court held that Mr Ainsworth's replica stormtrooper costumes infringed Lucasfilm's US copyrights, and ruled that those rights are enforceable in the United Kingdom with respect to activities outside of the United Kingdom.

85 *Sony Dadc Austria AG & Anor v Digital Digest Pty Ltd & Anor* Federal Court Proceeding No NSD417/2007.

primary person to whom a patent will be granted is the inventor and all other rights to the invention are derived from the inventor.[86] There is no statutory definition of 'inventor', but it is clear from case law that the word has its ordinary English meaning.[87]

In identifying the inventor, a court will identify the inventive concept of the invention by reference to the terms of the whole of the relevant specification, including the claims; and ascertain the timing of the invention and the identity of the inventor by reference to that inventive concept.[88]

The timing of the invention will be assessed by reference to the date when the inventive idea or concept was developed, rather than the date of its reduction to practice.[89] There may be multiple contributors to the inventive concept and multiple inventive concepts in a single invention.[90] The Full Court has in recent times held that failure to correctly nominate all inventors will not invalidate a patent.[91]

5. Copyright protection

5.1 Copyright protection of software and databases

The Copyright Act 1968 (Cth) protects original literary works, dramatic works, musical works, artistic works, sound recordings and films.[92] A work is original if it has not been wholly copied from another work and originates with the author.[93] Copyright protection is automatic and does not require registration. Literary, dramatic, musical and artistic works are protected for the life of the author plus 70 years.[94]

'Computer programs' – "a set of statements or instructions to be used directly or indirectly in a computer in order to bring about a certain result" – are protected as a type of 'literary work'.[95] Copyright protection extends to software that is embedded in firmware and built into devices.[96]

---

86 *Davies Shepard Pty Ltd v Stack* (2001) 51 IPR 513.

87 *Atlantis Corp v Schindler* (1997) 39 IPR 29.

88 *University of Western Australia v Gray* [2009] FCAFC 116, drawing on *Polwood Pty Ltd v Foxworth Pty Ltd* [2008] FCAFC 9 and *Kafataris v Davis* [2016] FCAFC 134.

89 *UWA, ibid* at [249].

90 *Polwood Pty Ltd v Foxworth Pty Ltd* [2008] FCAFC 9.

91 *University of British Columbia & Anor v Conor Medsystems Inc* (2006) 70 IPR 248 *per* Emmett, Stone and Bennett JJ, overturning the decision of Finkelstein J that the patent ought be revoked due to a failure to name all inventors in *Conor Medsystems, Inc v The University of British Columbia (No 2)* [2006] FCA 32; see also *Sigma Pharmaceuticals (Australia) Pty Ltd v Wyeth* [2011] FCAFC 132 *per* Yates J at [270]–[291], Bennett J agreeing at [125]–[126].

92 Section 32 of the Copyright Act. Also see Section 10(1) definitions for literary, artistic and dramatic works.

93 *Data Access Corp v Powerflex Services Pty Ltd* (1999) 202 CLR 1; *Ice TV v Nine Network Australia* (2009) 239 CLR 458

94 Section 33(2)–(3) of the Copyright Act.

95 Sections 10 (definition of computer program) 184 and 185 of the Copyright Act 1968. The definition of 'computer program' covers programs written in source code, object code, microcode or declarative programming language.

96 Subject only to an important exception in relation to circuit layouts discussed in the next section of this chapter.

Other material protected by copyright includes artistic works such as the design of graphical user interfaces, multimedia elements, computer and app icons, written material, tables and compilations. There is no *sui generis* statutory protection in respect of databases. Claims in respect of copying of data and databases have often been unsuccessful in Australia.[97] Although facts and data are not protected by copyright, a collection of data, a dataset, metadata, or a compilation of numerous items of data or metadata records may be protected by copyright if sufficiently original. Originality is determined having regard to application of independent intellectual effort by the authors, which may involve the exercise of skill, judgement, or creativity in selecting, presenting, or arranging the information.[98] Literary works are defined as including "a table, or a compilation, expressed in words, figures or symbols (whether or not in a visible form)".[99] Data compilations may therefore be 'literary works'.[100]

Where a data compilation is sufficiently original to be protected by copyright, copyright applies to the form in which the information has been compiled, arranged, or presented, not to the data itself.[101] Any computer program underlying a database may also be protected by copyright as a (separate) literary work. The Full Federal Court in *Acohs v Ucorp*[102] noted that it was possible that even an information tag, if it is original, sufficiently substantial, and functionally separate from the program of which it forms a part, might constitute a separate copyright work. In *CA, Inc v ISI Pty Ltd*[103] macros developed by the respondents to replace the applicant's macros were found to reproduce substantial parts of the applicant's database management system.

In addition to the requirement of originality that the relevant work not be copied and result from independent intellectual effort directed to expressing the work in its final form, the recent Australian cases as to copyright in data compilations and computer-generated works have placed significant stress upon

97 See *Data Access Corp v Powerflex Services Pty Ltd* (1999) 202 CLR 1; *Desktop Marketing v Telstra Corporation* (2002) 119 FCR 491; *Nominet UK v Diverse Internet* (2004) 63 IPR 543; *IceTV Pty Ltd v Nine Network Pty Ltd* (2009) 239 CLR 458, (2009) HCA 14; *Telstra Corporation Limited v Phone Directories Company Pty Ltd & Ors* [2010] FCA 44; *Acohs Pty Ltd v Ucorp Pty Ltd* [2010] FCA 577; *Acohs Pty Ltd v Ucorp Pty Ltd (No 2)* [2012] FCAFC 16; *Dynamic Supplies Pty Ltd v Tonnex Pty Ltd* [2011] FCA 362; *CA, Inc v ISI Pty Ltd* [2012] FCA 35; *JR Consulting & Drafting Pty Limited v Cummings* [2016] FCAFC 20.

98 See in particular *IceTV Pty Ltd v Nine Network Pty Ltd* (2009) 239 CLR 458; (2009) HCA 14 *per* French CJ, Crennan and Kiefel JJ at [33] and [48], Gummow, Hayne and Heydon JJ at [99]; also *Telstra Corporation Limited v Phone Directories Company Pty Ltd* [2010] FCAFC 149 *per* Keane CJ at [57], Perram J at [99] and Yates J at [133].

99 Section 10(1) of the Copyright Act.

100 The 'originality' requirement frequently arises in the context of compilations and is often problematic. See *University of London Press Ltd v University Tutorial Press Ltd* [1916] 2 Ch 607; *Desktop Marketing Systems Pty Ltd v Telstra Corp Ltd* (2002) 55 IPR 1; *IceTv Pty Ltd v Nine Network Australia Pty Ltd* (2009) 239 CLR 458.

101 *Victoria Park Racing & Recreation Grounds Co Ltd v Taylor* (1937) 58 CLR 479; *Autodesk Inc v Dyason* (1992) 173 CLR 330.

102 *Acohs Pty Ltd v Ucorp Pty Ltd (No 2)* [2012] FCAFC 16.

103 *CA, Inc v ISI Pty Ltd* [2012] FCA 35. Comment lines to source code that instruct the operator and are not converted to the object code have been held within the definition of 'computer program': *CCH Australia Ltd & Anor* v *Accounting Systems 2000 (Developments) Pty Ltd & Ors* (1991) AIPC 90–801.

the requirement for a plaintiff to prove that the work originated from an identifiable human author.[104] Authorship may include establishing a framework for the contents of a database, selecting certain data, assembling the contents, making decisions about defining fields and indexes, devising the schema of the database, and perhaps adapting software that provides the access and search capability. It is clear that more subjective selections, such as listing in order of importance, selecting sub-topics after exercise of creativity, or settling a directory where the selected universe is not obvious, may involve sufficient intellectual effort.[105] However, the relevant contributions of each human author may need to be proven, creating significant evidentiary difficulties for plaintiffs in many software and databases cases.

If an author produces software during the course of his or her employment,[106] any copyright vests in the author's employer.[107] If there is an agreement[108] about the creation of software (for example, a commissioned software development agreement), ownership will be determined by reference to the agreement. Assignment of copyright under an agreement may be express or inferred[109] but will not be inferred merely from making a payment for a commissioned work.[110] In many cases where an assignment cannot be so inferred, a licence in favour of the commissioning party will be inferred for use of the created work for the purpose for which it was commissioned.[111]

The author or authors of a compilation work are the persons who gather or organise the collection of material and who select, order or arrange for its fixation in material form, not those who individually created or supplied the individual parts.[112] Separate copyrights may subsist in individual parts of the whole from the compilation or collection copyright in the assembly of those individual parts to create the whole.

Where there is more than one creator of a copyright work the work will, in

---

104 *Telstra Corp v Phone Directories* (2010) 194 FCR 142; *Acohs Pty Ltd v Ucorp Pty Ltd* (2012) 95 IPR 117.

105 See further Andrews and Cameron, "Copyright in computer-generated work in Australia post-IceTV: Time for the Commonwealth to act" (2011) 22 AIPJ 29; Anne Fitzgerald and Natasha Dwyer, *Copyright in Databases in Australia*, 2012, available at http://eprints.qut.edu.au/50425/; Peter Knight, "Copyright in databases and computer programs: Why is it so hard to understand?" (2010) 21 AIPJ 118; Jani McCutcheon, "When sweat turns to ice: The originality threshold for compilations following IceTV and Phone Directories" (2011) 22 AIPJ 87.

106 "... in pursuance of the terms of his or her employment by another person under a contract of service or apprenticeship": Section 35(6). Note that this statutory vesting does not apply in respect of commissioned software developed by independent contractors or works created by an employee outside their scope of employment. See *EdSonic Pty Ltd v Cassidy* [2010] FCA 1008 at [41] and *Insight SRC IP Holdings Pty Ltd v The Australian Council for Educational Research Limited* [2012] FCA 779 at [40]. Whether a person is an employee is determined applying general principles of employment law.

107 *Redrock Holdings Pty Ltd v Hinkley* (2001) 50 IPR 565.

108 Section 35(4) of the Copyright Act.

109 Section 196(3) of the Copyright Act. See *Murray v King* (1984) 4 FCR 1.

110 An intention to effect an assignment of copyright must be evident. See *Larrikin Music Publishing Pty Ltd v EMI Songs Australia Pty Ltd* (2009) 81 IPR 584.

111 *Beck v Montana Constructions Pty Ltd* (1963) 5 FLR 298; *Achos Pty Ltd v R A Bashford Consulting Pty Ltd* (1997) 37 IPR 542.

112 *IceTV Pty Limited v Nine Network Australia Pty Limited* (2009) 239 CLR 458 and (2009) HCA 14 *per* Gummow J and *per* Gummow, Hayne and Heydon JJ at [99].

the absence of agreement to the contrary, be a 'work of joint authorship' if the contribution of each creator to the collaboration cannot be distinguished from the contribution of the other creator or creators.[113] In these circumstances, each creator will usually own an indivisible share of the copyright in the work, with the consequence that no one creator may exercise the rights without the permission of the other creator or creators.[114]

5.2 Exclusive rights and infringement

Owners of copyright in software have a number of exclusive rights,[115] including the right to:

- reproduce the software in a material form, such as copying the program to storage devices or typing the source code of the program;
- publish the software, such as offering the program to the public for the first time;
- make an adaptation of the software, such as making versions in different languages or code; and
- communicate the software to the public, such as making it available for download.

'Adaptation' in relation to a computer program is defined to mean "a version of the work (whether or not in the language, code or notation in which the work was originally expressed) not being a reproduction of the work".[116]

The rights of a copyright owner are infringed where an infringing act is done in relation to 'a substantial part' of the work or other subject matter.[117] In determining whether any reproduction is of a 'substantial part', greater weight is given to the quality of the taken part than to the quantity.[118] The courts have adopted a qualitative as well as a quantitative approach which is to be applied in determining whether a 'substantial' part of a copyright work has been copied. In one case a court held that a 'look-up table' which formed part of a piece of software was a substantial part of that software: the fact that the look-up table was not of itself software did not mean it could not form a substantial part of a computer program. In another case, a court held that an error text table in a piece of software was not a substantial part, apparently on the basis that the error table was not an essential element of the program and the software could function without an error text table.

There will be a direct infringement only if a person, not being the owner and

---

113 Section 10(1) (definition of 'work of joint authorship').
114 *Seven Network (Operations) Ltd v TCN Channel Nine Pty Ltd* (2005) 66 IPR 101.
115 Section 31(1) of the Copyright Act.
116 Section 10(1) of the Copyright Act.
117 Section 14(1) of the Copyright Act.
118 *Data Access Corp v Powerflex Services Pty Ltd* (1999) 202 CLR 1; *SW Hart & Co Pty Ltd v Edwards Hot Water Systems* [1985] HCA 59. The originality expressed in the part taken is also often a key factor.

without a licence, does or authorises the doing of any act that the copyright owner has the exclusive right to do.[119] It is also an infringement for a person other than the copyright owner to 'authorise' the doing of an act comprised in the copyright. 'Authorise' has been interpreted as "sanction, approve, countenance".[120] In determining whether a person is liable for 'authorising', the court considers the extent (if any) of the person's power to prevent the doing of the act concerned; the nature of the relationship existing between the person and the person who did the act concerned; and whether the person took any reasonable steps to prevent or avoid the doing of the act, including whether the person complied with any relevant industry codes of practice. In *JR Consulting & Drafting Pty Limited v Cummings*[121] the alleged infringers included an individual (Gianni Pacione) and two companies of which Pacione was a director and shareholder. The courts considered whether Pacione was liable for authorising the copyright infringing acts of those companies. The courts found he was liable by reason of his position as a director and the economic owner of the companies. The companies were under the control of his guiding mind, he knew they had no right to engage in the infringing conduct and while he could have stopped them from engaging in that conduct, he took no reasonable steps to do so and in fact consciously facilitated the infringing conduct and obfuscated having done so.

A licensee cannot pursue an infringement action unless the licensee is an exclusive licensee.[122]

5.3 Exceptions

Copyright in computer programs is not infringed by any of the following:

- making reproductions (copies) for normal use or study;[123]
- making back-up copies;
- making copies to obtain information to make interoperable products;
- making copies to correct errors; and
- making copies for security testing.[124]

---

119 Sections 13(1), 36(1) and 101(1) of the Copyright Act.

120 See *The University of New South Wales v Moorhouse & Anor* (1975) 133 CLR 1 *per* Gibbs J at 12–13, also *Roadshow Films Pty Ltd v iiNet Pty Ltd [*2012] HCA 16 *per* French CJ, Crennan and Kiefel JJ at [42] and *JR Consulting & Drafting Pty Limited v Cummings* [2016] FCAFC 20.

121 [2016] FCAFC 20. See in particular the review of relevant authorities at [320]–[361].

122 An assignee is able to bring an action for infringement and a partial assignee may also do so as far as the assignee's part. However, an assignment of copyright "does not, without more, carry with it accrued rights of action for infringement": *Insight SRC IP Holdings Pty Ltd v The Australian Council for Educational Research Limited* [2012] FCA 779 at [104]. Section 120(1) provides that if the copyright owner and exclusive licensee have concurrent rights of action in relation to an infringement, neither the copyright owner nor the exclusive licensee is entitled, except with leave of the court, to proceed with the action, unless the other party (the copyright owner or the exclusive licensee) is joined as a plaintiff or added as a defendant. Where the other party is not joined as a plaintiff but added as a defendant, the other party is not liable for any costs in the action unless he enters an appearance and takes part in the proceedings: Section 120(1)(b) of the Copyright Act. This does not affect the granting of an interlocutory judgment on the application of the copyright owner or the exclusive licensee: Section 120(2). See further Sections 119(a) and 120(1).

123 Sections 40 and 103C of the Copyright Act.

124 Sections 47B–47F of the Copyright Act.

The copies must be made by, or on behalf of, the owner or licensee of a non-infringing copy of the computer program. In addition, if the copy is used for a purpose other than in accordance with these provisions, the defence does not apply.[125] For the purposes of the defences, 'computer program' includes any literary work that is incorporated in, or associated with, the program so long as it is essential to its effective operation.[126]

The defence of 'normal use or study' is confined to making a reproduction incidentally and automatically as part of the technical process of running a program for the purpose for which it was designed. In this context, and the back-up copy defence, 'reproduction' has a special narrow meaning and does not include a version of the work in source code that has been derived from object code by any process, including decompilation.[127] By contrast, a broader definition of 'reproduction' of computer programs applies for other defences.[128]

In specific circumstances it is not an infringement of copyright in a computer program to make a reproduction, or an adaptation of, a program for the purpose of obtaining information necessary to make independently another program or article, to connect to and be used together with, or otherwise integrate with, the original program. However, the reproduction or adaptation may be made only to the extent necessary to obtain the relevant information which is not readily available from another source.[129]

A reproduction or adaptation may be made for the purpose of correcting an error in the original copy of a program that prevents it from operating as intended by its author, or in accordance with any specifications or documentation supplied with it. The reproduction or adaptation must be made only to the extent reasonably necessary to correct the error. The defence does not apply if another copy of the program without error is available within a reasonable time at an ordinary commercial price.

A reproduction or adaptation may also be made for the purpose of testing in good faith, the security of the original copy of the program or of a computer system or network of which it is a part or investigating or correcting a security flaw. However, the defence is limited to testing of the original to ascertain its security from unauthorised access or against hacking or other electronic invasion, and is not available where the testing process was for disaster recovery and business continuity purposes.[130]

In 2016 the Productivity Commission[131] recommended the introduction of a

125 Section 47G of the Copyright Act.
126 Section 47AB of the Copyright Act.
127 Section 21(5)(b) of the Copyright Act.
128 Section 21(5) of the Copyright Act.
129 Section 47D of the Copyright Act, considered in *CA, Inc v ISI Pty Ltd* [2012] FCA 35.
130 *Software AG (Australia) Pty Ltd v Racing & Wagering Western Australia* (2009) AIPC 92–334.
131 Productivity Commission Inquiry Report, Intellectual Property Arrangements, No 78 September 2016, available at https://www.pc.gov.au/inquiries/completed/intellectual-property/report.

US-style fair use exception into Australian law. The government noted that recommendation but indicated that further consultation is required before the government will consider amending the suite of existing copyright exceptions.[132]

5.4 Technological protection measures

Technological protection measures (TPMs) are technical locks that copyright owners may deploy to stop their material being accessed or copied. Examples include passwords, encryption software and access codes. Access control TPMs may prevent a person from being able to read, listen to or watch material. Copy control TPMs may allow a person to read, listen to or watch material but prevent a person from making a copy of the material.

The Copyright Act supports the use of TPMs as a means of practical protection of copyright owners. There are also provisions in the Copyright Act that give copyright owners the right to take legal action against people who make, supply, distribute or import devices to circumvent TPMs. In some cases, distribution of devices or services to circumvent TPMs is a criminal offence.

In particular, there are sanctions against:

- circumventing an access-control TPM;
- manufacturing or supplying a device to circumvent an access-control TPM; and
- providing a service to circumvent an access-control TPM.[133]

Under the Copyright Act, actions may be brought by the copyright owner or exclusive licensee against a person who makes or commercially deals with a device capable of circumventing, or facilitating the circumvention of, a TPM. The provisions are complex and have proven difficult to interpret and apply. In the leading case in the area, *Stevens v Kabushiki Kaisha Sony Computer Entertainment*,[134] on ultimate appeal the High Court of Australia accepted the trial judge's construction that "a 'technological protection measure', as defined, must be a device or product which utilises technological means to deny a person access to a copyright work [or other subject matter], or which limits a person's capacity to make copies of a work [or other subject matter] to which access has been gained, and thereby 'physically' prevents or inhibits the person from undertaking acts which, if carried out, would or might infringe copyright in the work [or other subject matter]". An access code system which had the

---

132 Australian Government Response to the Productivity Commission Inquiry, August 2017, response to recommendation 6.1, available at https://www.pc.gov.au/inquiries/completed/intellectual-property/intellectual-property-government-response.pdf.

133 Division 2A of Pt V of the Copyright Act 1968.

134 *Kabushiki Kaisha Sony Computer Entertainment* v *Stevens* (2002) AIPC 91-814 (Sackville J at first instance) at pp 37, 953; in the High Court *Stevens v Kabushiki Kaisha Sony Computer Entertainment* (2005) AIPC 92–140

purpose of discouraging copying but did not effectively prevent it did not satisfy the requirement that "the protective devices are designed to function, by their own processes or mechanisms, to prevent or hinder acts that might otherwise constitute an infringement of copyright".

Both civil remedies and criminal penalties are available where a person circumvents a TPM. A TPM is only protected by the scheme where it is used:

- by, or with the permission of, the owner or exclusive licensee of the copyright;
- on material which is protected by copyright (ie, TPMs on public domain material are not protected);
- in connection with the copyright owner's exclusive rights (ie, to reproduce, communicate etc); and
- to control access to the work (eg, by requiring a password or some other process for access).

The scheme specifically excludes TPMs that control geographic market segmentation. As a result, consumers can bypass region coding measures to play overseas purchased DVDs or computer programs on Australian devices. It also excludes TPMs which restrict the use of after-market goods or services, such as supply of spare parts or repair or maintenance services by third parties.

Additional exceptions are provided for in Schedule 10A to the Copyright Regulations. These additional exceptions are subject to review and relevantly include:

- reproduction of computer programs to make interoperable products;
- reproduction and communication of copyright material by educational and other institutions assisting people with disabilities;
- reproduction and communication of copyright material by libraries, archives and cultural institutions for certain purposes;
- access where a TPM is not operating normally and a replacement TPM is not reasonably available; and
- access where a TPM damages a product, or where circumvention is necessary to repair a product.

The Productivity Commission recently recommended amending the TPM provisions of the Copyright Act to:

- make unenforceable any part of an agreement restricting or preventing a use of copyright material that is permitted by a copyright exception; and
- permit consumers to circumvent technological protection measures for legitimate uses of copyright material.

The government has indicated some support for these recommendations and is presently consulting further with stakeholders.[135]

5.5 Electronic Rights Management Information (ERMI)

ERMI is information that has been embedded or attached to copyright-protected material. ERMI can include details about the material, the copyright owner and related data: for example, details embedded in the metadata of an MP3 file or the watermarking and other data embedded into an image or video file.

A copyright owner can take action against someone who alters or removes such data with the aim of enabling, concealing or facilitating infringement of that material.

6. Confidential information

Although there are jurisprudential views in Australia as to the basis for an action for wrongful disclosure of confidential information or trade secret information that significantly differ from the English cases, in practice the formulation of the basic principles and their application is usually similar to the English *Coco v Clark* formulation. This requires the establishment of three elements:[136]

- the information must have the necessary quality of confidence;
- the information must have been imparted in circumstances identifying an obligation of confidence; and
- there must be an unauthorised use of that information to the detriment of the person who claims the confidence.

To be confidential, information must be capable of being given the attribute or quality of confidence; that is, it must be something which is not within the public knowledge.[137] Disclosure under conditions of confidentiality may preserve its character as confidential information, "even if the dissemination was via mass media or the internet. It is a question of degree in each case".[138] In *Medic-Care,*[139] Finn J noted at [631] that the basis of the equitable jurisdiction attaching to abuse of confidential information in Australian law lies "in the notion of an obligation of conscience arising from the circumstances in or through which the information was communicated or obtained".

In the High Court of Australia, the dissenting judgment of Gummow J in *Corrs Pavey Whiting & Byrne v Collector of Customs* summarised the relevant law and proposed additional elements to the test:

---

135 Australian Government Response to the Productivity Commission Inquiry, August 2017, response to recommendation 5.1.

136 *Coco v AN Clark (Engineers) Ltd* (1968) 1A IPR 587 as applied, in particular, in *Commonwealth v John Fairfax & Sons Ltd* [1980] HCA 44; (1980) 147 CLR 39 at 51. In the context of computer programs, see in particular *CA, Inc v ISI Pty Ltd* [2012] FCA 35.

137 *Saltman Engineering Co Ltd v Campbell Engineering Co Ltd* [1963] 3 All ER 413 at 415.

138 *CA, Inc v ISI Pty Ltd* [2012] FCA 35 per Bennett J at [376]. The term 'relative secrecy' has been used to describe this concept: see *Franchi v Franchi* [1967] RPC 149 at 153; *Australian Medic-Care Co Ltd v Hamilton Pharmaceutical Pty Ltd* [2009] FCA 1220 at [633] *per* Finn J; and *Interfirm Comparison (Australia) Pty Ltd v Law Society of New South Wales* [1975] 2 NSWLR 104 at 119.

139 [2009] FCA 1220 at [631] *per* Finn J, citing *Moorgate Tobacco Co Ltd v Phillip Morris Ltd (No 2)* (1984) 196 CLR 414 at 438.

It is now well settled that in order to make out a case for protection in equity of allegedly confidential information, a plaintiff must satisfy certain criteria. The plaintiff: (i) must be able to identify with specificity, and not merely in global terms, that which is said to be the information in question; and must also be able to show that (ii) the information has the necessary quality of confidence (and is not for example, common or public knowledge); (iii) the information was received by the defendant in such circumstances as to import an obligation of confidence; and (iv) there is actual or threatened misuse of the information ... It may also be necessary ... that unauthorised use would be to the detriment of the plaintiff.[140]

Notwithstanding this judgment being a dissent, the judgment has thereafter been applied and accepted in Australia as defining the elements of a breach of confidence action.[141]

Actions for breach of confidential information may be brought in any court in Australia exercising equitable jurisdiction, including the State and Territory Supreme Courts and the Federal Court of Australia. Equitable relief may include injunctions restraining further use or disclosure of the information, orders for search and seizure of documents in preservation of evidence, account of profits or compensatory damages.[142]

6.1 Importance of specificity in identifying what is confidential

Justice McMeekin in *Deeson Heavy Haulage Pty Ltd v Cox*[143] stressed that "the plaintiff must be able to identify with specificity, and not merely in global terms, that which is said to be the information in question".[144] There has been consistent application of this 'fourth element' in Australian breach of confidence claims.[145] In the context of commercial disputes the courts have been particularly reluctant to extend the protection of the law of confidentiality to a broad suite of non-specific information.[146] The policy position behind this trend is the maintenance of free enterprise and competition.

In the 2013 decision of *Streetscape Projects (Australia) Pty Ltd and Another v*

140 *Corrs Pavey Whiting & Byrne v Collector of Customs (Vic)* (1987) 74 ALR 428 at [137]. See also *Moorgate Tobacco Co Ltd v Philip Morris (No 2)* (1984) 156 CLR 414 at 438 *per* Deane J; 59 ALJR 77; 56 ALR 193.

141 See eg, *Rapid Metal Developments (Australia) Pty Ltd v Anderson Formrite Pty Ltd* [2005] WASC 255; *Ekaton Corporation Pty Ltd v Chapman* [2010] SADC 150 *per* Brebner J at [17].

142 The High Court of Australia has applied equitable principles in a number of important decisions to require wrongdoing fiduciaries to disgorge all profits causally related to the breach of fiduciary duty, and not only the direct benefits or gains from the wrongdoing. See *Warman International Ltd v Dwyer* [1995] HCA 18; (1995) 182 CLR 544 particularly at 557–558; and *Ancient Order of Foresters in Victoria Friendly Society Limited v Lifeplan Australia Friendly Society Limited* [2018] HCA 43. Where there is shown to exist a causal connection between a fiduciary's breach of fiduciary obligation and a benefit or gain to the fiduciary or knowing participant, the onus shifts to the defendant to establish that it is inequitable to order that the defendant account for the value of the whole of the identified benefit or gain.

143 (2009) 82 IPR 521.

144 *Id*, at 556; See also *Corrs Pavey Whiting & Byrne v Collector of Customs (Vic)* (1987) 14 FCR 434 at 443.

145 See also *Smith Kline & French Laboratories (Aust) Ltd v Secretary, Dept of Community Services & Health* (1990) 22 FCR 73 at 86–7 *per* Gummow J in the Federal Court; and *Del Casale v Artedomus (Aust) Pty Ltd* (2007) 73 IPR 326 *per* Hodgson, McColl and Campbell JJA in NSW Court of Appeal.

146 *Ibid*.

City of Sydney[147] the New South Wales Court of Appeal made the following comments at [159] about determining whether the information on the subject of the claim had the requisite quality of confidence:

> *The need for specificity in the identification of the information said to be confidential in respect of which relief is sought comes from the fact that the court must make an assessment of the quality of that information, that is, whether it is in truth of a confidential nature. An aspect of that inquiry may turn on whether the whole or some part has become the subject of general disclosure or notoriety. Precise delineation of the subject matter is accordingly essential. The task of a plaintiff, in this respect, is, in the words of Gummow J in* Smith Kline & French Laboratories (Australia) Ltd v Department of Community Services and Health *(1990) 22 FCR 73 at 87; 95 ALR 87 at 102, "to identify with specificity, and not merely in global terms, that which is said to be the information in question".*

The consistent emphasis on specificity has significantly raised the threshold in this jurisdiction for satisfying the elements in a breach of confidence action.

6.2 Breach of confidence relating to software

Application of the principles in a software context is well illustrated by the case of *CA, Inc v ISI Pty Ltd* [2012] FCA 35. In that case CA's claim was based not on a readily identifiable single letter, client list or technical document. Rather, the case concerned complex programming information about complex systems that existed not only in the source code but also in the information about Datacom software, the way in which it works and in the information given to licensees to enable them and their employees and/or contractors to work with Datacom. Bennett J concluded that having regard to the nature of the information and the context of complex computer systems, she was satisfied that CA had sufficiently identified the confidential information: "the question arises whether, in the circumstances, a reasonable person standing in the shoes of [the relevant defendant's employees] would have realised that the information that they were receiving while working on ISI's behalf for Macquarie, Beta and other licensees was being given to them in confidence or, alternatively, whether the information was imparted for what was known or ought reasonably to have been known to be only for a particular purpose".[148]

As to the necessary quality of confidence, CA submitted that the CA Information:

- was created by research, the considerable expenditure of time and money and the application of skill and ingenuity;
- was inherently confidential by its nature;

---

147 (2013) 295 ALR 760 *per* Meagher, Barrett and Ward JJA.

148 *CA, Inc v ISI Pty Ltd* [2012] FCA 35 *per* Bennett J at [466], applying *Medic-Care per* Finn J at [636]–[637].

- was intrinsically valuable to CA and its customers (and would be to CA's competitors);
- was subject to stringent security measures limiting relevant disclosures, including various internal security measures and also external security measures implemented in respect of Datacom licensees, including licence notices on Datacom tapes, discs and the Key CA Manuals and the entering into of SLAs (or other forms of licence agreement) before a licensee is supplied Datacom and the Key CA Manuals.

Bennett J concluded that CA had:

... taken all reasonable and practical steps to ensure that the CA Information was created in confidence, maintained in confidence and understood to be confidential by those granted access to it. There is, by the nature of the CA Information, the need to disseminate it. However, as that dissemination is to persons engaged in an industry and in a section of that industry where it is generally understood that programs and certain information about them are confidential, this does not alter the inherent confidentiality of the CA Information.[149]

In then considering whether the information was imparted in circumstances of confidentiality, the judge noted that there is no single test for determining when the communication of confidential information will import an obligation of confidence: however, the "obligation may come into existence by reason of the terms of an agreement, or what is implicit in them, by reason of the nature of the relationship between persons, or by reason of the subject-matter and the circumstances in which the subject-matter has come into the hands of the person charged with the breach".[150]

7. Other means of protection

7.1 Registered design protection

The Designs Act 2003 (Cth) enables the owner of a 'new and distinctive'[151] design to apply for the exclusive right[152] to use the design through a registration system.

'New and distinctive' essentially means that a design must look different to the eye when compared with other products available on the market and that it must not be publicly disclosed or advertised before an application is filed.[153] The relevant prior art base[154] is:

- designs publicly used in Australia;

149 *Ibid*, at [463].
150 *Ansell Rubber Co Pty v Allied Rubber Industries Pty Ltd* [1967] VR 37 at 40 *per* Gowans J.
151 Section 15 of the Designs Act.
152 Section 10(1) of the Designs Act.
153 Section 15(1) of the Designs Act.
154 Section 16 of the Designs Act.

- designs published in a document within or outside Australia; and
- designs in relation to which each of the following apply, being: the design is disclosed in a design application, the design has an earlier priority date than the designated design and the first time that documents disclosing the design are made available for public inspection in Australia is on or after the priority date of the designated design. The priority date may be in another signatory country to the Paris Convention.

Design law protects the 'overall appearance' through 'visual features' of products, including shape, pattern, configuration and ornamentation.[155] The visual appearance can also serve a functional purpose: however, protection is not given for the functional aspect.[156] Design protection is not available for methods or principles of construction (although those aspects may be protected by patent law).

The Designs Act protects the visual appearance (not the function) of the design for up to 10 years.[157]

(a) *Copyright/design overlap*

Designs that could have been registered in the Designs Act will often be denied copyright protection due to contentious copyright/design overlap provisions. The overlap provisions are designed to prevent dual protection under the Designs Act and the Copyright Act: that is, copyright protection is lost where the artistic work is registered or capable of being registered as a design. The provisions also limit copyright protection for utilitarian three-dimensional products that are incapable of registration (eg, because they are not 'new' or 'distinctive').

Copyright in an artistic work in two dimensions may be infringed by an unauthorised reproduction of that work in three dimensions, and vice versa.[158] However, Part III Division 8 of the Copyright Act prevents the owner of copyright in an artistic work that has been exploited as a three-dimensional design from relying on copyright protection to prevent another person making an identical or a substantially identical product, if the design is:

- registered as a design;[159] or
- not registered but more than 50 products made to the design have been produced.[160]

---

155 Sections 5 and 7 of the Designs Act.
156 *Kestos Ltd v Kempat Ltd* (1935) 53 RPC 139; *Dalgety Australia Operations v F F Seeley Nominees Pty Ltd* (1985) 5 IPR 97.
157 Section 46 of the Designs Act.
158 Section 21(3) of the Copyright Act.
159 Section 75 of the Copyright Act.
160 Section 77 of the Copyright Act and Regulation 17 of the Copyright Regulations 1969 (Cth).

In other words, the overlap provisions provide a defence to infringement where the making of a product (ie, the 'non-infringing product') would otherwise involve an unauthorised reproduction of the artistic work. The overlap provisions also provide a defence to copyright infringement of the artistic work by reproducing the work in two dimensions provided the reproduction is derived from a three-dimensional product and was made in the course of making the non-infringing product.[161] This is intended to cover reverse engineering situations where a product manufacturer takes a product and uses it to create a plan drawing which is then used to manufacture their own product. It does not protect against an infringement which involves direct copying of an artistic work.

7.2 Semiconductor topologies

The Circuit Layouts Act 1989 provides a *sui generis* scheme of protection in Australia for original computer chip layouts, such as those used in audio players and games consoles, digital watches and calculators, computers and tablets, white goods and motor cars.

To be protected by the act in Australia the circuit must be an 'eligible layout' as defined by Section 5 of the Circuit Layouts Act, being:

- a circuit layout within the meaning of the act;
- original (ie, resulting from the designer's own creativity, skill and labour);
- the creator was either an Australian citizen, national or resident or an Australian corporation, or the circuit layout was first commercially exploited in Australia or in an eligible foreign country.[162]

Eligible foreign countries include all member countries of the World Trade Organisation (WTO).[163]

'Circuit layout' means "a representation, fixed in any material form, of the three-dimensional location of the active and passive elements and interconnections making up an integrated circuit". 'Material form', "includes any form of storage (whether visible or not) from which the layout, or a substantial part of it, can be reproduced". 'Integrated circuit' means "a circuit, whether in a final form or an intermediate form, the purpose, or one of the purposes, of which is to perform an electronic function, being a circuit in which the active and passive elements, and any of the interconnections, are integrally formed in or on a piece of material". 'Original' is not defined but interpretation of the term is affected by Section 11, which provides that a circuit layout shall

161 Section 77A of the Copyright Act.
162 Sections 5 and 10(a) of the Circuit Layouts Act 1989.
163 Regulation 3 (Eligible Foreign Countries) of the Circuit Layout Regulations.

be taken not to be original if its making involved no creative contribution by the maker or it was commonplace at the time it was made.

A circuit layout is taken to have been 'commercially exploited' if the layout or an integrated circuit made in accordance with it is by way of trade imported for the purpose of sale, hire or other distribution, sold, hired or otherwise distributed, or imported for the purpose of sale, hire or other distribution, in or into Australia or a WTO member country.[164] Accordingly, sale in a WTO member country of an electronic device incorporating the eligible circuit layout will be a relevant commercial exploitation that satisfies this condition.

The act grants automatically to the owner of an original circuit layout, for the duration of the 'protection period',[165] certain rights to:

- copy the layout, directly or indirectly, in a material form;
- make an integrated circuit in accordance with the layout (that is, a three-dimensional copy of the layout); and
- exploit the layout commercially in Australia.[166]

'Material form' covers masks but may not cover computer code used to fabricate an integrated circuit by direct laser.

The protection period is 10 years from first commercial exploitation. The first commercial exploitation must occur within 10 years of creation of the layout, or 10 years from when it was made. It follows that the maximum possible protection period is 20 years from the year of making an eligible layout.

7.3 Trademarks and unfair competition

The Trademarks Act 1995 (Cth) enables any person to register a mark or sign[167] used in connection with their goods or services in the course of trade.[168] The owner of a registered trademark has the exclusive right to use, and authorise others to use,[169] the trademark for 10 years.[170] Trademarks can then be renewed for potentially indefinite periods of 10 years.[171]

Although Australian law includes torts of unfair competition and in particular the tort of passing off, there has been limited development of jurisprudence as to these torts in the last few decades. To the extent that unfair competition is by way of a competitor taking action or otherwise facilitating or creating confusion by any means of its products or business with the business, products or the business activities of a competitor, generally the unfair practices

164 Section 8(1) of the Circuit Layouts Act 1989.
165 Section 5 of the Circuit Layouts Act 1989.
166 Section 17 of the Circuit Layouts Act 1989.
167 Section 6 of the Trademarks Act defines a 'sign'.
168 Section 17 of the Trademarks Act.
169 Section 7(3)–(5) of the Trademarks Act.
170 Section 72 of the Trademarks Act.
171 Section 77 of the Trademarks Act.

provisions in Part 3-1 of the Australian Consumer Law[172] may be invoked by a plaintiff against a commercial competitor.

For example, Section 29 of the Australian Consumer Law prohibits any person, in trade or commerce, in connection with the supply or possible supply of goods or services or in connection with the promotion by any means of the supply or use of goods or services from making a false or misleading representation:

- that goods are of a particular standard, quality, value, grade, composition, style or model or have had a particular history or particular previous use;
- that services are of a particular standard, quality, value or grade;
- that goods are new;
- that a particular person has agreed to acquire goods or services;
- that purports to be a testimonial by any person relating to goods or services;
- concerning a testimonial by any person, or a representation that purports to be such a testimonial relating to goods or services;
- that goods or services have sponsorship, approval, performance characteristics, accessories, uses or benefits;
- that the person making the representation has a sponsorship, approval or affiliation;
- with respect to the price of goods or services;
- concerning the availability of facilities for the repair of goods or of spare parts for goods;
- concerning the place of origin of goods;
- concerning the need for any goods or services;
- concerning the existence, exclusion or effect of any condition, warranty, guarantee, right or remedy; or
- make a false or misleading representation concerning a requirement to pay for a contractual right that is equivalent to a statutory right that a person has under Federal or State or Territory law or an automatic condition, warranty, guarantee, right or remedy.

Section 29 was adapted and re-enacted to replace Sections 52 and 53 of the Trade Practices Act 1974 (Cth) and the very extensive body of case law developed under those provisions continues to be relevant and applied in relation to Section 29.[173] As a result, it is generally not necessary to invoke the traditional common law torts of unfair competition, which relevantly protect a trader against:

---

172 The full text of the Australian Consumer Law is set out in Schedule 2 to the Competition and Consumer Act 2010 which is the new name of the Trade Practices Act 1974.

173 See eg, the many cases cited in Ray Steinwall, *Annotated Competition and Consumer Act 2010*, LexisNexis Butterworths, 2012.

- false allegations in the course of trade of such a nature as to discredit the business, products, or commercial activities of a competitor;
- indications or allegations the use of which, in the course of trade, is likely to mislead the public as to the nature, the manufacturing process, the characteristics and the suitability for their purpose;
- passing off, being a representation in the course of trade by one trader which damages the goodwill of another trader. To succeed in an action for passing off, a plaintiff must show three things: goodwill in its business; a misrepresentation by the other trader; damage to its goodwill as a result of the misrepresentation.

8. Counterfeiting

Under the Trademarks Act 1995 and the Copyright Act 1968, a trademark owner, an authorised user of a trademark, a copyright owner, or the owner's exclusive licensee may lodge a Notice of Objection with the Australian Customs Service objecting to the importation of:

- products that have a trademark applied to them which is substantially identical with or deceptively similar to an Australian registered trade mark that is nominated in the Notice of Objection;[174]
- nominated copyright material, if those copies would have constituted an infringement of copyright if the copies had been made in Australia by the importer.[175]

A Notice of Objection provides Customs with correct details of trademarks or copyright material and details of who owns the trademarks and/or copyright material.[176] Where a product or its packaging includes both an Australian registered trademark and copyright material, both Notices of Objection may be given.

A Notice of Objection is valid for four years, but may then be re-lodged.[177] Each notice may refer to multiple trademarks and/or copyright material.

If a Notice of Objection has been lodged and Customs becomes aware of a product shipment that Customs suspects is counterfeit and that it believes is intended for commercial use (sale, lease or hire), Customs may seize and hold the goods for 10 working days, enabling the objector to either commence legal proceedings or come to an arrangement with the importer. At the end of 10

---

174 Section 132 of the Trademarks Act.

175 Section 135 of the Copyright Act.

176 Sections 37 and 135(7) of the Copyright Act; Sections 132, 133, 133A and 141 of the Trademarks Act and Australian Customs Notice No 95/31. Importation of counterfeit products for sale, hire or distribution with knowledge that they are counterfeit is a criminal offence: Section 132AH of the Copyright Act and Section 148 of the Trademarks Act. An example of successful use of the seizure provisions is *Jemella Australia Pty Ltd v Vosiliatis* [2010] FCA 331 (23 March 2010). Many recipients of seizure notices voluntarily forfeit infringing products without legal action.

177 Section 132(4)–(5) of the Trademarks Act; Section 135(5) of the Copyright Act.

working days the products will be released to the importer unless the objector has initiated legal proceedings and has obtained an appropriate order from the court.[178]

If counterfeit products are voluntarily forfeited, Customs will dispose of the goods by destruction or donation to a charity or in accordance with an order of the court. Alternatively, if proceedings are commenced, the court will make an order regarding the goods at the conclusion of any legal proceedings that may be initiated.

Generally, a written undertaking to repay the Commonwealth's expenses of seizure will be required.[179] Customs may require payment of security sufficient to meet the expenses of seizing goods incurred by Customs. If Customs requires any payment for its expenses, Customs will not be obliged to seize any goods until payment has been made.[180]

Of course, to the extent that parallel importation provisions permit importation of some categories of products that were legitimately produced in the place of origin,[181] Customs is not able to seize such categories of products.

9. Enforcement

9.1 Obtaining information

The identity of a patent infringer may not be known until access is obtained to relevant information and documents not in the patentee's possession. In such circumstances, an order for pre-trial or 'preliminary' discovery can be sought that requires a person to give information about possible infringers to a patentee. The person concerned may not end up being a respondent, but merely someone who has relevant information or documents.

Preliminary discovery is available under the Federal Court Rules (2011) (Cth).[182] Division 7.3 of the Rules provides for two forms of preliminary discovery:

- to ascertain the identity or whereabouts of a prospective respondent;[183] and
- to assess whether or not to commence proceedings against that other person.[184]

The first type of application is available where a patentee may have a right to relief and seeks information from a person who is likely to know the identity

178 Sections 133–140 of the Trademarks Act; Section 135 of the Copyright Act; Regulations 22, 22A and 23 of the Copyright Regulations.

179 Section 133(3) of the Trademarks Act.

180 Section 135AJ of the Copyright Act.

181 Section 123 of the Trademarks Act; *Montana Tyres Rims & Tubes Pty Ltd v Transport Tyre Sales Pty Ltd* (1998) 41 IPR 301; Sections 37, 38, 102 and 103 of the Copyright Act; *Interstate Parcel Express Co Pty Ltd v Time-Life International (Nederlands) BV* (1977) 138 CLR 534.

182 Preliminary discovery is similarly available under rules for the Supreme Court and under Rules 5.2–5.3 of the Uniform Civil Procedure Rules (UCPR).

183 Rule 7.22.

184 Rule 7.23.

of the prospective respondent or have documents in his or her control that would help ascertain the prospective respondent's description.[185]

The second type is designed to enable a patentee to assess prospects and take a decision whether to sue the prospective respondent. To obtain such a discovery order, it must first be demonstrated to the court in accordance with Rule 7.3, that the applicant:

(a) reasonably believes that he or she may have the right to obtain relief in the Court from a prospective respondent whose description has been ascertained, and

(b) after making reasonable inquiries, does not have sufficient information to decide whether to start a proceeding in the Court to obtain that relief, and

(c) reasonably believes that:

(i) the prospective respondent has or is likely to have or has had or is likely to have had in the prospective respondent's control documents directly relevant to the question whether the prospective applicant has a right to obtain the relief; and

(ii) inspection of the documents by the prospective applicant would assist in making the decision.

Where trade competitors seek such orders, the court may make an order for preliminary discovery subject to an undertaking by the applicant to respect the confidentiality of discovered information.[186]

9.2 Availability of interim relief

Interim injunctions restraining infringing conduct can be obtained by a patentee through the Australian Courts, either with or without notice to the infringer. Interim injunctions are usually ordered for short periods of time, often just a few days, after which the parties are required to attend court to determine whether or not the injunction should be continued at an 'interlocutory' hearing.

The relevant principles governing the exercise of the court's discretion to grant interlocutory relief are well established. The applicant must demonstrate that:

- there is a serious question to be tried or a *prima facie* case such that if the evidence remains the same there is a probability that at a final hearing it would be entitled to relief;
- it will suffer irreparable harm for which damages will not be adequate compensation, unless an injunction is granted; and
- the balance of convenience favours the granting of an injunction.[187]

The above conditions should not be considered in isolation of each other but should be seen to apply concurrently with greater weight focused on one or

185 See eg, *Pfizer Ireland Pharmaceuticals v Samsung Bioepis AU Pty Ltd* [2017] FCAFC 193.
186 *Smith Kline Beecham plc v Alphapharm Pty Ltd* [2001] FCA 271.
187 *Hexal Australia Pty Ltd v Roche Therapeutics Inc* (2005) 66 IPR 325 at [17].

more conditions depending on each of the circumstances of the case.[188] In a patent case, the first factor requires the applicant to establish that there is a serious question to be tried on the issue of infringement despite whatever evidence has been led on the issue of invalidity.

Australian courts traditionally demonstrated reluctance to impose interlocutory restraints on an alleged infringer unless the patent in suit had either been judicially held to be valid or stood unchallenged for many years.[189] However, the recent approach of Australian courts in granting interlocutory injunctions have widened significantly to allow for greater application of this remedy, especially in relation to pharmaceutical-related patents.[190]

While interlocutory injunctions remain infrequently granted in technology patent cases, in the high-profile decision between Apple and Samsung,[191] the Federal Court continued its trend of granting interlocutory injunctive relief. In that case Apple was able to satisfy the above criteria and restrain Samsung's Galaxy Tab 10.1 shortly before Christmas. The injunction was subsequently overturned, however, by the Full Court on appeal six weeks later when the Full Court reassessed the balance of convenience.[192] The Full Court's decision was carefully reasoned and provides a timely reminder to applicants that such extraordinary remedies are to be reserved for extraordinary circumstances. It is also important to note that in such cases acquiescence towards infringement or delay in instituting an application for an interlocutory injunction may be a decisive factor in a court's refusal to exercise its discretion.[193]

Finally, it is possible in extreme cases for interim injunctions to be ordered on an *ex parte* application (ie, without notice) to either prevent a respondent removing assets or documents from Australia (sometimes referred to as a *Mareva* injunction) or to obtain civil search and seizure orders (sometimes referred to as an *Anton Pillar* order).

9.3 Unjustified threats

All Australian IP statutory regimes provide that it is unlawful to improperly threaten to initiate proceedings for infringement, even though it may be honestly believed that infringement is occurring.[194] By contrast, the mere notification of the existence of a patent, or a patent application, will not of itself constitute a threat of infringement proceedings.[195]

188 *Tidy Tea Ltd v Unilever Australia Ltd* (1995) 32 IPR 405 at 416.

189 See eg, *Beecham Group Ltd v Bristol Laboratories Pty Ltd* (1968) 118 CLR 618 at [6] (High Court).

190 See *Sigma Pharmaceuticals (Australia) Pty Ltd v Wyeth* (2009) 81 IPR 339; *Interpharma Pty Ltd v Commissioner of Patents* (2008) 79 IPR 261; *GenRx Pty Ltd v Sanofi-Aventis* [2007] FCA 1485; *Merck & Co inc v GenRx Pty Ltd* (2006) 70 IPR 286; and *Pharmacia Italia SpA v Interpharma Pty Ltd* (2005) 67 IPR 397.

191 *Apple Inc v Samsung Electronics Co Limited* [2011] FCA 1164.

192 *Samsung Electronics Co Limited v Apple Inc* [2011] FCAFC 156 *per* Foster, Dowsett and Yates JJ.

193 See eg, *Apple Inc v Samsung Electronics Co Ltd* [2011] FCA 1164 at [219]–[225].

194 See Sections 128–132 of the Patents Act; Section 202 of the Copyright Act 1968; Section 129 of the Trademarks Act 1995; Sections 77–81 of the Designs Act 2003; and Section 46 of the Circuit Layouts Act 1989. Also see *Skinner & Co v Perry* (1893) 10 RPC 1.

A person aggrieved by a threat of infringement proceeding may apply to a court for a declaration that the threat is unjustifiable, an injunction against the continuance of the threat, and the recovery of any damages sustained as a result of the threat.[196] A prerequisite to the operation of the unjustified threats provisions is that a person be 'threatened' with an action or proceedings which have not yet been brought.[197] Once infringement proceedings have actually commenced, the threat to commence proceedings no longer exists and the provisions do not generally apply.[198]

The following key considerations are relevant to establishing liability under the Patents Act's unjustified threats provisions:

- The 'aggrieved party' must establish that a threat of legal proceeding has been made in relation to an Australian patent and that the conduct complained of occurred within Australia.[199]
- A threat arises where the language used directly or impliedly conveys to a reasonable person that the author of the letter intends to bring infringement proceedings against the person threatened.[200]
- It is not, however, a threat to make:
  - a communication merely notifying a person of the existence of a patent or patent application;
  - a statement that any suggestion that the recipient is entitled to replicate the invention is not maintainable; or
  - a communication seeking confirmation that no improper or wrongful use or infringement of the patent has occurred.[201]
- Once it is established that a threat has been made, it is *prima facie* unjustifiable unless the person making it establishes that it was justifiable.[202] In this context, threats are only justifiable if the relevant conduct is shown to be actually infringing.[203]
- The grant of relief is discretionary.[204]

---

195 Section 131 of the Patents Act. Ultimately whether a particular communication amounts to a threat will depend on whether the language would convey to a reasonable person that the author intended to bring proceedings for infringement. See *JMVB Enterprises Pty Ltd v Camoflag Pty Ltd* (2005) 67 IPR 68; *Occupational and Medical Innovations Ltd v Retractable Technologies Inc* (2007) 73 IPR 312.

196 Section 128 of the Patents Act.

197 Any threat of infringement proceedings will automatically be regarded as unjustifiable where an innovation patent application has not been certified under Sections 128 and 129A(1)–(2) of the Patents Act.

198 *Nine Network Australia Pty Ltd v IceTV Pty Ltd* [2007] FCA 1172 at [230]–[231]. Similarly, whilst damages compensate for any loss that a person threatened has sustained as a result of those threats, such damage must flow from the making of the threats, not from any legal proceedings that may be subsequently commenced. See also *Australian Mud Company Pty Ltd v Coretell Pty Ltd* [2017] FCAFC 44 at [27].

199 *Australian Steel Co (Operations) Pty Ltd v Steel Foundations Ltd* (2003) 58 IPR 69 at [16]; *Townsend Controls Pty Ltd v Gilead* (1989) 16 IPR 469 at 474.

200 *U & I Global Trading (Australia) Pty Ltd v Tasman-Warajay Pty Ltd* (1995) 60 FCR 26 at 31.

201 Section 31 of the Patents Act; *Australian Steel Co (Operations) Pty Ltd v Steel Foundations Ltd* (2003) 58 IPR 69 at [17].

202 The onus is now on the defendant to prove that the acts complained of amounted or would amount to an infringement of a patent claim whose validity the applicant has not been able to challenge *per* Sections 129 and 129A(3) of the Patents Act.

In the decision of *Bell v Steele (No 3)*,[205] Collier J of the Federal Court provided examples of compensable damages flowing from threats, including: losses arising from the readiness of third parties to do business with the applicant;[206] losses arising from threats to the applicant's customers;[207] lost sales or lost potential sales;[208] and the cost to the applicant of instituting relief – such as a declaration – following receipt of the threat.[209] Collier J also added that the following were not compensable damages: damage to the threatened party flowing from the actual commencement of legal action by the party who made the threats; and rumours which may spring up but are not attributable to the party making the threat.

The Full Court most recently considered unjustified threats in a patent case[210] and emphasised the importance of demonstrating that the threats are causative of any loss claimed rather than as a result of legal proceedings.[211] The Full Court in that case reversed a multi-million-dollar award of damages and costs.

9.4 Remedies

Section 122 of the Patents Act provides for relief for infringement in the following terms:

> *(1) The relief which a court may grant for infringement of a patent includes an injunction (subject to such terms, if any, as the court thinks fit) and, at the option of the plaintiff, either damages or an account of profits.*
>
> *(1A) A court may include an additional amount in an assessment of damages for an infringement of a patent, if the court considers it appropriate to do so having regard to:*
>
> *(a) the flagrancy of the infringement; and*
>
> *(b) the need to deter similar infringements of patents; and*
>
> *(c) the conduct of the party that infringed the patent that occurred:*
>
> *(i) after the act constituting the infringement; or*
>
> *(ii) after that party was informed that it had allegedly infringed the patent; and*
>
> *(d) any benefit shown to have accrued to that party because of the infringement; and*
>
> *(e) all other relevant matters.*

---

203 An action for threats will therefore depend on whether the defendant can ultimately establish infringement, or whether the applicant may be able to preclude such a finding by showing the patent's invalidity. See Section 130 of the Patents Act for counterclaim for infringement and counter-counterclaim for revocation. Also see *U & I Global* at 32–33; *Doric v Lockwood* [2001] FCA 1977.

204 *Townsend Controls Pty Ltd v Gilead* (1989) 16 IPR 469 at 475; *Occupational and Medical Innovations Ltd v Retractable Technologies Inc* (2007) 73 IPR 312.

205 [2012] FCA 246.

206 *World of Technologies (Aust) Pty Ltd v Tempo (Aust) Pty Ltd* [2007] FCA 114 at [124].

207 *Avel Pty Ltd v Intercontinental Grain Exports Pty Ltd* (1996) 143 ALR 533 at 542.

208 *Cowan, B S & Anor v Avel Pty Ltd* [1995] FCA 546 at [25].

209 *Nine Network Australia Pty Ltd v IceTV Pty Ltd* [2007] FCA 1172 at [234].

210 *Australian Mud Company Pty Ltd v Coretell Pty Ltd* [2017] FCAFC 44.

211 *Id* at [27].

(2) On the application of either party, the court may make such order for the inspection of any thing in or on any vehicle, vessel, aircraft or premises, and may impose such terms and give such directions about the inspection, as the court thinks fit.

An injunction is the principal relief available to a patentee to protect its monopoly. The infringer cannot then conduct infringing activities without first obtaining a licence from the patentee. Damages compensate the patentee for the loss suffered as a result of the infringement,[212] whereas an account of profits will require an infringer that has sold or benefited financially from infringing goods, to pay to the patentee a sum equivalent to the profits wrongfully made.[213]

Although additional damages are available in Australia under Section 122(1A), there is scant authority for the grant of such an award.[214] By contrast, additional damages are routinely awarded in copyright cases[215] where flagrancy is demonstrated. Additional damages have also been awarded in a designs case.[216]

10. Hot topics

10.1 Computer-implemented business methods

Patents for business methods remain a vexed issue under Australian law. While business methods that are implemented in a computer or other physical environment are patentable,[217] simply interpolating the operation of a computer into a process will not necessarily make a business method patentable.

There is an entire spectrum of possible uses of technology in the implementation of new processes, including business methods. Whether a computer-implemented process is eligible for patent protection will therefore depend where it falls within that continuum between, on the one hand, being an incidental or insignificant use of technology, and on the other, it having a "concrete effect or phenomenon or manifestation or transformation".

See further our detailed discussion in Section 2.1 above.

10.2 Productivity Commission Review

Following the reforms introduced under "Raising the Bar" in 2013, the

212 *General Tire & Rubber Co v Firestone Tyre & Rubber Co Ltd* [1976] RPC 197.

213 Profits made by the defendant as a result of the infringement will be assessed according to the principles in *Colbeam Palmer Ltd v Stock Affiliates Pty Ltd* (1968) 122 CLR 25 and *Dart Industries Inc v Décor Corp Pty Ltd* (1993) 179 CLR 101.

214 *Pacific Enterprises (Aust) Pty Ltd v Bernen Pty Ltd* [2014] FCA 1372; 321 ALR 715 at [13]–[14] and *Gram Engineering Pty Ltd v Oxworks Pty Ltd* [2019] FCA 689 [113]–[124].

215 Under Section 115(4) of the Copyright Act (1968) (Cth); see eg, *Raben Footwear Pty Ltd v Polygram Records Inc* (1997) 75 FCR 88; *Aristocrat Technologies Australia Pty Ltd v DAP Services (Kempsey) Pty Ltd (in liq)* [2007] FCAFC 40; *Allam v Aristocrat Technologies Australia Pty Ltd* [2012] FCAFC 34 (Special Leave granted); *Facton Ltd v Rifai Fashions Pty Ltd* [2012] FCAFC 9.

216 Under Section 75(3) of the Designs Act 2003 (Cth); see *GM Holden Ltd v Paine* [2011] FCA 569.

217 *Welcome Real-Time SA v Catuity Inc* [2001] FCA 445.

Australian government asked the Productivity Commission in August 2015 to conduct an inquiry into Australia's IP system. In particular, the commission was asked to consider whether the current arrangements provide an appropriate balance between access to ideas and products, and encouraging innovation, investment and the production of creative works.

In December 2016, the Productivity Commission published its Inquiry Report on "Intellectual Property Arrangements".[218] The Inquiry Report came after extensive consultation with various stakeholders including IP professionals, economists and IP owners and users.

The Productivity Commission's key recommendations to patents[219] included:

- incorporating an objects clause in the Patents Act 1990;[220]
- a further raising of the inventive step standard to equal or above the European patent system and including a statement in the Explanatory Memorandum that a 'scintilla' of invention is an insufficient threshold. The Productivity Commission has suggested Australia adopt the European 'obvious to try' test;[221]
- abolishing the innovation patent system in its entirety;[222]
- a requirement to identify the technical features of the invention;[223]
- raising selected patent fees to discourage maintenance of 'low-value' patents, including specific recommendations that renewal fees should rise each year at an increasing rate, and to increase fees for applications with many claims.[224]

The Productivity Commission considered that the above recommendations would increase the quality of Australian patents such that it is on a par with the European patent system and reduce 'low-value' patents. From a patentee's perspective, the likely effect of these recommendations is that it will make patents more difficult and costly to obtain, enforce and maintain. Despite submissions from the profession[225] to keep the innovation patent system and that the change to inventive step was premature so soon after the Raising the Bar changes, the Productivity Commission has maintained its position and it is

218 The Australian Productivity Commission's final report on its Inquiry into "Australia's Intellectual Property Arrangements" was publicly released on 20 December 2016 and is available at: https://www.pc.gov.au/inquiries/completed/intellectual-property/report (Inquiry Report).

219 See chapters 7 and 8 of the Inquiry Report.

220 Recommendation 7.1 of the Inquiry Report.

221 Recommendation 7.2 of the Inquiry Report.

222 Recommendation 8.1 of the Inquiry Report.

223 Recommendation 7.3 of the Inquiry Report. Section 40 of the Patents Act presently requires that claims must define the invention, be clear and succinct and supported by the matter disclosed in the specification, not rely on references to descriptions or drawings unless absolutely necessary to define the invention, and relate to one invention only.

224 Recommendation 7.4 of the Inquiry Report.

225 For example, Institute of Patent and Trademark Attorneys (IPTA) of Australia.

anticipated that the Australian government will soon introduce draft legislation to this effect.

Regarding software and business methods patents, the Productivity Commission did not recommend any blanket exclusion to patentability as possibly thought but rather considered that the combined effect of the recommendations (inclusion of an objects clause, raising the inventive step requirement, and requiring the identification of technical features to be present in the claims) will better balance the patent rights of software innovators and users.

Canada

Kendrick Lo
Law Office of Kendrick Lo
Joanna Ma
Bereskin & Parr LLP

Canada is the world's second largest country by total area, and shares the largest land border in the world with its neighbour to the south, the United States. Canada is the United States' top trading partner, and the recipient of almost one-fifth of the United States' total goods exports. The annual total trade in goods between Canada and the United States has been estimated at over US$500 billion.[1]

Canada's economy is one of the world's largest, and while it is highly reliant on the country's natural resources, the information and communications technologies (ICT) sector also plays a significant role in the country's economic growth. According to Industry Canada, the ICT sector is made up of nearly 40,000 companies, approximately 90% of which are identified as being in the software and computer services industries.[2] In 2016, the annual growth of this sector outpaced that of the overall Canadian economy by a factor of two. This has been driven largely by service industries, which include software and computer services, and communications services. The sector also continues to be the largest performer of research and development (R&D), accounting for 30.8% of all private sector R&D expenditures in the country.[3]

Canada is probably best known by IP practitioners in the electronics and software arena as the home of BlackBerry[4] and, more recently, as a hotspot for pioneering research and talent in the burgeoning field of artificial intelligence (AI).

Significant changes to the Canadian Patent Act and Rules took effect on 30 October 2019.[5] This revised edition discusses the major changes and the practical effect of these changes.

1 US Department of Commerce International Trade Administration, Top US Trade Partners – Ranked by 2017 US Total Export Value for Goods, www.trade.gov/mas/ian/build/groups/public/@tg_ian/documents/webcontent/tg_ian_003364.pdf.

2 Industry Canada, Canadian ICT Sector Profile 2016, www.ic.gc.ca/eic/site/ict-tic.nsf/eng/h_it07229.html.

3 *Ibid.*

4 The trademark Blackberry® is owned by BlackBerry Limited and is registered in the United States and may be pending or registered in other countries. The author/publisher is not endorsed by, sponsored by, affiliated with or otherwise authorised by BlackBerry Limited.

5 Canada Gazette, Part II, Volume 153, Number 14, Patent Rules: SOR/2019-251 (10 July 2019), www.gazette.gc.ca/rp-pr/p2/2019/2019-07-10/html/sor-dors251-eng.html.

1. National patent procedure

1.1 Filing

In Canada, patents are granted on a 'first-to-file' basis. Canada is an 'absolute novelty' jurisdiction; therefore, an application for an invention must be filed prior to any public disclosures of the invention by third parties. There is, however, a one-year grace period that may be applicable with respect to an inventor's own prior disclosures.[6]

It is expected that most non-Canadian applicants wishing to obtain a Canadian patent will either submit an application claiming priority directly from an earlier-filed application under the Paris Convention, or enter the national phase in Canada from a Patent Cooperation Treaty (PCT) application. Different procedural considerations will apply depending on how applicants elect to proceed in Canada.

(a) Canadian national phase entry for Patent Cooperation Treaty applications

In general, the time limit for entering the national phase in Canada is 30 months from the earliest priority date of a PCT application designating Canada, regardless of whether or not a demand for international preliminary examination has been filed during the international phase. However, Canada is one of a small number of PCT signatories that allow potential applicants to enter the national phase 'late' if certain criteria are met.

For PCT applications filed before 30 October 2019,[7] applicants may enter the national phase up to 12 months beyond the default 30-month deadline upon payment of a nominal late entry surcharge (C$200, 2019),[8] effectively extending the deadline for national phase entry in Canada to 42 months from the earliest priority date of the PCT application.[9] For PCT applications filed on or after 30 October 2019, applicants may still enter the national phase up to 42 months from the earliest priority date of the PCT application with the payment of the late entry surcharge, but applicants are also required to state that the failure to timely enter the national phase was unintentional. No evidence is required at filing but this statement is reviewable by the Federal Court. As of the publication date of this edition, the circumstances that constitute 'unintentional' have yet to be clearly defined. Great caution, therefore, must be applied when deciding whether to delay Canadian national phase entries beyond the 30-month deadline for PCT applications filed on or after 30 October 2019.

Restoration of priority is available for national phase entries of PCT

---

6 See 2.2 below.
7 Patent Rules, SOR/2019-251, Sections 154(3), 210(3) and 234.
8 Patent Rules, SOR/2019-251, Tariff of Fees – Patents, Part 4 – Fees in Respect of International Applications, Item 22.
9 World Intellectual Property Organization (WIPO), Time Limits for Entering National/Regional Phase under Patents Convention Treaty Chapters I and II, www.wipo.int/pct/en/texts/time_limits.html.

applications filed on or after 30 October 2019. This change brings Canada into compliance with the Patent Law Treaty. Restoration of priority will be available if the priority was restored during the international phase,[10] or at the earlier of 16 months from the earliest priority date, or one month from the national phase entry date, along with a statement indicating that the failure to timely file the application was unintentional.[11] However, this statement is reviewable by the Federal Court and the courts could later deny the restoration of priority.

The effective Canadian filing date of a PCT national phase application is the international filing date, and not the date on which the national phase is entered in Canada. Where an applicant enters the national phase in Canada more than two years after the international filing date, at least one maintenance fee will also be due at the time of national phase entry.

To enter the national phase, an applicant must provide,[12] before expiry of the applicable time limit (no later than 42 months from the earliest priority date): a copy of the international application if the international application has not yet been published, a translation of the international application into English or French if the international application is not in English or French, the prescribed fee for national phase entry (C$400, 2019),[13] and any late entry or maintenance fees where applicable.[14] For PCT applications filed on or after 30 October 2019, a certified copy of each priority document is required to be submitted to the Canadian Intellectual Property Office (CIPO) if the certified copy was not submitted to WIPO during the International Phase.[15]

Although there are advantages to having a transfer document or assignment (eg, as executed by the inventor) registered against the application, it is not a requirement to enter the national phase in Canada. For PCT applications with a filing date before 30 October 2019, a declaration (which need not be signed by the applicant or the inventor) that states that the applicant is the legal representative of the inventor will typically be prepared where the applicant is not the inventor,[16] and this declaration may be filed at the time of national phase entry. For PCT applications filed on or after 30 October 2019, a statement of entitlement or inventorship is required.[17]

In any event, entering the national phase of a PCT application in Canada is a relatively quick and efficient process, as it is not necessary to have any documents executed by the applicant or inventor(s) in order to effect national phase entry.

---

10 Patent Rules, SOR/2019-251, Section 162.
11 Patent Rules, SOR/2019-251, Section 77(1).
12 Patent Rules, SOR/2019-251, Section 154(1).
13 Patent Rules, SOR/2019-251, Tariff of Fees – Patents, Part 4 – Fees in Respect of International Applications, Item 21.
14 Patent Rules, SOR/2019-251, Section 154(2) for 30-month entries; Section 154(3) for late entries.
15 Patent Rules, SOR/2019-251, Section 74(1).
16 Patent Rules, SOR/2019-251, Sections 225–227.
17 Patent Rules, SOR/2019-251, Section 54(2).

(b) ***Direct national patent applications***

The requirements for obtaining a filing date for direct filings or filings claiming Paris Convention priority were simplified by the Patent Rules that came into force on 30 October 2019. The applicant can secure a filing date by providing:

- an indication that the granting of a Canadian patent is sought;
- the name of the applicant;
- contact information of the applicant; and
- a document, in any language, that appears to be a description.[18]

Under the new Patent Rules, the applicant can initially obtain a filing date even without payment of the filing fee and also without submitting an English or French description. However, an English or French translation must be submitted within two months of a notice issued by CIPO requiring the translation.[19] The translation would then replace the originally filed description.[20] Despite the more flexible filing requirements, applicants should note that a timely and accurate translation is critical since no extension of time is available for submitting the translation[21] and any subsequent amendments must be reasonably inferred from both the original description and the translation.[22] If the filing fee was not paid with the original filing submission, the filing fee (C$400, 2019),[23] along with a late fee (C$150, 2019),[24] are due within three months of a notice issued by CIPO;[25] otherwise, the application is considered to be withdrawn.[26]

In order to claim priority to an earlier-filed application under the Paris Convention, the Canadian application must be filed within 12 months of the filing date of the earlier-filed application. For patent applications filed on or after 30 October 2019, it is possible to simply refer to the earlier-filed application (eg, no specification needs to be submitted), along with a statement that the application is being filed by reference, to obtain a filing date.[27] Details of the Convention priority claim, including the country or office of filing, filing date and serial number of the application are required for securing the priority

18 Patent Rules, SOR/2019-251, Section 71. Note that prior to 30 October 2019, to obtain a filing date for direct Canadian patent applications, the applicant must provide: an indication in English or French that the granting of a Canadian patent is sought; the name of the applicant; the address of the applicant or of a patent agent representing the applicant; a document in English or French that on its face appears to describe an invention; and the prescribed filing fee (see Patent Rules, SOR/96–423, Section 27.1(1)(a), which was repealed when Patent Rules, SOR/2019-251 came into force on 30 October 2019).

19 Patent Rules, SOR/2019-251, Section 15(3) and (4).

20 Patent Rules, SOR/2019-251, Section 15(5).

21 Patent Rules, SOR/2019-251, Section 15(7).

22 Patent Act, RSC 1985, c P–4, Section 38.2(2) as amended 2015, c.36, Section 58, in force 30 October 2019.

23 Patent Rules, SOR/2019-251, Tariff of Fees – Patents, Part 3 – Fees in Respect of Applications for a Patent, Item 6.

24 Patent Rules, SOR/2019-251, Tariff of Fees – Patents, Part 3 – Fees in Respect of Applications for a Patent, Item 7.

25 Patent Rules, SOR/2019-251, Section 66(1).

26 Patent Rules, SOR/2019-251, Section 66(2).

27 Patent Act, RSC 1985, c P–4, Section 27.01(1) as amended 2014, c.39, Section 121, in force 30 October 2019; Patent Rules, SOR/2019-251, Section 67.

claim. The applicant must provide the details of the earlier-filed application by the earlier of 16 months from the earliest priority date or four months after the Canadian filing date (whichever date is later); or the date on which the applicant requests early publication.[28]

Under the old regime, it was not necessary to file a certified copy or other evidence of the filing of the earlier-filed application. However, for any application on which a priority claim is made on or after 30 October 2019, CIPO will require a certified copy of the earlier-filed application by the later of 16 months from the earliest priority date or four months after the Canadian filing date.[29]

Another significant change resulting from the new Patent Rules is that restoration of priority is now available for Canadian applications filed on or after 30 October 2019. Like with the restoration of priority available to national phase entries of PCT applications filed on or after 30 October 2019, this change places Canadian patent law in compliance with the Patent Law Treaty. Restoration of priority will be available for two months after the 12-month period has expired. To restore priority, the applicant must submit a statement indicating that the failure to timely file the application was unintentional.[30] However, this statement will be reviewable by the Federal Court. As will be described in 2.2 below, the availability of restoration of priority in Canada results in a limited extension to the one-year grace period.

Again, although there are advantages to having a transfer document or an assignment (eg, as executed by the inventors) registered against the application, it is not a requirement to obtain a filing date. Under the new Patent Rules, a statement of entitlement or inventorship is required for applications filed on or after 30 October 2019.[31] It is not necessary to have documents executed by the applicant or inventor(s) in order to obtain a filing date in Canada.

1.2 Examination system

Canada operates under a full examination system. Examination by CIPO will not be conducted until an explicit request for examination is made and the requisite fee is paid (C$800, 2019).[32] Any person, even a third party, may request examination of an application.[33]

---

28 Patent Rules, SOR/2019-251, Section 73(1) and (2). For applications filed before 30 October 2019, the applicant must provide the details of the earlier-filed application by 16 months from the earliest priority date (Patent Rules, SOR/2019-251, Section 188(1)).

29 Patent Rules, SOR/2019-251, Section 74(1) and (2).

30 Patent Rules, SOR/2019-251, Section 77(2); Patent Act, RSC 1985, c P–4, Section 28.4(6), 2014, c.39, Section 125, as amended by 2018, c.27, Section 206, in force 30 October 2019.

31 Patent Rules, SOR/2019-251, Section 54(2). For applications filed before 30 October 2019, a declaration that states that the applicant is the legal representative of the inventor will typically be prepared where the applicant is not the inventor, and this declaration is generally provided at the time of filing (SOR/2019-251, Sections 225–227).

32 Patent Rules, SOR/2019-251, Tariff of Fees – Patents, Part 3 – Fees in Respect of Applications for a Patent, Item 10,

33 Patent Act, RSC 1985, c P–4, Section 35(1).

The new Patent Rules shortened the timeframe for requesting examination of applications from five years to four years for PCT patent applications and direct Canadian patent applications filed on or after 30 October 2019.[34] The Patent Office may require that the request be made earlier.[35] Once examination is requested, a first office action can typically be expected within two to three years, although response times from the Patent Office will vary.

The new Patent Rules also introduce a more complex reinstatement procedure if the deadline for requesting examination is missed. When examination requests due on or after 30 October 2019 are missed, CIPO will issue a notice requiring the examination fee along with a late fee within two months of the notice.[36] However, when the date set by the notice is missed, the application becomes abandoned.[37] In the previous regime, the application could be reinstated within 12 months of abandonment as of right, but that may not be the case under the new Patent Rules. Under the new Patent Rules, reinstatement of an application at or after six months from when the examination request was due requires a statement outlining the reasons for the failure to file the request along with a showing of due care.[38] The Commissioner would determine whether the failure satisfies the due care standard. CIPO has indicated that they plan to adopt a practice consistent with the PCT Receiving Office Guidelines[39] which indicate that the 'due care' standard requires "all measures which a reasonably prudent applicant would have taken". It is likely that the standard will vary with how sophisticated the applicant is, and likely also whether the applicant retained an agent.

Third-party rights also begin to be available starting at six months from the examination request due date.[40] Third-party rights provide exemptions for acts that would otherwise constitute infringement during prescribed periods.

The effect of missing the request for examination due date under the new Patent Rules can, therefore, be quite detrimental – great care should be taken to ensure the request for examination due date, and related due dates, are properly docketed.

During examination, the examiner will assess the patentability of the claimed subject matter. The claimed subject matter must be novel, useful, non-obvious, and eligible for patent protection. The examiner will raise any

34 Patent Rules, SOR/2019-251, Sections 81(1)(a), 188(1), 197.

35 Patent Act, RSC 1985, c P–4, Section 35(6), as amended 2014, c.39, Section 128, in force 30 October 2019.

36 Patent Act, RSC 1985, c P–4, Section 35(3), as amended 2014, c.39, Section 128, in force 30 October 2019; Patent Rules, SOR/2019-251, Section 82 and Tariff of Fees – Patents, Part 3 – Fees in Respect of Applications for a Patent, Item 11.

37 Patent Act, RSC 1985, c P–4, Section 73(1)(d), as amended 2014, c.39, Section 137, in force 30 October 2019.

38 Patent Act, RSC 1985, c P–4, Section 73(3)(b), as amended 2014, c.39, Section 137, in force 30 October 2019; Patent Rules, SOR/2019-251, Section 135(1)(b).

39 Patent Cooperation Treaty (PCT) receiving Office Guidelines, Paragraph 166J, www.wipo.int/export/sites/www/pct/en/texts/pdf/ro.pdf.

40 Patent Rules, SOR/2019-251, Section 128(b).

objections in an examination report, along with any informalities or other issues that the applicant will be required to address in order to advance the prosecution of the application or to put the application in order for grant.

(a) *Advanced and accelerated examination*

If a request for examination has been filed and the application has been published, any person may request 'advanced' examination upon payment of a fee (C$500, 2019).[41] The request will be granted upon a determination by the commissioner that failure to advance the application is likely to prejudice the rights of the person making the request.[42] In the current practice, it does not appear that much is required for the commissioner to make this determination beyond receiving a broad submission that "a failure to advance the application for examination is likely to prejudice the rights of the [person making the request]". No other supporting documentation (eg, search results, prior art listings, details of corresponding applications) is required to accompany the request for advanced examination. A first office action in respect of an application subject to advanced examination may be expected to issue in a matter of months, rather than years.

An applicant can also request advanced examination of an application by declaring that the patent application relates to 'green' technology. Specifically, the applicant must file a declaration indicating that the application "relates to technology the commercialization of which would help to resolve or mitigate environmental impacts or to conserve the natural environment and resources". There are no additional fees required for requesting advanced examination if the patent application relates to green technology.

For office actions issued on or after 30 October 2019 in respect of applications under advanced examination, the new Patent Rules introduced a longer response deadline (four months from the date of the office action, instead of the previous three months).[43]

CIPO is also a participant in a number of Patent Prosecution Highway pilot projects with various patent offices and organisations worldwide, currently including, for example: the Japan Patent Office, the National Intellectual Property Office Administration of the People's Republic of China, the United States Patent and Trademark Office, and the European Patent Office. When certain requirements are met, applicants can request an accelerated examination based on an indication of patentable subject matter that has been previously made by a partner office[44] in respect of a corresponding application.

---

41 Patent Rules, SOR/2019-251, Tariff of Fees – Patents, Part 3 – Fees in Respect of Applications for a Patent, Item 12.

42 Patent Rules, SOR/2019-251, Section 84(1) and (2).

43 Patent Rules, SOR/2019-251, Sections 86(2) and 131(1).

44 CIPO, Patent Prosecution Highway, www.ic.gc.ca/eic/site/cipointernet-internetopic.nsf/eng/h_wr02160.html.

(b) Corresponding applications

While Canadian examiners are entitled to perform their own independent prior art searches, it is extremely common for documents that have been cited in the prosecution of corresponding applications filed outside of Canada – particularly corresponding applications filed in the United States Patent and Trademark Office or the European Patent Office – to be relied upon by examiners in Canadian office actions.

The examiner may also formally require an applicant to provide details of prior art documents cited in the prosecution of corresponding foreign applications. It is extremely important that the requisition be acknowledged in reply even if no such documents are available (eg, even in situations where no office actions have yet to issue in a foreign jurisdiction or no corresponding applications have been filed). Failure to explicitly respond to such a requisition during the prosecution of an application is a ground for invalidating any patent that may subsequently issue from the application.[45]

Furthermore, in practice, allowances of claims by the United States or European Patent Offices in particular are often looked upon favourably by CIPO. Therefore, it may be advisable to bring such allowances to the attention of the Canadian examiner at the earliest possible opportunity during examination of the Canadian application. That said, more recently, CIPO has taken a relatively stricter stance when it comes to subject matter rejections, relative to the United States (see 2.1 below).

(c) Time limits for responding to office actions

The new Patent Rules shortened the timeframe for replying to office actions issued by CIPO. The due date for replying to office actions issued on or after 30 October 2019 is four months from the issue date.[46] The due date can be extended up to six months from the date of the office action but a request must be made prior to the original due date,[47] along with payment of an extension fee (C$200, 2019).[48] Similar to the previous Patent Rules, the application will be abandoned when the due date is not met and a request for reinstatement along with the reinstatement fee are required to reinstate the application.[49]

Furthermore, while it is not uncommon for examiners in jurisdictions outside of Canada to raise 'final' rejections after a handful of office actions, in

---

45 *DBC Marine Safety Systems Ltd v Canada (Commissioner of Patents)*, 2008 FCA 256, 69 CPR (4th) 189.

46 Patent Rules, SOR/2019-251, Sections 86(2) and 131(1). Note that, for office actions issued before 30 October 2019, the typical deadline is six months from the date of the office action. If a response is not filed within six months, the application will be abandoned, but can be reinstated as a matter of right as long as a response is filed with a request for reinstatement and a modest reinstatement fee within one year of the date of abandonment. This effectively gives applicants an 18-month period in which to respond to Canadian office actions.

47 Patent Rules, SOR/2019-251, Section 3(1).

48 Patent Rules, SOR/2019-251, Tariff of Fees – Patents, Part 1 – Fee in Respect of Extension of Time, Item 1.

49 Patent Rules, SOR/2019-251, Section 135(1)(a).

Canada it is relatively rare for a final action to be issued unless it seems abundantly clear that the applicant and the examiner have reached an impasse.

In view of the above factors, applicants are able to exercise a reasonable degree of control over how quickly they would like the prosecution of their Canadian application to proceed. This allows for considerable flexibility, particularly if applicants expect that a favourable outcome in respect of a corresponding application is imminent.

(d) No continuation or terminal disclaimer practice in Canada

Unlike the United States, in which applicants may file continuation applications, and where applicants can typically overcome any subsequent obviousness-type double patenting objections in respect of a parent application or patent by filing a terminal disclaimer, there is no continuation or terminal disclaimer practice in Canada. Furthermore, the double patenting rules in Canada are relatively strict.

Therefore, applicants are generally advised to include any continuation-type or divisional-type claims in a single Canadian application before the application is allowed; should the examiner choose to raise a unity of invention objection, non-elected claims may then be pursued in a divisional application, and the risks of facing a subsequent double patenting rejection can be minimised. Pursuing claims in a 'voluntary' divisional application (ie, not in response to an explicit unity objection) is not recommended.

There is also no equivalent to a US-style 'continuation-in-part' application in Canada. A new Canadian application needs to be filed if the applicant wishes to add new matter to an existing Canadian application. The existing application may then be citable against the new application, depending on the filing date of the new application.

1.3 Costs

The following chart illustrates an example of official fees payable during the prosecution of a typical Canadian application. This listing illustrates government fees only (2019),[50] and does not include patent agent or professional fees that may be assessed to prepare the application, official documents, office action responses, or other supporting documentation, for example.

Filing fee or national phase entry fee	C$400
Registration of transfer/document (eg, assignment)	C$100
Requesting examination	C$800
Final fee	C$300
Maintenance fees	C$100–450 per year (varies with each anniversary date).

50 Patent Rules, SOR/2019-251, Tariff of Fees – Patents.

In theory, individual inventors, certain small businesses,[51] and universities may be entitled to reduced fees as a 'small entity'.[52] However, a number of court cases[53] illustrate the difficulties involved with accurately assessing whether or not a given applicant is eligible for small entity status. An erroneous prior claim of small entity status during the prosecution of an application in Canada is a ground for invalidating any patent that issues from the application.

Previous amendments to the Patent Rules attempted to address concerns that even a claim of small entity status made in good faith might nevertheless be fatal to a patent if the status is later determined to have been made in error. Despite those amendments, the risks associated with an incorrect assessment are still generally considered by Canadian practitioners to be extremely high. Most patent agents continue to strongly advise applicants against claiming small entity status, particularly since the potential cost savings afforded to small entities in Canada are relatively small.

1.4 Privileged communications

Recent amendments to the Patent Act introduced statutory protection over communications between a patent agent and their clients. Any communication between a patent agent and the client that is intended to be confidential and is made for the purpose of seeking or giving advice regarding the protection of an invention is deemed to be privileged.[54] This protection is intended to be similar to that afforded between a lawyer and his or her client. The agent-client privilege can extend to communications between foreign patent agents and their clients provided that the country of the foreign patent agent also recognises such privileged communication.

2. Examination of software patents

2.1 Patentable subject matter

Section 2 of the Patent Act defines 'invention':

> *'invention' means any new and useful art, process, machine, manufacture or composition of matter, or any new and useful improvement in any art, process, machine, manufacture or composition of matter*
>
> and Section 27(8) of the Patent Act states:
>
> *No patent shall be granted for any mere scientific principle or abstract theorem.*

---

51 A 'small entity' is a business that employs 50 or fewer employees, or is a university, and is not controlled by an entity (other than a university) that employs more than 50 employees, or has transferred or licensed (or are under an obligation to do so) to an entity (other than a university) that employs more than 50 employees. This is different from the definition of small entity in the United States, for example.

52 Patent Rules, SOR/96–423, Section 3.01.

53 See eg, *Barton No-Till and Flexi-Coil v Dutch Industries*, 2003 FCA 121, 24 CPR (4th) 157, reversing in part 2001 FCT 879, 14 CPR (4th) 499; *Johnson & Johnson Inc v Boston Scientific Ltd*, 2004 FC 1672, 37 CPR (4th) 385.

54 Patent Act, RSC 1985, c P–4, Section 16.1(1).

A list of specifically excluded subject matter is not provided in the Patent Act or the Patent Rules. Traditionally, it has been left to the Commissioner of Patents to determine what he or she believes is to be explicitly excluded from the definition of 'invention', taking judicial interpretation into account. Guidelines as to what may be excluded are then outlined in documents published by CIPO.[55]

The law regarding the patentability of computer software, and computer-implemented 'business methods', remains in a state of flux in Canada.

In November 2011, the Federal Court of Appeal had delivered a judgment[56] in respect of the application filed in Canada in 1998 by Amazon.com Inc (Amazon) for what has now become well known as the 'one-click' online purchasing process.[57] The Commissioner of Patents initially refused to grant the patent. This decision was appealed to the Federal Court, which quashed the commissioner's decision and effectively directed the commissioner to grant the patent.

Upon further appeal to the Federal Court of Appeal, the judgment of the lower court was set aside. However, the appellate court refused to specifically include in its ruling an opinion on whether the patent ought to be granted or refused. Instead, the Federal Court of Appeal directed the commissioner to re-examine the patent application in accordance with guidance as provided in its decision. Within one month, CIPO then issued a notice of allowance for Amazon's application, and the application subsequently issued to patent. Notably, no explicit reasons for grant were given.

The allowance of Amazon's application for the one-click process would, on its face, appear to have been a favourable outcome to prospective applicants wishing to protect computer software inventions, and even to what might be characterised as computer-implemented 'business methods' in Canada. In fact, the Federal Court of Appeal explicitly indicated that a purposive construction of claims[58] is to be undertaken, with a mind open to the possibility that a novel business method may be an essential element of a valid patent claim.[59]

However, given the paucity of jurisprudence in this area, and the fact that the Federal Court of Appeal did not take the opportunity in *Amazon* to provide a clear analytical framework, patent practitioners, examiners and prospective applicants were left with little guidance on how to determine whether a given software or business method invention (computer-implemented or otherwise) is

---

55 See eg, CIPO, Manual of Patent Office Practice (MOPOP), online: www.ic.gc.ca/eic/site/cipointernet-internetopic.nsf/eng/h_wr00720.html; Practice Notices, online: www.ic.gc.ca/eic/site/cipointernet-internetopic.nsf/eng/h_wr00292.html; and Practice Guidelines, eg, online: www.ic.gc.ca/eic/site/cipointernet-internetopic.nsf/eng/wr03627.html.

56 *Canada (Attorney General) v Amazon.com, Inc*, 2011 FCA 328, 97 CPR (4th) 171.

57 Canadian Application No 2,246,933: *Method and system for placing a purchase order via a communication network.*

58 Purposive construction is discussed further at 2.3 below.

59 *Canada (Attorney General) v Amazon.com, Inc*, 2011 FCA 328 at paragraph 63, 97 CPR (4th) 171.

patent-eligible subject matter. The Patent Act states that no patent shall be granted for any mere scientific principle or abstract theorem,[60] but the question of exactly what would or would not constitute a 'mere' scientific principle or abstract theorem remains open.[61]

CIPO attempted to provide greater certainty by issuing a number of draft practice notices in April 2012 that, if accepted, would serve to provide official guidance to patent examiners on what constitutes patent-eligible subject matter. The Commissioner of Patents requested public comment on these. With few exceptions, the submissions (largely constituting letters from the profession) were generally critical of the draft practice notices, asserting that the proposed guidelines represented an approach that was entirely inconsistent not only with the leading Supreme Court of Canada case in respect of patentable subject matter,[62] but also with the Federal Court of Appeal's decision in *Amazon*[63] to which the practice notices were responding.[64]

In March 2013, CIPO issued revised examination guidelines ("Examination Practice Respecting Computer-Implemented Inventions – PN 2013-03"), which currently govern the approach used by CIPO in assessing subject matter patent eligibility of computer software inventions. CIPO has also since updated its *Manual of Patent Office Practice* (MOPOP) chapter on subject matter[65] to refer to specific passages in the Federal Court of Appeal's *Amazon* decision. For instance, the updated chapter now states that "[a]n invention is a solution to a practical problem. In order to solve a practical problem, the solution must be something with physical existence, or something that manifests a discernible effect or change and, hence, that will itself enable a person skilled in the art to obtain the intended result or benefit. Such a form is referred to herein as a 'practical form' or a 'practicable form'".[66] Some practitioners have observed an increasing number of subject matter objections being raised in the wake of these guidelines and updated MOPOP chapter; depending on the nature of the particular invention, these objections may be quite challenging to overcome at present.

The authors have observed that not all software-implemented methods have

---

60 Patent Act, RSC 1985, c P–4, Section 27(8).

61 The Federal Court of Appeal suggested that an abstract idea cannot become patentable subject matter merely because it has a practical embodiment or a practical application. See *Canada (Attorney General) v Amazon.com, Inc*, 2011 FCA 328 at paragraph 61, 97 CPR (4th) 171. However, the Federal Court of Appeal did not provide a test as to what may or may not constitute a mere abstract idea.

62 *Shell Oil Co v Commissioner of Patents* [1982] 2 SCR 536 at 554: "A disembodied idea is not *per se* patentable. But it will be patentable if it has a method of practical application."

63 It has been suggested that despite the fact that a test for what constitutes an 'art' has been provided in *Canada (Attorney General) v Amazon.com, Inc*, 2011 FCA 328 at paragraph 50, 97 CPR (4th) 171, the test has not been acknowledged in the practice notices: "There are thus three important elements in the test for art ... : (i) it must not be a disembodied idea but have a method or practical application; (ii) it must be a new and inventive method of applying skill and knowledge; and (iii) it must have a commercially useful result."

64 The reader is directed to the finalised guidelines, online: www.ic.gc.ca/eic/site/cipointernet-internetopic.nsf/eng/wr03627.html.

65 MOPOP, Chapter 12.

66 MOPOP, Chapter 12, Section 12.02 – Inventions must not be disembodied – November 2017.

been subject to the same degree of intense scrutiny by Canadian examiners. Applications concerning methods executed on portable electronic devices (eg, relating to user interface features, data communication, cryptographic functions, etc) and more complex systems (eg, image processing, data processing, data storage techniques), for example, appear to be regarded by examiners somewhat more favourably (perhaps regarded as being more 'scientific' in nature) and seem less likely to attract subject matter rejections.

On the other hand, applications that might be classified as relating to a method for doing business (eg, software with relevance to the financial services industry in particular) or a method of playing a game[67] are more likely to be considered controversial, and one can expect that applications for such inventions will continue to attract greater scrutiny by Canadian examiners.

2.2 Novelty and non-obviousness

(a) *Public disclosures originating from the applicant*

Canada provides a grace period of one year for filing an application for an invention that has been disclosed to the public anywhere by the applicant, or by a person who obtained knowledge, directly or indirectly, from the applicant.[68] It is very important to note, however, that this one-year grace period extends back from the actual Canadian filing date and not the priority date where priority is claimed.

For example, suppose a potential applicant first disclosed an invention to the public on 2 January 2018. At that time, no patent applications had been filed anywhere in the world. In most jurisdictions governed by an absolute novelty standard, this self-disclosure would be a bar to patentability should the applicant subsequently file an application for a patent. However, much like the United States, Canada offers potential applicants a one-year grace period to file a patent application after such self-disclosure of the invention.

Now assume, in the above example, that the applicant files a patent application in the United States six months following the date of public disclosure, on 2 July 2018. It would be wrong for the applicant to assume that the Canadian application may be validly filed any time before 2 July 2019, claiming Convention priority from the filing of the application in the United States one year earlier.

If the actual filing date of the Canadian application is more than one year from the date of the applicant's public disclosure of the invention, then claims to that invention will no longer be considered to be directed to novel subject matter. In this example, the Canadian application must be filed before 2 January

---

67 MOPOP, Chapter 12, Section 12.03.09 – Games.
68 Patent Act, RSC 1985, c P-4, Section 28.2(1)(a).

2019 (ie, within one year of the first public disclosure), notwithstanding the fact that an application in the United States was timely filed for which the period for claiming Convention priority extends beyond the one-year anniversary of the public disclosure.

In addition, an invention must not be obvious as of the earlier date (either the filing date or any applicable priority date), having regard to any public disclosure anywhere made more than one year before the filing date by the applicant, or by a person who obtained knowledge, directly or indirectly, from the applicant.[69]

It is interesting to note that, as a result of the restoration of priority provisions brought in by the new Patent Rules, there may be occasions where the grace period can be further extended. For example, in a case where a disclosure to the public was made on 14 September 2020, a Canadian application would normally need to be filed by 14 September 2021 (one-year grace period). Under the new Canadian patent regime,[70] if a priority application (eg, US patent application) was filed before the public disclosure date (14 September 2020), such as 14 August 2020, the Canadian application could be filed up to 14 October 2021 (assuming the request for restoring the priority claim to the US patent application was properly made, see 1.1(b) above).

(b) Third-party public disclosures

There must be no public disclosure of the invention anywhere by a third party (ie, someone other than the applicant, or someone other than a person who obtained knowledge, directly or indirectly, from the applicant) before the Canadian filing date, or before the priority date of the application if priority is claimed.[71]

The invention must also not be obvious as of the earlier of the filing date and any applicable priority date, having regard to any public disclosure anywhere made by a third party.[72]

2.3 Tests for anticipation and obviousness

In order for a prior art document to be anticipatory, it must meet a two-part test of 'disclosure' and 'enablement'. More specifically, to anticipate, the single prior art document must disclose the invention claimed in the patent at issue, and no trial-and-error is permitted in the analysis. Further, the prior art document must provide sufficient information to enable the skilled person to perform or make the invention without undue burden[73] in order to arrive at a finding of anticipation.

---

69 Patent Act, RSC 1985, c P–4, Section 28.3(a).

70 Patent Act, RSC 1985, c P-4, Sections 28.2(1)(a) and 28.3(a), as amended by 2014, c.39, Sections 123 and 124; in force 30 October 2019.

71 Patent Act, RSC 1985, c P–4, Section 28.2(b).

72 Patent Act, RSC 1985, c P–4, Section 28.3(b).

73 *Apotex Inc v Sanofi-Synthelabo Canada Inc*, 2008 SCC 61, [2008] 3 SCR 265.

The Canadian test for obviousness does not involve an assessment of 'inventive step' *per se*, despite the fact that the current four-step approach to an obviousness enquiry set out below is based on UK law:

- Identify the notional 'person skilled in the art'; and the relevant common general knowledge of that person.
- Identify the inventive concept of the claim in question or, if that cannot readily be done, construe it.
- Identify what, if any, differences exist between the matter cited as forming part of the 'state of the art' and the inventive concept of the claim or the claim as construed.
- Viewed without any knowledge of the alleged invention as claimed, determine whether those differences constitute steps which would have been obvious to the person skilled in the art or whether they require any degree of invention.[74]

With respect to the fourth step of this approach, the court indicated that application of an 'obvious to try' test might be warranted in "areas of endeavour where advances are often won by experimentation";[75] practically, such considerations are generally more appropriate when dealing with inventions in the pharmaceutical and chemical arts, and less so with inventions relating to electronics or computer software.

3. Infringement

3.1 Claim interpretation

The claims of a patent should be given a purposive construction. The principles of purposive construction in Canada are elucidated in the Supreme Court of Canada cases of *Free World Trust* and *Whirlpool*.[76]

The act of construing the claims is antecedent to the consideration of both validity and infringement issues, and claims should receive the same interpretation for both purposes. A purposive construction of the claims involves identifying, with the assistance of the skilled reader, particular words or phrases in the claims that describe what the inventor considered to be the 'essential' elements of his or her invention. What constitutes an essential element is to be interpreted in light of the knowledge of the art at the date of publication of the patent specification.

Purposive construction involves interpreting words of the claims in the context of the specification as a whole. It has been held that a 'dictionary

74 *Ibid.*

75 *Ibid* at paragraph 68.

76 *Free World Trust v Électro Santé Inc*, 2000 SCC 66, [2000] 2 SCR 1024; and *Whirlpool Corp v Camco Inc*, 2000 SCC 67, [2000] 2 SCR 1067.

approach' to the construction of claim terms is improper; it is permissible to look at the rest of the specification, including the drawings, to understand what is meant by the terms in the claims. Recent changes to Canadian law establish that prosecution history may be relevant to claim construction in any action or proceeding that has not been finally disposed of as at 13 December 2018.[77] This can include any communication between the applicant and CIPO, and also any communication in respect of a parent application in the case of a divisional application. This change in Canadian law is similar to the 'file wrapper estoppel' practice in the United States.

3.2 Acts of infringement

Section 42 of the Patent Act grants to a patentee the "exclusive right, privilege and liberty of making, constructing, and using the invention and selling it to others to be used".[78] The act does not specifically define 'infringement', but generally any act that interferes with the full enjoyment of those rights will be considered to be an infringement.[79] Deprivation of the full enjoyment of an invention occurs when another person, without licence or permission from the patentee, uses the invention to further a business interest.[80] The burden is on the patentee to prove on the balance of probabilities that infringement has occurred.[81]

Despite the broad definition of 'infringement' generally accepted by the courts, determining whether a given act constitutes an infringement can be a challenge when the circumstances are somewhat unusual. Selling parts of a disassembled machine in Canada to be sold to recipients abroad[82] and importation of products made abroad that are the subject of patented process claims in Canada[83] may constitute infringement, whereas a *bona fide* experimental use without intention to sell or make use of a product in the commercial market[84] or a repackaging of a patented substance into a commercially usable form[85] may not constitute infringement. Prior use before the claim date of a patent may exempt use on or after that date from being an infringing act. Any person who commits an act that would otherwise constitute an infringement of the patent, or made serious and effective preparations to

---

77 Patent Act, RSC 1985, c P–4, Section 53.1 (1).

78 Patent Act, RSC 1985, c P–4, Section 42.

79 See eg, *Monsanto Canada Inc v Schmeiser*, 2004 SCC 34 at paragraphs 34–7, 31 CPR (4th) 161.

80 *Ibid* at paragraphs 34–7, 31 CPR (4th) 161 at paragraph 37.

81 *Lishman v Erom Roche Inc* (1996), 68 CPR (3d) 72 at 77 (FC), aff'd (1996), 71 CPR (3d) 146 (FCA).

82 *Beloit Canada Ltd v Valmet-Dominion Inc* [1997] 3 FC 497, 73 CPR (3d) 321 (*sub nom J M Voith GmbH v Beloit Corp*).

83 *Eli Lilly and Co v Apotex Inc*, 2009 FC 991, 80 CPR (4th) 1, aff'd 2010 FCA 240, 90 CPR (4th) 327.

84 *Merck & Co v Apotex Inc*, 2006 FCA 323, 55 CPR (4th) 1. Newly introduced Section 55.3(1) of the Patent Act excludes an act committed for the purpose of experimentation as infringement of the patent, and applies to any action or proceeding that has not been finally disposed of as of 13 December 2018. Further regulations are expected to be issued to clarify what constitutes experimental use.

85 *Eli Lilly & Co v Apotex Inc*, [1998] SCJ No 59 at paragraph 101, 80 CPR (3d) 321.

commit such an act, would not infringe a patent if the act previously occurred, in good faith, before the claim date.[86] However, this exemption would not apply if the knowledge of the claimed invention was obtained directly or indirectly from the applicant of the patent application from which the patent was granted, and the user knew that the applicant was the source of the knowledge.

The authors are not aware of any recent Canadian jurisprudence of significant note on the issue of infringement with respect to software patents specifically.

3.3 Contributory infringement and Infringement by Common Design

Strictly speaking, there is no cause of action in Canada for contributory infringement of a patent.[87] However, it is well established that one who induces or procures another to infringe a patent is guilty of infringement of the patent.[88] In Canada, this is known as infringement by inducement.

A determination of inducement requires the application of a three-prong test:

- the act of infringement must have been completed by the direct infringer;
- the completion of the acts of infringement must be influenced by the acts of the alleged inducer to the point that, without the influence, direct infringement would not take place; and
- the influence must knowingly be exercised by the inducer (ie, the inducer knows that this influence will result in the completion of the act of infringement).

The Federal Court of Appeal has commented that the test is a difficult one to meet.[89] By way of example, selling parts as unassembled components in a kit for making a patented product[90] or supplying a product in a manner that constitutes inviting users to infringe the method claims of a patent[91] have been considered to constitute infringement by inducement.

Canadian jurisprudence also appears to support the concept of infringement by common design, which originates from the United Kingdom.[92] Under this concept, a patent can be infringed by the acts of multiple parties when those parties acted in concert – even if only based on a tacit agreement. The Canadian courts have yet to make a finding of infringement by common design but the Federal Court of Canada[93] has repeatedly endorsed this concept.

---

86 Patent Act, RSC 1985, c P-4, Section 56(1).

87 *Nycomed Canada Inc et al v Teva Canada Limited*, 2012 FCA 195.

88 *Corlac Inc et al v Weatherford Canada Ltd et al*, 2011 FCA 228, 95 CPR (4th) 101.

89 *Ibid* at paragraph 162, 95 CPR (4th) 101.

90 *Windsurfing International Inc v Bic Sports Inc* [1985] FCJ No 1147, 8 CPR (3d) 241.

91 *Procter & Gamble Co v Bristol-Meyers Canada Ltd* (1978), 39 CPR (2d) 145, [1978] FCJ No 812, aff'd 42 CPR (2d) 33, [1979] FCJ No 405; aff'd 39 CPR (2d) 171, [1978] FCJ No 1014.

92 *Unilever plc v Gillette (UK) Limited*, [1989] RPC 583 (UKCA) at 609.

93 See eg, *Bauer Hockey Corp v Easton Sports Canada Inc*, 366 FTR 242; *Packers Plus Energy Services Inc v Essential Energy Services Ltd* [2017] FCJ No 1200; *Georgetown Rail Equipment Company v Rail Radar Inc*, 2018 FC 70.

3.4 Cross-border issues

The high-profile litigation involving RIM and NTP Inc in the United States brought the issue of cross-border infringement to the forefront of the minds of patent practitioners worldwide. The United States Court of Appeals for the Federal Circuit held that US patents in respect of an email system were infringed, despite the 'control point' of RIM's system being located outside of the United States, in Canada.

The issue of whether a Canadian patent might be infringed when certain claimed acts are performed by network components located outside of Canada has not yet been tested before Canadian courts. While it is generally accepted that the exclusive rights conferred by a Canadian patent extend only to acts performed within Canada, the broad definition of infringement as set out in *Monsanto*[94] – that any act that interferes with the full enjoyment of the rights is considered to be an infringement. In fact, the Federal Court was unable to locate any Canadian jurisprudence to address the "specific question of whether infringement of a claim to a method (or a sever [sic] arrangement or a computer) can be avoided by locating one component thereof (a server) outside Canada".[95] This leaves undecided the question of whether all steps of a computer-based method need to be carried out in Canada or all components of a computer system need to be located in Canada to constitute infringement.

More recently, the effects of a number of 2017 Supreme Court decisions[96] in both Canada and the United States on the issue of litigation forum selection are being watched closely by litigators from both countries; those parties whose activities are subject to cross-border commercialisation risk are well advised to seek updated advice from North American counsel.

4. Ownership, inventorship and compensation

4.1 Ownership issues

The Patent Act provides for certain requirements relating to the assignment and licensing of patent rights. For example, the Patent Act explicitly states that every patent issued for an invention is assignable in law, either as to the whole interest or as to any part thereof, by an instrument in writing.[97] More particularly, in Canada, the rights to the invention (even before the filing of an application), to the interest in an application once filed, and to issued patents may be assigned.

The Patent Act, however, is silent on the question of ownership of rights in

94 *Monsanto Canada Inc v Schmeiser*, 2004 SCC 34 at paragraphs 34–7, 31 CPR (4th) 161.

95 *Canadian National Railway Company v BNSF Railway Company*, 2018 FC 614, [2018] FCJ No 624, 156 CPR (4th) 1.

96 See eg, *Douez v Facebook, Inc*, 2017 SCC 33; *Google Inc v Equustek Solutions Inc*, 2017 SCC 34; *BNSF Railway Co v Tyrrell*, 137 S Ct 1549 (2017).

97 Patent Act, RSC 1985, c P–4, Section 50(1). Under the proposed amendments to the Patent Act and Rules, the actual transfer document may not be necessary to register ownership at CIPO (not yet in force as of early 2019).

a patent or a patent application in the absence of a written agreement. Such ownership disputes must apparently be resolved by applying applicable law, typically provincial law, in Canada.

For example, unlike many other jurisdictions, the general principle in Canada is that the invention of an employee belongs to the employee, and not the employer, unless there is an overriding contractual or fiduciary obligation requiring the employee to assign applicable rights to the employer.[98]

4.2 Inventorship issues

The Patent Act itself does not explicitly define the terms 'inventor' or 'inventorship'. However, the Supreme Court of Canada has held that the inventor is the person responsible for the inventive concept;[99] and the concept, or the ingenious idea, must be "reduced ... to a definite and practical shape".[100]

The Patent Act provides that:

A patent is void if any material allegation in the petition of the applicant in respect of the patent is untrue, or if the specification and drawings contain more or less than is necessary for obtaining the end for which they purport to be made, and the omission or addition is wilfully made for the purpose of misleading.[101]

Therefore, in general, an improper identification of inventors is a ground for invalidating a patent, if the incorrect information is given wilfully for the purpose of misleading. Whether there has been a 'material' allegation (practically, a misstatement that makes a difference to the issuance of the patent or the grant of rights for the invention in the eyes of the Commissioner of Patents) will depend on the facts of each particular case;[102] however, the courts do not appear to be quick to invalidate a patent on this ground.

4.3 Compensation (as between co-patentees)

Disputes between co-owners of a patent may also arise, particularly when one co-owner wishes to perform some act in respect of the patent without the consent of the other co-owners and without compensating the other co-owners. In general, in the absence of contractual provisions to the contrary, the co-owner of a patent can use the patent without accounting to the other co-owners. However, any act of a co-owner that acts to dilute the rights of the other co-owners will amount to an infringement of the patent.[103]

---

98 *Comstock Canada et al v Electec Ltd et al* (1991), 38 CPR (3d) 29, [1991] FCJ No 987.
99 *Apotex Inc v Wellcome Foundation Ltd* (2002), 21 CPR (4th) 499 at paragraph 96, [2002] SCJ No 78.
100 *Ibid* at paragraph 97, [2002] SCJ No 78, citing *Rice v Christiani & Nielson* [1930] SCR 443.
101 Patent Act, RSC 1985, c P–4, Section 53(1).
102 See eg, *671905 Alberta Inc v Q'Max Solutions Inc*, 2003 FCA 241 at paragraph 31, [2003] FCJ No 873; *Proctor & Gamble Co v Bristol-Myers Canada Ltd* (1978), 39 CPR (2d) 145, [1978] FCJ No 812, aff'd 42 CPR (2d) 33, [1979] FCJ No 405; *Weatherford Canada Ltd v Corlac Inc*, 2010 FC 602, [2010] FCJ No 721 rev'd in part 2011 FCA 228, [2011] FCJ No 1090, leave to appeal to SCC refused, 2012 CarswellNat 846 (WL Can).
103 See eg, *Forget v Specialty Tools of Canada Inc* (1993), 48 CPR (3d) 323, [1993] BCJ No 822. The Quebec Civil Code may present additional restrictions where applicable.

For example, a co-owner is permitted to sell his or her entire interest in a patent without the consent of the other co-owners, and the purchaser will then become a co-owner of the patent. It is not an infringement when one co-owner, without the consent of the other, makes or sells in Canada the patented invention.

On the other hand, a co-owner may not sell a partial interest in the patent without the consent of the other co-owners, as increasing the number of co-owners may be seen as a dilution of the rights of the other co-owners. Furthermore, a co-owner may not grant a licence to work the patent to a third party without the consent of the other co-owners, and any such licence will be invalid absent the consent of the other co-owners. Also, if one co-owner, without obtaining the necessary consent from the other co-owners, licenses the right to use the patent to a licensee, and the licensee manufactures the patented invention without the consent of the other co-owners, this will constitute infringement.

5. Copyright protection

5.1 Copyright protection of software programs

The copyright protection of computer programs is explicitly provided for in the Copyright Act.[104] Section 5(1) provides that copyright subsists in every original literary work, and a 'literary work' includes computer programs.[105] The incorporation of a computer program into an integrated circuit product, for example, may therefore constitute an infringement of the copyright or moral rights in a work.[106]

It is well accepted that copyright can protect the form of expression of computer programs, but not the 'ideas' embodied therein. Infringement of copyright will not be found merely because an allegedly infringing product performs the same functions as a computer program for which copyright is asserted.

In addition to the literal copying of a computer program, 'copying' the source code of a program by rewriting the code in a different programing language may still constitute copyright infringement; in an analogous manner, an "original literary work protected by copyright and written in English is just as much copied if the plagiarist copying it translates it into German".[107] However, it would appear that the greater the departure from a literal copying of the source code for a program, the less likely courts will find copyright to be

104 Copyright Act, RSC 1985, c C–42.
105 Copyright Act, RSC 1985, c C–42, Section 2.
106 Copyright Act, RSC 1985, c C–42, Section 64.2(2).
107 *Delrina Corp v Triolet Systems Inc* (1993), 47 CPR (3d) 1, aff'd 156 OAC 166 (2002) (Ont CA), further aff'd 165 OAC 160 (2002) (Ont CA).

infringed, particularly when it is determined that parts of the program with genuine originality have not been copied, recognising that certain portions of the program may not be protectable by copyright.[108]

5.2 No copyright protection for consumer electronics

The Copyright Act does not specifically extend protection to consumer electronics.

Where the Copyright Act leaves gaps – particularly with respect to protecting ornamental designs of electronic devices, for example, and integrated circuits embedded within consumer electronics[109] – it can be helpful to look to the Industrial Design Act as well as the Integrated Circuit Topography Act.[110]

5.3 Strong anti-circumvention provisions

In 2012, the Copyright Act underwent significant reform; one of the most controversial changes relates to the introduction of anti-circumvention provisions (Sections 41 to 41.22) in Canadian copyright law. Generally, anyone who circumvents a technological protection measure (eg, 'digital lock') that controls access to a work, or provides services or devices that enable others to do so, may be liable for copyright infringement, although the act provides for a number of exceptions. Notably, a 2017 Federal Court decision awarded Nintendo of America Inc a record-breaking C$12.7 million in damages for circumvention of technological protection measures and copyright infringement, suggesting Canadian anti-circumvention laws may be among the most restrictive in the world.

6. Confidential information

In Canada, there is no specific statutory protection for what is generally known as 'confidential information'. Correspondingly, there is no statutory definition of 'confidential information', which is also sometimes referred to, among other names, as trade secrets.

Trade secrets can take the form of, for example, customer lists, a technical process, a chemical formula, a tool, or any other valuable information that is not generally known to or accessible by the public at large.

If an owner takes adequate steps to ensure that the information is kept secret (eg, by treating the information as confidential, and through the use of contractual barriers and physical barriers where possible), the owner may then be able to recover damages from a party who breaches the confidentiality.

---

108 See eg, *Delrina Corp v Triolet Systems Inc* (1993), 47 CPR (3d) 1, [1993] OJ No 319, aff'd 156 OAC 166 (2002) (Ont CA), further aff'd 165 OAC 160 (2002) (Ont CA).

109 See Section 64 of the Copyright Act, RSC 1985, c C–42, where it exempts some uses of designs applied to a useful article and reproduced in a quantity of 50 or more, as well as topographies or part of topographies.

110 See 7 below.

This protection, however, only lasts for as long as the information remains secret; once the secret is disclosed to the public – whether accidentally or not – protection as a trade secret will cease (although a claim against those responsible for the loss of trade secret protection may still be available). In particular, with respect to subject matter that is reverse-engineered, independently discovered, or independently developed, the trade secrets laws offer no protection.

As for the remedy for a breach of confidentiality, a proprietary remedy will not automatically follow, although there may be cases where it is appropriate. In general, compensation for breaches of confidentiality is to be distinguished from remedies available for patent infringement; otherwise, plaintiffs would be "obtaining the benefit of patent remedies without establishing that their invention meets the statutory criteria for the issuance of a patent, or paying the price of public disclosure of their secret".[111]

7. Other IP rights

7.1 Industrial design protection

The Industrial Design Act[112] defines 'design' or 'industrial design' as "features of shape, configuration, pattern or ornament and any combination of those features that, in a finished article, appeal to and are judged solely by the eye".[113] Once registered, the owner obtains an exclusive right to the use of that design in Canada for 10 years from the date of registration (for applications filed before 5 November 2018), or the later of 10 years from the date of registration and 15 years after the filing date of the application (for applications filed on or after 5 November 2018).[114]

In the absence of a licence from the proprietor of a design, any act constituting a making, importing for the purpose of trade or business, selling, renting, or offering or exposing for sale or rent, of any article in respect of which the design is registered and to which the design "or a design not differing substantially therefrom" has been applied will constitute an infringement. When considering whether certain differences are 'substantial', the extent to which the registered design differs from previously published designs can be taken into account.[115]

7.2 Semiconductor topographies

The Integrated Circuit Topography Act[116] provides protection for an integrated circuit topography, by registration, for a period of up to 10 years.[117] In practice, we see few registrations under this Act.

111 *Cadbury Schweppes Inc v FBI Foods Ltd* [1999] 1 SCR 142, 83 CPR (3d) 289.
112 Industrial Design Act, RSC 1985, c I–9.
113 *Ibid*, Section 2.
114 *Ibid*, Sections 10(1) and 34.01.
115 *Ibid*, Section 11.
116 Integrated Circuit Topography Act, SC 1990, c 37.
117 *Ibid*, Section 5.

8. Counterfeiting

According to the World Trade Organisation (WTO), pirated copyright goods include copies of an article made without the consent of a copyright holder, where the making of the copies would have constituted a copyright infringement.[118] On the other hand, 'counterfeit trademark goods' generally include any goods, including packaging, bearing, without authorisation, a trademark which is identical to the trademark validly registered in respect of such goods, or which cannot be distinguished in its essential aspects from such a trademark, and which thereby infringes the rights of the trademark owner.[119] More generally, both piracy and counterfeiting typically involve the manufacture and sale of goods that are made in imitation, to be passed off as copies of authentic goods, but which are not in fact genuine. Piracy and counterfeiting are serious concerns, especially for computer software developers and manufacturers of consumer electronic devices.

With respect to counterfeit goods, a trademark owner may bring an action for the common law tort of passing off without prior registration of a mark. The Trademarks Act[120] codifies the tort of passing off and creates additional rights for registered trademark owners to sue for trademark infringements. Remedies for trademark infringement and passing off can include damages or an accounting of a defendant's profits, the delivery up of infringing goods, and injunctive relief against further unauthorised use of an infringing mark. Suspected infringing goods may also be temporarily detained by Canadian customs officials at their point of entry into Canada.

It is not uncommon in trademark matters, particularly with operations of a temporary nature involving individuals selling counterfeit merchandise, to seek an Anton Piller order that allows the trademark owner to seize infringing articles without prior warning and deposit them for safekeeping at the court pending a determination of litigation.

9. Enforcement

9.1 Court structure

Both federal and provincial court systems exist in Canada. Most commercial litigation in Canada is commenced in the superior courts of the provinces, which have jurisdiction over most civil and criminal proceedings, including practically all commercial matters. The Federal Court of Canada, in contrast, is a statutory court with narrower jurisdiction. This jurisdiction is restricted to

118 Agreement on Trade-Related Aspects of Intellectual Property Rights, 15 April 1994; Marrakesh Agreement Establishing the World Trade Organization, Annex 1C, 1869 UNTS 299, 33 ILM 1197 (1994), Article 51, fn 14.

119 *Ibid.*

120 Trademarks Act, RSC 1985, c T–13, Section 7.

specified federal matters such as immigration law, maritime law and some intellectual property.

With respect to IP matters, the Federal Court may have exclusive or concurrent jurisdiction with the superior courts of the provinces, depending on the nature of the particular matter being litigated. The Federal Court has exclusive jurisdiction in cases of conflicting applications for any patent of invention, or for the registration of any copyright, trademark, industrial design or topography within the meaning of the Integrated Circuit Topography Act;[121] in all other cases, the Federal Court has concurrent jurisdiction in which a remedy is sought "under the authority of Act of Parliament or at law or in equity" respecting any patent, copyright, trademark, industrial design or topography.[122] In certain situations, a choice between the two court systems will be required; at other times, multiple actions may proceed separately in both systems. However, the Federal Court is generally considered to have much more experience than the provincial superior courts in IP matters.

9.2 Obtaining information

In the Federal Court, every party must serve an affidavit of documents on every other party within 30 days of the close of pleadings.[123] The affidavit is in standard form and should contain separate lists and descriptions of all relevant documents that:

- are in the possession, power or control of the party and for which no privilege is claimed;
- are or were in the possession, power or control of the party and for which privilege is claimed;
- were but are no longer in the possession, power or control of the party and for which no privileged is claimed; and
- the party believes are in the possession, power or control of a person who is not a party to the action.[124]

A document is considered 'relevant' if a party intends to rely on it or if the document tends to adversely affect the party's case or to support another party's case.[125]

The affidavit of documents must also include a statement of the grounds for each claim of privilege in respect of a document,[126] a description of how the party lost possession, power or control of any document and its current

121 Federal Courts Act, RSC 1985, c F-7, Section 20(1)(a).
122 *Ibid*, Section 20(1)(b).
123 Federal Courts Rules, SOR /98-106, Rule 223(1).
124 *Ibid*, Rule 223(2)(a). Requirements for the affidavit of documents may differ in provincial superior court proceedings: see eg, Ontario Rules of Civil Procedure, O Reg 575/07, Section 6(1), Rule 30.
125 *Ibid*, Rule 222(2).
126 *Ibid*, Rule 223(2)(b).

location,[127] and the identity of each third party who may possess relevant documents.[128] The deponent of the affidavit is required to make all reasonable inquiries of any officer, servant, agent or employee of the party, including any outside Canada, who might reasonably be expected to have knowledge relating to any matter in question in the action.[129]

Examinations for discovery are examinations conducted under oath, typically orally. Unless the parties agree, a discovery is not held until each party has served an affidavit of documents.[130] Except by leave of the court, each party may only examine a single representative of each adverse party.[131] The scope of an examination for discovery can potentially be quite broad; any question 'relevant' to any unadmitted allegation of fact in a pleading filed by the party being examined or by the examining party must be answered.[132] However, the test for relevance is subject to the "overriding discretion of a prothonotary or judge to control abuses of the discovery process".[133] In particular, if a question is determined to be relevant, the relevance of that question will be weighed against several factors, including, for example, the degree of relevance, the generality or breadth of the question, the onerousness of providing an answer, and whether the question is fair or abusive.[134]

A motion may be brought for leave to examine for discovery a non-party who might have information on an issue in the action. However, for leave to be granted, the court must be satisfied of the following:

- that the non-party may have information on an issue in the action;
- the party has been unable to obtain the information informally from the non-party or from another source by any other reasonable means;
- it would be unfair not to allow the party an opportunity to question the non-party before trial; and
- the questioning will not cause undue delay, inconvenience or expense to the non-party or to the other parties in the action.[135]

These may be difficult tests to satisfy.

---

127 *Ibid*, Rule 223(2)(c).

128 *Ibid*, Rule 223(2)(d).

129 *Ibid*, Rule 224(2).

130 *Ibid*, Rule 236(1).

131 *Ibid*, Rule 235.

132 *Ibid*, Rule 240(a). Some provincial courts may permit an even broader scope. See eg, Ontario Rules of Civil Procedure, O Reg 575/07, Section 6(1), Rule 31.06: "A person examined for discovery shall answer ... any proper question relevant to any matter in issue in the action."

133 *Sanofi-Aventis Canada Inc v Laboratoire Riva Inc*, 2011 FC 1469 at paragraph 13 (citing *AstraZeneca Canada Inc v Apotex Inc*, 2008 FC 1301 at paragraphs 12 and 17, [2009] 4 FCR 243; *Apotex Inc v Bristol-Myers Squibb Co*, 2007 FCA 379 at paragraph 30, [2007] FCJ No 1597).

134 *Ibid* at paragraph 14 (citing *AstraZeneca Canada Inc v Apotex Inc*, 2008 FC 1301 at paragraphs 16 and 18, [2009] 4 FCR 243; *GSC Technologies Corp v Pelican International Inc*, 2009 FC 223, [2009] FCJ No 274 at paragraphs 9, 11–12; *T-Mobile USA, Inc v Telus Corporation*, 2010 FC 455, [2010] FCJ No 539 at paragraphs 4–5).

135 Federal Courts Rules, SOR /98–106, Rules 238(1) and 238(3).

9.3 Availability of interim relief

A party to a pending or intended proceeding may obtain an interlocutory injunction by way of a motion to a judge.[136] A judge may grant an interim injunction (pending an interlocutory injunction) on an *ex parte* motion for a period of not more than 14 days where the judge is satisfied, in a case of urgency, that no notice is possible or that to give notice would defeat the purpose of the motion.[137]

The test[138] for granting an interlocutory injunction, in the Federal Court and in the provincial courts, consists of three parts:

- First, the applicant for interlocutory relief must demonstrate a serious question to be tried. A determination of whether this prong of the test has been satisfied is based on common sense and an extremely limited review of the case on the merits. Going beyond a preliminary investigation into the merits is generally warranted only when the result of the interlocutory motion effectively amounts to a final determination of the action.
- Second, the applicant must demonstrate that irreparable harm will result if the relief is not granted. 'Irreparable' refers to the nature of the harm rather than its magnitude. This prong of the test is generally considered the most difficult to meet. The court requires clear evidence, which is not speculative, that the applicant has or will suffer harm that cannot be compensated in monetary damages. Furthermore, losses of market share that are quantifiable and recoverable will not amount to irreparable harm.[139] Accordingly, many applications for interlocutory injunctions in IP cases (but particularly in patent cases) fail given the inherent difficulty of showing that any harm caused would not be compensable by money.
- Third, an assessment of the balance of inconvenience to the parties will be made, considering the harm that each party will suffer from the granting or refusal of the interlocutory injunction.

An Anton Piller order is an exceptional order that permits pre-trial search and seizure of evidence without prior warning, as a means of ensuring that the evidence will not be destroyed and will be preserved for trial. The Federal Court has granted 'rolling' Anton Piller orders, which are not defendant-specific, and are typically executed against street vendors and transient flea market vendors.[140] The court must be convinced that the applicant has a very strong *prima facie* case before it will grant an Anton Piller order. These orders are rare in patent cases.

---

136 *Ibid*, Rule 373(1). Similar considerations will generally apply in provincial court systems.
137 Federal Courts Rules, SOR /98–106, Rule 374(1).
138 See *RJR MacDonald Inc v Canada (Attorney General)* [1994] 1 SCR 311, [1994] SCJ No 17.
139 See eg, *Bayer HealthCare AG v Sandoz Canada Inc*, 2007 FC 352, [2007] FCJ No 585.
140 See eg, *Fila Canada Inc v Doe* [1996] 3 FC 493, [1996] FCJ No 736.

9.4 Springboard relief and relief for post-expiry losses

Relief based specifically on the application of the 'springboard' doctrine in patent infringement cases is not particularly well developed in Canada. It has been considered to some extent in cases involving trade secrets and confidential information. Courts have granted interlocutory injunctive relief in respect of alleged acts of infringement occurring just prior to a patent's expiry, that would presumably allow a defendant to gain an early foothold in the market after expiry of the patent.[141] However, provided there is a reasonable method for calculating a patentee's losses due to an infringer's head-start in the marketplace (prior to the patent expiration), courts have been generally reluctant to grant interlocutory injunctions to prevent infringers from springboarding. Applicants face difficulties in demonstrating that any permanent loss of market share that may result from the defendant's early entry into market will amount to irreparable harm.[142]

In Canada, historically, the prospects of obtaining relief specifically for post-expiry patent losses would not have been considered to be strong. The Federal Court of Appeal, in 1980, had explicitly remarked that "protection beyond the expiry of the patent is not something to which the patentee is entitled under the patent".[143] However, in subsequent decisions, the Federal Court held that post-expiry damages may be quantifiable, and that "damages will be calculated in reasonable fashion, providing a normal remedy for infringement, if the trial [sic] finds that to have occurred, whether those are caused before or after expiry of the plaintiffs' patent".[144] In this regard, it would appear that damages attributed to losses occurring after the expiry of a patent may be available to successful plaintiffs.

9.5 Competition Act

The Competition Act[145] governs most business conduct in Canada. It contains both criminal and civil provisions aimed at preventing anti-competitive practices in the marketplace. The Competition Bureau is responsible for the administration and enforcement of the Competition Act.

With respect to IP rights, the Competition Bureau recognises that, in general, the mere exercise of the owner's right to unilaterally exclude others from using

---

141 See eg, *Baker Hughes Inc v Galvanic Analytical Systems Ltd* [1991] FCJ No 826 at p3 (FCTD); *Proctor & Gamble Inc v Colgate-Palmolive Canada Inc* [1995] FCJ No 719 at paragraph 73 (FCTD); *ICI Americas Inc et al v Ireco Canada Inc et al* [1985] FCJ No 1229 at paragraph 57 (FCTD).

142 See eg, *Bayer HealthCare AG v Sandoz Canada Inc*, 2007 FC 352, [2007] FCJ No 585.

143 *Cutter Ltd v Baxter Travenol Laboratories of Canada, Ltd et al* (1980), 47 CPR (2d) 53 at 57 (FCA).

144 *Bayer HealthCare AG v Sandoz Canada Inc*, 2007 FC 352, [2007] FCJ No 585 at paragraph 57. More recently, see *Dow Chemical Company v Nova Chemicals Corporation*, 2017 FC 350 at paragraph 124: "[S]pringboard damages are nothing more than a type of loss to be proven with evidence, and I see no reason why this principle should operate differently to a plaintiff's gains in the context of an accounting of profits".

145 Competition Act, RSC 1985, c C–34.

intellectual property is not cause for concern under the general provisions of the Competition Act.[146] The Competition Act also explicitly provides that:

> *... an act engaged in pursuant only to the exercise of any right or enjoyment of any interest derived under the Copyright Act, Industrial Design Act, Integrated Circuit Topography Act, Patent Act, Trademarks Act or any other Act of Parliament pertaining to intellectual or industrial property is not an anti-competitive act.*[147]

However, the Competition Act also recognises that the mere exercise of an IP right may have anti-competitive consequences. Section 32 of the Competition Act provides that the Federal Court may make an order for various 'special remedies' in cases where the use of patent, trademark, or copyright rights serves to:

- limit unduly the facilities for transporting, producing, manufacturing, supplying, storing or dealing in any article or commodity that may be a subject of trade or commerce;
- restrain or injure, unduly, trade or commerce in relation to any such article or commodity;
- prevent, limit or lessen, unduly, the manufacture or production of any such article or commodity or unreasonably enhance the price thereof; or
- prevent or lessen, unduly, competition in the production, manufacture, purchase, barter, sale, transportation or supply of any such article or commodity.[148]

9.6 Threats

A mere threat to litigate for patent infringement can be sufficient to cause significant commercial loss in certain situations, and such a threat may be considered unjustified or groundless where there is, in fact, no intent to litigate, or no basis for the allegation of infringement. With respect to the subject of potential relief for harm caused by an unjustified threat of patent litigation, there is very little Canadian jurisprudence. In 2018, the Patent Act was amended to include provisions mandating that demand letters comply with certain prescribed requirements,[149] and giving the recipient of a non-compliant written demand (or any other person aggrieved by such a demand) the right to bring a proceeding to seek relief in the Federal Court.[150] If the Federal Court is satisfied that the written demand does not comply with the prescribed requirements, it may grant any relief that it considers appropriate, including by way of recovery of damages, punitive damages, an injunction, a declaration or an award of costs.[151]

---

146 Competition Bureau, Intellectual Property Enforcement Guidelines, Item 4.2.1, online: www.competitionbureau.gc.ca/eic/site/cb-bc.nsf/eng/04031.html.
147 Competition Act, RSC 1985, c C-34, Section 79(5).
148 Competition Act, RSC 1985, c C-34, Section 32(1).
149 Patent Act, RSC 1985, c P-4, Section 76.2(1).
150 Patent Act, RSC 1985, c P-4, Section 76.2(2).

Notably, the Patent Act[152] lacks provisions analogous to the unfair competition provisions found in the Trademarks Act[153] which prevent persons from threatening litigation for infringement of trademarks. As a result, cases regarding threatened litigation for patent infringement have been occasionally framed as claims under the Trademarks Act.[154]

9.7 Remedies

(a) *Monetary remedies*

Damages, including punitive damages, are available from the issue date of a patent. A person who infringes a patent is liable to the patentee and to all persons claiming under the patentee for all damages sustained by the patentee or by any such person, after the grant of the patent, by reason of the infringement. No remedy can be awarded for an act of infringement committed more than six years before the commencement of an infringement action.[155]

Upon a determination of liability, the basic compensatory remedies are generally an election between damages, which are available as of right subject to any equitable bar, or an accounting of profits, which is a discretionary, equitable remedy. Damages are intended to put the patentee in the position they would have been in had the infringement not occurred. General categories of compensatory damages include, for example:

- lost profits due to lost sales;
- a reasonable royalty; and
- lost profits due to price suppression.[156]

On the other hand, an accounting of profits is measured by the profits made by the infringer, rather than the amount lost by the patentee. If the patentee elects to seek an accounting of profits, the patentee is only entitled to that portion of the infringer's profit which is causally attributable to the invention and may be limited to that over a non-infringing alternative. These two types of remedy are alternatives; the patentee may elect one or the other, not both.[157]

---

151 Patent Act, RSC 1985, c P–4, Section 76.2(3). A special provision also exists to attach joint and several, or solidary liability, to a corporation's officers, directors, agents or mandataries that directed, authorised, assented to, acquiesced in, or participated in the sending of a non-compliant demand, for which defects were not rectified within a reasonable time after receiving notice of those defects, see Section 76.2(5).

152 Patent Act, RSC 1985, c P–4.

153 Trademarks Act, RSC 1985, c T–13, Section 7(a).

154 See eg, *S & S Industries v Rowell* (1966), 48 CPR 193, [1966] SCJ No 22; *Riello Canada, Inc v Lambert* (1986), 9 CPR (3d) 324, [1986] ACF No 243.

155 Patent Act, RSC 1985, c P–4, Sections 54, 55(1) and 55.01.

156 *Allied Signal v Du Pont Canada Inc* (1998), 78 CPR (3d) 129 at 139, [1998] FCJ No 190.

157 *Monsanto Canada Inc v Schmeiser*, 2004 SCC 34 at paragraphs 34–7, 31 CPR (4th) 161 at paragraphs 100–02. See also, on the issue of accounting of profits, *Monsanto Canada Inc v Rivett*, 2009 FC 317, [2010] FCR 93, aff'd 2010 FCA 207, 87 CPR (4th) 444.

The quantification of damages and profits is typically done on a reference, following a trial at which liability is determined.

A 'reasonable compensation' may be recoverable for damage sustained by the patentee after the publication of the application and before the grant of the patent, as a result of any act that would have constituted an infringement of the patent if the patent had been granted on the day the application was published.[158] This generally requires that the claims of the patent determined to have been infringed be the same or substantially similar to the corresponding claims as published. It is regarded as providing for lower compensation than damages associated with infringement after the grant of the patent, and what is 'reasonable' will depend on the facts of the given case.[159]

Punitive damages may be assessed to address "high-handed, malicious, arbitrary, or highly reprehensible conduct that departs to a marked degree from ordinary standards of decent behaviour",[160] but are rarely awarded.

Most recently, an award in *Dow Chemical Company v Nova Chemicals Corporation* (see note in 9.4 above) representing the largest monetary award for patent infringement in Canada to date was granted in a Federal Court judgment, in the amount of C$645 million. This case may be regarded as affirming the expansive nature of monetary remedies available for patent infringement in Canada.

(b) *Permanent injunctions*

In an infringement action, the court generally will enjoin the unsuccessful defendant from further use, manufacture or sale of the subject matter of a patent, and if need be, consequentially provide for his or her punishment in the event of disobedience of that order.[161] This is a discretionary remedy. The court will typically provide for a notice period before a permanent injunction takes effect, and may in exceptional cases, deny a plaintiff injunctive relief depending on the facts (eg, delay). An order for destruction or delivery up of infringing product normally follows a grant of a permanent injunction.

The authors would like to acknowledge the input of Mike Charles, Ian McMillan, Nicholas Aitken, Christopher Bury, Michael Canniffe, Rita Gao and Julian Ho in respect of the first edition of this chapter, and the kind assistance of Shuo Xing, Laura MacDonald and Wynnie Chan in the preparation of the second edition of this chapter.

---

158 Patent Act, RSC 1985, c P–4, Section 55(2).

159 See eg, *Jay-Lor International Inc v Penta Farm Systems Ltd*, 2007 FC 358, 59 CPR (4th) 228.

160 *Apotex Inc v Merck & Co*, 2003 FCA 291, 26 CPR (4th) 278 (FCA). See also, eg, *Lubrizol v Imperial Oil* (1994), 58 CPR (3d) 167, [1994] FCJ No 1441; *Eurocopter v Bell Helicopter Textron Canada Ltée*, 2012 FC 113, 100 CPR (4th) 87.

161 Based on the Patent Act, RSC 1985, c P–4, Section 57(1).

China

Handong Ran
Maucher Jenkins
Min Xu
Huahe IP Limited

China has become the largest patent filing country in the world and an important arena for patent litigation. While in the early days the Chinese patent system was to a large extent modelled on European and German counterparts, its subsequent development has been subject to influence from many external factors as well as growth of the domestic industries. This is clearly reflected by the evolution of the Chinese patent practice in the electronics and software sectors.

1. Outline of national patent procedure

1.1 Filing

A patent application can either be filed directly at the Patent Office within the Chinese Intellectual Property Administration (CNIPA) or alternatively derived from an international patent application filed under the Patent Cooperation Treaty (PCT).

China has a first-to-file patent system. The earliest patent application for an invention will be granted, provided that the application satisfies other relevant requirements for grant stipulated by the Chinese Patent Law.

1.2 Examination system

Invention patent applications[1] are subject to a full examination by the Chinese Patent Office.

When a patent application is filed or Chinese national phase entry of a PCT application is effected, the Patent Office conducts a formalities examination of the application and, if applicable, invites the applicant to correct any formal defects in the application. Once the formalities examination has been completed, and unless the applicant requests early publication, the application will be published 18 months from the filing date or, if applicable, the earliest priority date or, in the case of Chinese national phase entry of a PCT application after international publication, as soon as technical preparation for publication has been completed.

1 As opposed to applications for Utility Models which are discussed in section 6 below.

Search and examination of a patent application are conducted at the applicant's request, provided that such request is made and relevant fees are paid within the prescribed period, ie, 36 months from the filing date or, if applicable, the earliest priority date. Substantive examination may commence earlier upon explicit request of the applicant. In rare cases, search and substantive examination may be conducted at the Patent Office's own volition.

When an examination report issues, the applicant is given an extendable period of time (usually four or two months) to file any observations and, if necessary, make amendments to the application. Once the application is found to be in an allowable form, a Notice of Allowance will issue, which sets a two-month period of time for the applicant to pay relevant grant and renewal fees. When the fees have been paid, the application will be sent to grant and a corresponding patent will be published in due course.

If after observations have been filed and/or, where appropriate, amendments have been made to the application, the application still does not meet with the requirements for grant as stipulated by the Patent Law, the Patent Office may issue a decision refusing the application. Such a decision is issued without giving the applicant an opportunity for hearing. Where such a decision issues, the applicant is entitled to request re-examination of the application by the Re-examination and Invalidation Department of the Patent Office.

Divisional applications can be filed while a parent application is pending. Second and later generations of divisional applications may also be filed while an original application is pending. However, no cascading of divisional applications is allowed after an original (ie, the first filed parent) application is granted or refused.

1.3 Amendments

There are several opportunities for an applicant to make voluntary amendments to a patent application at the Patent Office. For a patent application filed directly at the Chinese Patent Office, voluntary amendments may be filed at the time of requesting substantive examination or within three months after notification of entry into the substantive examination stage. For a PCT application which has entered the Chinese national phase, voluntary amendments may also be filed at the time of national phase entry.

It is also possible to amend the patent application in response to examination reports, but at these later stages amendments are admissible only if they are made for the purpose of addressing objections raised in the examination reports. Accordingly, the following types of amendments are normally not admissible at these stages:

- amendments that seek to broaden the scope of independent claims by way of deleting or altering technical features;
- amendments that introduce into claims any subject matter that is

present in the description but does not have unity with the pending claims; and
- addition of claims the subject matter of which was not presented in the original claims.

Amendments to a patent application may not introduce matter that was not disclosed in the application as filed.[2] Matter disclosed by a patent application as filed is limited to what is directly and unambiguously derivable from the description, claims and drawings of the application as originally on file. In practice, the test is often applied in a rather strict manner and any deviation from the original wording of the description or claims may risk attracting objections under Article 33 of the Patent Law.

1.4 Timescales

Typically, patent applications are searched and examined within the first year after entry into the substantive examination stage. In most cases, examination can be completed within three years from entry into the substantive examination stage.

Several routes are available for requesting accelerated prosecution of Chinese patent applications. The Patent Office has signed up bilateral Patent Prosecution Highway (PPH) agreements with Patent Offices of a number of countries and participates in the IP5 PPH pilot. The Patent Office also implements the "Guidelines for Accelerated Examination of Patent Applications",[3] which allow accelerated examination of patent applications:

- relating to inventions in selected technical fields encouraged by the state, provincial or municipal governments, eg, energy saving, environmental protection, new generation of information technology, biology, new materials, etc;
- relating to inventions that the applicants have started to implement or made serious preparations to do so, or have evidence showing that third parties are using;
- relating to inventions of great importance to national or public interests; or
- which serves as a basis for foreign applications.

Where requests for accelerated examination are allowed, the Patent Office is obliged to, where appropriate, issue the first examination reports within 45 days and, if applicable, complete the examination and grant procedures within 12 months. Applicants will be given shorter time period, typically 2 months, for responding to examination reports.

---

2 Article 33 of the Chinese Patent Law.
3 Effective from 1 August 2017.

1.5 Outline of costs

The main official fees for filing, prosecuting and grant of an application are as follows:

- filing and publication fee RMB 950;
- examination fee RMB 2,500; and
- stamp duty payable at the grant stage RMB 5.

Renewal fees are due at the time of grant, for the patent year within which the grant date falls, and each year after grant and range from RMB 900 for the first year to RMB 8,000 for the 20th year.

2. Examination of software patents

Issues of patentability for software inventions[4] can be raised pre-grant in substantive examination proceedings or post-grant in invalidation proceedings.

During substantive examination, a Chinese Examiner may refuse to assess novelty and inventive merit of certain claims, if he or she believes that the claims relate to computer programs as such, or exclude computer programs from such assessment if the claims include technical features in addition to the computer programs.

2.1 Inventions

Article 2 of the Chinese Patent Law defines inventions for the purposes of invention patents[5] as "new technical solutions proposed for a product, a process or the improvement thereof". A "technical solution" is an "aggregation of technical means applying the laws of nature to solve a technical problem".[6]

In addition, the Patent Law[7] stipulates that, in order to qualify for patent protection, an invention needs to be novel, involve an inventive step and be capable of industrial application.

For software patent applications, determination of technical solutions, either when assessing inventive merit of a claimed subject matter (discussed in section 2.4 below), or separately when assessing whether the claimed subject matter lacks a technical solution,[8] normally follows a three-step test:

- whether the claimed subject matter involves execution of computer programs in order to solve a technical problem;
- whether the computer programs are executed by a computer so as to

---

4 A software invention is defined in the Guidelines for Patent Examination (hereinafter referred to as the 'Guidelines') as "a solution to an identified problem, which is wholly or partly based on processes of computer programs, for controlling or processing external or internal objects of a computer via execution of the computer programs by a computer".

5 As opposed to utility model patents and design patents which are discussed in sections 6 and 7 below.

6 Guidelines, Part II, Chapter 1, Section 2.

7 Article 22 of the Chinese Patent Law.

8 Article 2(2) of the Chinese Patent Law.

control or process internal or external objects of the computer in accordance with the laws of nature;
- whether any technical effect is achieved in accordance with the laws of nature by execution of the computer programs.

Failing any of these three steps, the claimed subject matter will not be regarded as a technical solution as required by the Chinese Patent Law. In other words, the claimed subject matter will not qualify as an 'invention' in the sense of the Chinese Patent law.

2.2 Exceptions to patentability

Several types of creations are excluded from patent protection. Such exceptions include, inter alia, scientific discovery and rules and methods for performing mental acts.[9]

The Patent Office Guidelines further clarify[10] that if a claim relates to a rule or a method for performing metal acts or for doing business, but includes technical features in addition to such rule or method, then the claimed subject matter as a whole should not be regarded as a rule or a method for performing mental acts or doing business and should not be excluded from patent protection under Article 25 of the Chinese Patent Law.

More specifically, the Guidelines state[11] that patentability of inventions relating to computer programs[12] should be determined following the principle set out for rules and methods for performing mental acts. Accordingly, if a claim relates to a mathematical method, a computer program as such, a computer program recorded in a medium as such, a rule or method for playing games as such, then the claimed subject matter falls within the scope of "rules and methods for performing mental acts" and is excluded from patent protection under Article 25 of the Chinese Patent Law. The other side of the coin is – if the claim includes a mathematical method, a computer program, a computer program recorded in a medium, or a rule or method for playing games, as well as other technical features, patentability cannot be ruled out for the claimed subject matter under Article 25(2) of the Chinese Patent Law. Of course, before the claimed subject matter is found to be allowable, there are other hurdles that need to be overcome, such as the requirements of novelty, inventive step and industrial applicability.

---

9 Article 25 of the Chinese Patent Law.

10 Guidelines, Part 2, Chapter 1, Section 4.2.

11 Guidelines, Part 2, Chapter 9, Section 2.

12 Computer programs are defined in the Guidelines as "coded instruction sequences executable by an information processing device, e.g., a computer, to obtain certain results", or "symbolized instruction sequences or symbolized statement sequences that can be automatically transformed into coded instruction sequences".

2.3 Novelty

Novelty means that the invention is neither a part of the state of the art, nor the subject of an intervening patent application filed at the Patent Office. With only a few exceptions (eg, disclosure at international exhibitions or disclosure resulting from breach of confidence) the state of the art encompasses technical information known to the public anywhere in the world prior to the filing date or, if applicable, the relevant priority date, of the patent application in question. An intervening patent application is a patent application for the same invention that was filed before but published after the filing date of the present patent application.

2.4 Inventive step

Inventive step means that, having regard to the state of the art, the invention includes prominent substantive features and represents remarkable improvements.

A problem-solution test is often adopted by the Patent Office and courts in order to assess whether an invention involves an inventive step. The test includes four steps:

- identifying the closest prior art;
- comparing the claimed subject matter against the closest prior art and identifying distinguishing features;
- defining a technical problem solved by the claimed subject matter based on technical effects achieved by the distinguishing technical features; and
- assessing whether the technical solution provided by the claim is obvious having regard to the closest prior art and in view of any other relevant disclosure.

Following this test, if the answer to the question in the final step is 'no', one can then conclude that the subject matter of the relevant claim involves an inventive step. If the answer is 'yes', if a technical solution cannot be properly defined, or if a technical problem cannot be identified in the third step, the subject matter of the relevant claim will be deemed to lack an inventive step.

2.5 Industrial applicability

Industrial applicability means that the invention can be used for production or be utilised, and may produce positive results. In practice, it is rare for patent applications to be refused in examination or found invalid in post-grant proceedings on this ground.

2.6 Drafting software patents

In order to obtain appropriate patent protection for software inventions, certain rules need to be respected when drafting applications for such inventions.

A claim drafted as a process for resolving a technical problem and reciting specific steps performed by way of executing computer programs to perform specific functions can normally meet the requirement of Article 2(2) and does not fall within the exception of Article 25(2) of the Chinese Patent Law.

A product claim normally needs to be drafted in a way that it spells out not only each specific component of an apparatus, but also connections between the components and describes how each function of the computer program is performed by a corresponding component or group of components.

If it is not practicable to define the components by their physical structures, it is not necessary to do so. The components can then be defined in means-plus-function language. If means-plus-function language is used for a product claim, it is important to ensure strict one-to-one correspondence between recited means and steps recited in the corresponding method claim. Although the Guidelines allow means-plus-function language to be used for product claims, in practice it is not uncommon for Chinese examiners to raise objections against means-plus-function claims on grounds of lack of support by the description.

The Chinese practice still does not allow certain types of claims. For example, claims to data structures are not patentable, but such claims can be amended to define methods for generating the relevant data structures which are potentially allowable. Claims to computer programs as such are normally not allowable either, but claims to computer-readable medium storing such computer programs are now allowable provided that a technical effect can be achieved as a result of execution of the computer programs. Note that claims to computer-readable media whose characteristics are dictated by the stored computer program as such are not allowable.

Therefore, terms such as 'data structure', 'signal', 'computer program product' etc should not be used as definition of inventions in claims if fast grant is desired. However, for Chinese patent applications based on earlier foreign applications (eg, Chinese national phase entry of PCT applications, or direct Chinese filing and claiming priority from earlier foreign applications), it may be worthwhile keeping such claims at the time of entry/filing in the event that the Chinese practice changes in the future to allow such types of claims. Often, amendments or new claims (eg, in divisional applications) based on originally filed claims are more likely to be accepted by Chinese examiners than those made based on the description.

Description of Chinese patent applications also needs to be carefully crafted to support amendments to the claims. Chinese examiners are often reluctant to allow claim amendments unless there is almost verbatim support in the specification. This is particularly relevant if the patent applications originate from jurisdictions which adopt less strict policies on added matter issues than China.

Apart from describing the solution as a whole, the description needs to clearly and completely describe the design and technical features involved in the computer programs that are necessary for the desired technical effects. It is advisable to include in the description technical problems that the prior art fails to solve and how the invention solves such problems specifically. Where appropriate, it is also advisable to include some description of technical effects achieved by each technical feature or group of technical features. Chinese examiners tend to be more prepared to accept that the invention is technical if they see such description. Arguments for technical effect not based on the description, even if presented with supporting evidence, are often not accepted by Chinese examiners.

Normally at least a main flow chart needs to be included in the description of a software invention. Where specific functions or branches of a computer program involve multiple steps, it is advisable to also include flow charts corresponding to the specific functions or branches. The description needs to clearly and completely describe each and every step shown in the flow charts. If particular steps are not essential, it is important that the description explicitly says so. Otherwise, the applicant may be forced to include corresponding steps and means in the independent claims.

Where execution of computer programs involves changes to the hardware of a computer, a diagram showing the structure of the computer should be included, and the description should describe the relevant hardware components and connections between them in a way that is clear and sufficient for a skilled reader to implement the invention.

3. Invalidation

After a patent has been granted (publication of grant), any person may request at any time a declaration of invalidity if he or she believes that the requirements for grant as set out in the Chinese Patent Law are not met.

Grounds for invalidation include:

- the claimed subject matter is not an invention;[13]
- the claimed subject matter does not meet with the requirements of novelty, inventive step and industrial applicability;[14]
- the claimed subject matter is excluded of patentability;[15]
- the invention is not sufficiently disclosed by the description so as to allow a person of ordinary skill in the art to work the invention;[16]
- the invention is not clearly and precisely defined by the claims and supported by the description;[17]

---

13 Article 2 of the Chinese Patent Law.
14 Article 22 of the Chinese Patent Law.
15 Article 25 of the Chinese Patent Law.
16 Article 26(3) of the Chinese Patent Law.
17 Article 26(4) of the Chinese Patent Law.

- the independent claim(s) lack(s) essential feature(s);[18]
- double patenting;[19]
- the invention violates relevant laws, public morality or is a detriment to the public interest;[20] and
- the patent contains matters not disclosed in the application as filed[21] or the patent as granted based on a divisional application contains matters not disclosed in the patent application as filed.[22]

Invalidation proceedings are handled by the Re-examination and Invalidation Department of the Patent Office. Decisions from the Re-examination and Invalidation Department are appealable to the Beijing IP Court.

Where a patent is declared invalid, it is deemed to have never existed. However, a declaration of invalidity does not have any retrospective effect with respect to court decisions, settlements, licensing agreements, and assignment agreements that have already been enforced or executed, unless damages are suffered by a party as a result of bad faith acts of the patentee or it is clearly against the fair and reasonable principle to not allow recovery of damages, licence fees or assignment fees.

Claims can be amended in post-grant invalidation proceedings, but allowable types of amendments at this stage are limited to:[23]

- deletion of claims;
- deletion of alternative solution(s) presented in a single claim;
- limitation of claims by way of adding technical features presented in other claims; and
- correction of obvious mistakes.

4. Infringement

4.1 Acts of infringement

After the grant of an invention patent or a utility model,[24] no entity or individual is entitled to, without permission of the patentee, exploit the patent, that is, to make, use, promise the sale of, sell or import the patented product, or use the patented process and use, promise the sale of, sell or import the product directly obtained from the patented process, for production or business purposes.[25]

After the grant of a patent for a design, no entity or individual shall, without

18 Article 22(2) Implementing Regulations of the Chinese Patent Law.
19 Article 9 of the Chinese Patent Law.
20 Article 5 of the Chinese Patent Law.
21 Article 33 of the Chinese Patent Law.
22 Article 43(1) of the Implementing Regulations of the Chinese Patent Law.
23 Guidelines, Part IV, Chapter 3, Section 4.6.2.
24 Utility models are discussed in section 6 below.
25 Article 11 of the Chinese Patent Law.

permission of the patentee, exploit the patent, that is to say, they shall not make, promise to sell, sell, or import the product incorporating its or his patented design, for production and business purposes.[26]

Definition of infringing acts in China is largely in line with most other counties.

4.2 Claim interpretation

Since China is a civil law jurisdiction, there is no Markman hearing or similar procedure. Patent claims are interpreted by courts at trials before determination of infringement.

In a patent trial, a Chinese court would first try to interpret the claim language with reference to the specification and drawings. If the specification and drawings are not sufficient to allow this task to be completed, extrinsic evidence, such as reference books, textbooks and the understanding of experts in this technical field, will be referred to.[27]

In addition, doctrine of equivalents is also adopted in determining the protection scope of a claim. If the accused technical features in the infringing products use substantially the same means, achieve substantially the same function(s), result in substantially the same effect(s), and this is obvious to the person skilled in the art after the application date of the patent but before the date of the infringement, the court shall determine that the accused features are equivalent to the claimed features.[28]

4.3 Contributory infringement

Until quite recently, China did not adopt the principle of contributory infringement like some other countries. Contributory infringement was tried by courts under the theory of joint infringement in civil law.

However, in a Judicial Interpretation from the Supreme Court, which became effective in 2016, contributory infringement and inducement were introduced into the legal system.

If a party, knowing that certain products are the materials, equipment, parts and components or intermediate items, provides such products to another party committing patent infringement, such activity is an act of contributory infringement. If a party, knowing that a product or process has been granted patent induces another party to commit patent infringement, such activity is an act of inducing another party to commit infringement.[29]

---

26 *Id.*

27 Articles 2 and 3 of the Interpretation of the Supreme People's Court on Several Issues concerning the Application of Law in the Trial of Patent Infringement Dispute Cases.

28 Several Provisions of the Supreme People's Court on Issues concerning the Application of Law in the Trial of Cases on Patent Disputes (2015 Amendment).

29 Article 21 of Interpretation (II) of the Supreme People's Court on Several Issues concerning the Application of Law in the Trial of Patent Infringement Dispute Cases.

5. Enforcement

5.1 Obtaining information

In civil litigation in China, the plaintiff bears the burden of proof. There is no discovery procedure in Chinese litigation proceedings, which increases the plaintiff's burden for evidence collection. On top of the already heavy burden, the court usually requires notarisation of the infringing evidence when their authenticity is challenged by the other party.

Notary purchases are fairly common in cases involving electronics products. In those cases that the infringing products are not purchased easily, a private investigation firm may be hired for collecting evidences. If infringement cannot be easily identified from the products, such as in software patent infringement cases, reverse engineering may become necessary.

5.2 Case docketing

As an initial matter in any infringement proceeding, a determination is made as to which court has jurisdiction over the matter. In principle, the court located in the place where the infringing act took place or where the defendant is domiciled has jurisdiction over an infringement proceeding.[30] In practice, the plaintiff often tries to ensure a case is heard at one the specialised IP courts based in Beijing, Shanghai and Shenzhen through forum shopping, where the judges are more experienced with patent cases.

To enable a case to be docketed, the plaintiff must have the following materials on hand:

- a civil complaint;
- preliminary infringing evidence; and
- formal documents, such as a Power of Attorney and Certificate of legal representation.

The civil complaint and preliminary evidence need not be comprehensive. The plaintiff will be given opportunities to supplement arguments and evidence during the proceeding.

Formal documents such as a Power of Attorney and Certificate of legal representation have to be notarised and legalised if they are produced overseas, which often takes time. Thus, it is advisable to have formality documents prepared before case docketing.

5.3 Interim relief

Interim relief is available for the plaintiff to prevent infringement before judgment is rendered.

30 Article 29 of the Civil Procedure Law.

Before any preliminary injunction (sometimes referred to as act preservation in China) is granted, the court considers the following factors:

- stability of the IP right and the likelihood of success;
- whether delay or refusal of injunctive relief will likely cause irreparable harm to the right owner;
- a balance of the parties' hardships in favour of injunction; and
- the public interest.[31]

The main challenge for obtaining preliminary injunction often lies in the task of proving irreparable harm. The following factors are usually considered by Chinese courts when deciding whether irreparable harm exists:

- where the behaviour of the defendant will infringe the plaintiff's personal rights such as the right to goodwill or publication right, privacy, etc and will cause irreparable harm;
- where the behaviour of the defendant will lead to uncontrollable infringement and significantly increase the loss suffered by the plaintiff;
- where the infringement of the defendant results in a significant reduction in the relevant market share of the plaintiff; and
- where other irreparable harm is suffered by the plaintiff.

A guarantee from the plaintiff is also required in case the injunction is wrongly granted.

5.4 Trial

Trials for patent infringement case usually last half-a-day to a day, and there may be more than one trial if issues remain after the first trial. There is no jury trial in China.

5.5 Remedies

(a) *Permanent injunction*

In most patent cases, a permanent injunction is granted if infringements are found. There are only exceptional cases where injunctions are not granted and normally permanent injunctions are only ever refused on public interest grounds.

(b) *Damages*

The plaintiff is entitled to damages if his case is supported by the court. The amount of compensation for a patent infringement shall be determined on the following basis:[32]

---

31 Provisions of the Supreme People's Court on Several Issues concerning the Application of Law in Examining Cases Involving Act Preservation in Intellectual Property Disputes.

32 Article 65 of the Chinese Patent law.

- the actual losses incurred to the patentee as a result of the infringement;
- if it is difficult to determine the actual losses, the actual losses may be determined on the basis of the gains which the infringer has obtained from the infringement;
- if it is difficult to determine the losses incurred to the patentee or the gains obtained by the infringer, the amount shall be reasonably determined by reference to the multiple of the royalties for this patent;
- if it is difficult to determine the losses incurred to the patentee, the gains obtained by the infringer as well as the royalty obtained for the patent, the court may, by taking into account such factors as the type of patent, nature and particulars of the infringement, etc, decide a compensation in the sum of not less than RMB 10,000 but not more than RMB 1 million; and
- in addition, the compensation shall cover reasonable costs that the patentee has incurred in order to stop the infringement.

6. Utility models

A Chinese utility model refers to any new technical solution relating to a product's shape, structure, or a combination thereof, which is fit for practical use.

The pros of protecting inventions by Chinese utility models include:

- utility model can be granted quickly, usually in 6–12 months, much shorter than invention patents, which is particularly suitable for fast-changing industries;
- there is no substantive examination on utility models, which means the claim scope can be fairly broad, thus, more threatening; and
- the cost for obtaining utility models is lower than invention patents.

That said, the cons of utility models are also obvious, which include:

- the stability of utility models are low due to the lack of substantive examination;
- an evaluation report is required if an infringement action is brought based on a utility model;
- the protection scope of a utility model is restricted to physical product; if a software invention were to be protected by utility model, claims need to be drafted in mean-plus-function language and may be rejected even then;
- the duration of a utility model is limited to 10 years as opposed to the 20-year term of a full innovation patent.

When granted a utility model can be enforced or invalidated in a similar way to an invention patent.

7. Designs

Design patents protect any new design of a product's shape, pattern or a combination thereof, as well as the combination of the colour and the shape or pattern of a product, which creates an aesthetic feeling and is fit for industrial application. Besides ordinary designs, many internet companies may like to protect the user interfaces of their apps and other software. GUI designs can help them obtain such protection.

Determining an infringement of a design patent can be subjective in some cases. The court shall determine whether the designs are the same or similar in accordance with the knowledge and discriminability of an ordinary consumer.[33]

The Supreme Court elaborated the level of an ordinary consumer in a recent Judicial Interpretation, which recites that when determining the level of knowledge and discriminability of an ordinary consumer to a design, the court shall normally consider the design space of the products in the same or similar category as the patented design at the time of infringement. Where the design space is relatively large, the court may determine that it is usually unlikely for an ordinary consumer to notice the minor differences between the compared designs; where the design space is relatively small, the court may determine that it is usually more likely for an ordinary consumer to notice the minor differences between two compared designs.[34]

8. Copyright protection

Copyright protection is available for computer software, but extends only to expressions rather than the underlining ideas *per se*. In order to establish copyright infringement in court proceedings, copyright owners need to prove copying acts by the alleged infringer, which is often difficult in the absence of discovery procedures.

China is a signatory of the copyright treaty known as 'The Berne Convention' and, accordingly, copyright protection for computer software does not depend on compliance with any formalities such as registration or deposit of copies. Nonetheless, China operates a copyright registration system under which right owners, often employers of software designers, can register their copyright by way of submitting, inter alia, copies of source codes of their software. Copyright registration relieves copyright owners of their burden to prove their ownership to the software concerned in copyright infringement proceedings in the first place. Of course, an alleged infringer can challenge the copyright owner's ownership if he or she is able to produce relevant evidence.

Copyright protection is also available for a database in China insofar as

---

33 Article 10 of the Interpretation of the Supreme People's Court on Several Issues concerning the Application of Law in the Trial of Patent Infringement Dispute Cases.

34 Article 14 of Interpretation (II) of the Supreme People's Court on Several Issues concerning the Application of Law in the Trial of Patent Infringement Dispute Cases.

there is originality in the selection and compilation of the content of the database.[35]

9. Trade secrets

Information relating to electronic or software innovations may also be protected as trade secrets. Protection of trade secrets is governed primarily by the Regulations Relating to Prevention of Infringement of Trade Secrets (the Regulations) in combination with the Law of the PRC against Unfair Competition.

Protection of trade secrets extends to the following infringing acts:

- obtaining trade secrets by way of theft, inducement, threats or any other unfair measures;
- disclosing, using or allowing others to use trade secrets obtained by any of the above-mentioned measures; and
- disclosing, using or allowing others to use trade secrets, by an employee or a party having business connections with the owner of the trade secrets, in breach of relevant contracts or confidentiality obligations.

It is also an infringement of trade secrets where a third party obtains, uses or discloses the trade secrets of others as a result of the above-listed infringing acts, when he or she knew, or could reasonably be assumed to know under the relevant circumstances, the infringing acts.

Trade secrets are defined as technical and business information which is unknown to the public, which is of practical use and may advance the economic interests of the owner, and for which measures have been put in place in order to keep the information confidential.

The Regulations clarify that 'unknown to the public' means that the information cannot be obtained via public channels. The requirement of practical use and for advancing economic interests of the owner means that the information can be used to achieve actual or potential economic benefits or improve the owner's competitive edge. Measures for keeping information confidential includes signing up non-disclosure agreements, setting up confidentiality schemes, and any other reasonable measures for keeping the information confidential.

10. Hot topics and recent changes

10.1 Setup of the Intellectual Property Tribunal in the Supreme People's Court

In accordance with a recent decision from the Standing Committee of the National People's Congress, the Supreme People's Court has set up a brand new

35 Article 14 of the Chinese Copyright Law.

Intellectual Property Tribunal (IP Tribunal).[36] The IP Tribunal started functioning on 1 January 2019.

The IP Tribunal of the Supreme Court will mainly handle second instance appeals of IP cases. More specifically, it takes responsibility over appeal cases of disputes involving invention patents, utility models, new variety of plants, layout design of integrated circuit, trade secrets, computer software and antitrust. Both civil and administrative IP appeals are now handled by the IP Tribunal of Supreme Court directly.

10.2 Establishment of the Internet court

Three specialised Internet courts were established in September 2018. These newly established Internet courts have jurisdiction over the following IP-related cases:

- disputes over the ownership of the copyrights or neighbouring rights of the works published on the Internet for the first time;
- disputes arising from infringements upon the copyrights or neighbouring rights of the works published or disseminated online through the Internet; and
- disputes over Internet domain name ownership, infringements and contracts.[37]

---

36 Decision of the Standing Committee of the National People's Congress on Several Issues concerning Judicial Procedures for Patents and Other Intellectual Property Cases and the Provisions of the Supreme People's Court on Several Issues concerning the Intellectual Property Tribunal.

37 Notice of the Supreme People's Court on Issuing the Plan for Adding the Beijing Internet Court and the Guangzhou Internet Court.

Hong Kong

Adelaide Yu
Rouse Legal Hong Kong

Hong Kong is a small market and often overlooked in terms of IP protection and enforcement. Nevertheless, it can be a strategically important jurisdiction given its status as a trading and shipping centre, particularly for high-tech goods made just across the border in mainland China, and its more recent emergence as a telecommunications hub. Significantly, the IP regimen in Hong Kong is completely separate from China and rights are strictly territorial in scope. A crucial difference lies in Hong Kong's patent system, which has not changed since handover from the United Kingdom in 1997. No substantive examination is carried out; while there is a useful but underutilised system of so-called 'short-term' patents. Standard patents consist merely of re-registrations of patents granted in China or the United Kingdom.[1]

1. National patent procedure

1.1 A registration system

Under the Patents Ordinance[2] there are two types of patent available: standard patents and short-term patents.

(a) *Standard patents*

The issuance of a standard patent in Hong Kong is based on re-registration of a patent granted by one of the designated offices, namely the State Intellectual Property Office (SIPO) in the People's Republic of China and the UK Patent Office or the European Patent Office (only for European patents designating the United Kingdom).

Patent specifications can be filed in either English or Chinese. Given the choice, for linguistic reasons and reasons of legal certainty, many applicants prefer to base their Hong Kong registrations on the corresponding European or UK patent.

The procedure is two-stage. First, an applicant must file a request to record

1 Including European patents designating the United Kingdom.
2 Cap 514, Laws of Hong Kong.

in respect of a published designated patent application. The request to record must be filed within six months of the designated application being published. Following a formalities examination, the request to record is published by the Hong Kong IP Office.

In the second stage, within six months of the designated application being granted, the patentee files a request for registration and grant. The Hong Kong patent is then granted as soon as practicable following another formalities examination. Fees are payable annually from the fourth year onwards.

Should the base patent be revoked, a copy of the revocation must be filed with the Hong Kong patent office, following which third parties may apply to have the Hong Kong patent revoked accordingly.[3] Third parties can also bring revocation proceedings[4] in court against the Hong Kong standard patent itself, without affecting the base patent, though in practice this is rare.

(b) *Short-term patents*

A short-term patent application may be granted in Hong Kong independently of any overseas patents and is filed directly with the Hong Kong Patents Registry. Short-term patents must conform to the same format as a standard patent, including in their specification a title, abstract, description, drawings and claims. Following a formalities examination, the short-term patent is granted as soon as practicable.

An application for a short-term patent must be accompanied by a search report obtained from an international search authority designated under the Patent Cooperation Treaty. However, it is possible to request that grant (and formalities examination) of the application be deferred for up to 12 months, which gives applicants additional time to obtain the search report following their filing.

A short-term patent lasts for a maximum period of eight years, subject to renewal from the expiry of the fourth year from the filing date of the application for a further period of four years.

Short-term patents may claim priority from another application under the Paris Convention. They are also a valid priority basis for other applications under the Paris Convention.

1.2 Timescales

Following completion of the formalities (for which no grace periods or extensions are available) both standard and short-term patents are usually registered within three months.

---

3 Section 44 of the Patents Ordinance.
4 Section 91 of the Patents Ordinance.

1.3 Outline of costs

(a) *Standard patent*

Request to record a standard patent	HK$380
Advertising fee for a request to record	HK$68
Request for registration and grant of a standard patent	HK$380
Advertising fee for a request for registration and grant	HK$68

Annual renewal fees of HK$540 are due for maintaining a standard patent from the third year and each successive year. Annual renewal fees of HK$270 are due on pending standard patent applications for the fifth and each successive year.

(b) *Short-term patents*

Filing application for grant for a short-term patent	HK$755
Advertising fee for grant of a short-term patent	HK$68

A single renewal fee of HK$1,080 is payable from the expiry of the fourth year from the filing date of the application to extend the term of the patent by a further period of four years.

2. Examination of software patents

As patents are not substantively examined, the following applies mainly to revocation proceedings, where non-patentability is a ground for revoking a patent.[5]

Section 93(1) of the Patents Ordinance provides that an invention is patentable if it is susceptible of industrial application, is new and involves an inventive step.

However, according to Section 93(2) of the Ordinance, the following shall not be regarded as inventions within the meaning of subsection (1):

- a discovery, scientific theory or mathematical method;
- an aesthetic creation;
- a scheme, rule or method for performing a mental act, playing a game or doing business, or a program for a computer; and
- the presentation of information.

As a computer program is not regarded as an invention, it is *prima facie* not patentable.

However, there is an exception under Section 93(3) of the Patents Ordinance, which provides that non-patentable subject matter under Section

5 Section 91 of the Patents Ordinance.

93(2) shall only be non-patentable to the extent that it relates to that subject matter as such.

In other words, if something further has been added to a computer program such that the claimed invention as a whole does not relate to excluded subject matter 'as such', it can still be patentable.

Since there has been no relevant case law in Hong Kong regarding the meaning of software 'as such', the Hong Kong IP Office, in its guidance to potential applicants, reiterates the positions of the UK IP Office and the European Patent Office (EPO) – that software can be patentable if it is a "computer related invention that produces a further technical effect ... software that merely automates a process that was previously done mentally or manually is not patentable. But if the software solves a technical problem it is patentable".[6]

What is meant by producing a further technical effect as would be allowed under Hong Kong law is illustrated in the EPO Technical Board of Appeal's decision in *Vicom's Application.*[7] In that case, the applicant used a mathematical method in a process for enhancing digital images. The board held that the claim to this process was not a claim to a mathematical method as such and should not be excluded. A mathematical method as such would not produce any direct technical result; on the other hand a technical process would produce a change in a physical entity by technical means. In modern times, said the board, the choice of whether to implement such a technical invention in hardware or software "is not of an essential nature, but is based on technical and economic considerations".

Similarly, in *Merrill Lynch's Application,*[8] decided under the equivalent of Section 93(2)(c),[9] the English courts were guided by *Vicom's Application*. The claimed invention in the case was an automated securities trading system. The English Court of Appeal held that the data processing system which included the computer program had to be considered as a whole. However, the court concluded that the system was not patentable as any technical result of the data processing system was simply a method of doing business. It was further held that the technical contribution in the computer program must be "some technical advance on the prior art in the form of a new result (eg, a substantial increase in processing speed as in *Vicom*)".

Similar reasoning was applied by the EPO Technical Board of Appeal in the case of *IBM/Card Reader,*[10] decided under the equivalent of Section 93(2)(c).[11] There, the subject matter of the application concerned an automatic self-service

6 FAQ: can a patent protect my computer software? See www.ipd.gov.hk/eng/faq/patents/pat_protection_inhk.htm.

7 EPO Decision T208/84; [1987] OJ EPO 14.

8 [1989] RPC 561.

9 Section 1(2)(c) of the Patents Act 1977.

10 EPO Decision T854/90 [1994] EPOR 89.

11 Article 52(2)(c) of the European Patent Convention.

machine operable upon presentation of a card carrying identification data, as well as a method of operating such a machine. In the board's view, when taken as a whole the claimed method is a method of deciding whether a card which is presented for the first time to a machine should be recognised as an authorised card by that machine. Such a method is part of a business operation. Of course, the claimed method may include steps which include a technical component (eg, using the machine to store and read information). However, the presence of such technical components does not alter the fact that the claimed method is a business method as such, rather than a technical method (just as the use of a typewriter to perform a business activity would not change such an activity into a technical method).

In the more recent UK case of *Aerotel Ltd v Telco Holdings Ltd*,[12] the UK Court of Appeal provided a new four-step approach, applied in subsequent cases, to assess whether inventions relate to a patentable matter. Again, it would be expected that the Hong Kong courts would find the UK Court of Appeal approach highly persuasive when considering technical effect.

Practically speaking, since Hong Kong is a re-registration system, applicants for software patents in one of the three base jurisdictions who wish to cover the territory should apply and should expect the cases to be upheld in Hong Kong if the base applications fulfil the requirements for patentability under local laws.

3. Patent infringement

3.1 Claims interpretation

If all elements of the alleged infringement fall entirely within the scope of one of the claims, the patent is literally infringed.

The scope of a claimed invention is set by Section 76 of the Patents Ordinance, which provides that it is specified by the claims as interpreted by the description and drawings. That section goes on to say that the scope must fall between the extremes of:

- the strict, literal meaning of the wording used in the claims, the description and drawings being employed only for the purpose of resolving an ambiguity found in the claims; and
- the claims serving only as a guideline, the actual protection conferred by a patent extending to what, from a consideration of the description and drawings by a person skilled in the art, the patentee has contemplated.

Instead, a patent is to be interpreted as defining a position between these extremes which combines a fair protection for the proprietor of the patent with a reasonable degree of certainty for third parties.

12 7 RPC [2007].

In the case of non-literal infringement where what is alleged to infringe is some variant of what is claimed, there would still be infringement if:

- the variant has no material effect on the way the invention works;
- it was obvious to a skilled reader, at the date of publication of the specification, that this was so; and
- it would be apparent to any skilled reader that the words of the claim from which the variant departs could not have been intended to exclude such minor variants.

The above was formulated by Lord Hoffmann in the UK House of Lords' decision in *Improver Corp v Remington Consumer Products Ltd*,[13] based on *Catnic Components Ltd v Hill & Smith Ltd*[14] (the basic principle established by *Catnic* was that patents are to be read in a 'purposive' manner) and approved and adopted by the Hong Kong Court of Appeal in *Improver Corporation v Raymond Industrial Ltd* [1991].[15]

In the UK Improver case, the patent covered a depilatory device having a curved 'helical spring' driven by a motor. The rotating spring gripped hairs between its coils and plucked them from the skin. The alleged infringement replaced the spring with a rubber rod with slits. The question was whether the rubber rod was 'a helical spring'. The judge answered the three questions as follows:

- the change to a rubber rod had no material effect on the way the invention worked; and
- it would have been obvious to a skilled reader that the rubber rod would work in the same way; but
- the expert would have understood from the patent that the patentee meant to confine his or her claim to a 'helical spring', in its primary meaning and not in a wide generic sense. As such, the rubber rod did not infringe.

Despite subsequent developments in the United Kingdom somewhat changing its approach to claim interpretation, *Improver* is still the leading case in Hong Kong.

3.2 Acts of infringement

Infringement can be in the form of direct or indirect use.[16] Direct use in relation to a patented product invention means making, putting on the market, using or importing the product, or stocking the product. Where the invention is a

---

13 [1990] FSR 181.
14 [1982] RPC 183 at 237.
15 1 HKLR 251, [1990] 2 HKC 28 (CA).
16 Sections 73 and 74 of the Patents Ordinance.

process, direct use is using the process or offering the process for use in Hong Kong when the third party knows, or it is obvious to a reasonable person in the circumstances, that the use of the process is prohibited without the consent of the proprietor of the patent. Process claims are also infringed by products obtained directly by means of that process, where the products are put on the market, used, imported, or stocked.

Indirect use (corresponding to the UK concept of contributory infringement) is supplying or offering to supply means for putting into effect an essential element of the invention, knowing, or it being obvious in the circumstances to a reasonable person, that the means are suitable and intended for putting that invention into effect in Hong Kong – unless the means are a staple commercial product and the supply or offer to supply is not an inducement to commit infringement.

Exceptions and exemptions to infringement include:

- prior users, including those who have made "in good faith effective and serious preparations" to do acts which would have constituted an infringement had the patent been in force;
- acts done privately for non-commercial purposes; and
- acts done for experimental purposes relating to the subject matter of the patent.[17]

3.3 Network claims

Where a patented invention can be implemented with a computer network, and part of the network is outside Hong Kong, the courts would likely still find infringement as use of the claimed invention is taking place in Hong Kong – it does not matter where the servers are situated.[18]

4. Ownership, inventors and compensation

The person who may apply for a patent on an invention is the person entitled to the property in that invention,[19] following normal principles of law.

Section 57 of the Patents Ordinance sets out special rules in relation to inventions made by employees. Unless agreed to the contrary, an invention shall belong to the employer if:

- it was made in the course of the normal duties of employees or in the course of duties falling outside their normal duties, but specifically assigned to them, and the circumstances in either case were such that an invention might reasonably be expected to result from the carrying out of their duties; or

17 Section 75 of the Patents Ordinance.

18 Following the UK courts in *Menashe Business Mercantile Ltd & Another v William Hill Organisation Ltd* [2002] EWCA Civ 1702.

19 Section 12 of the Patents Ordinance.

- the invention was made in the course of an employee's duties and, at the time of making the invention, because of the nature of those duties and the particular responsibilities arising from the nature of the duties, the employee had a special obligation to further the interests of the employer's undertaking.

Compensation is payable to employees if their invention results in a granted patent that produces an outstanding benefit to their employers. The amount should be a fair share of the benefit derived or reasonably expected to derive from the patent.[20]

5. Copyright protection

5.1 Requirements for protection

According to the Copyright Ordinance,[21] copyright subsists in (among other forms of creation) literary and artistic works. The definition of a 'literary work' includes computer programs and any preparatory design material, and "compilations of data or other material, in any form, which by reason of the selection or arrangement of its contents constitutes an intellectual creation, including but not limited to a table" (ie, databases). Artistic works would include circuit schematics and printed circuit board layouts.

There are no formalities required to obtain copyright protection for a work in Hong Kong. Copyright subsists in a literary or artistic work if it is recorded in writing or otherwise, and:[22]

- the author satisfies the qualification requirements (namely, is an individual domiciled, resident or having a right of domicile in Hong Kong or elsewhere; or is an incorporated entity[23]); or
- the work is published in Hong Kong or elsewhere.

There is a *de minimis* condition, namely input of a minimum level of so-called skill, judgement or labour on the part of the person making the work – although the work itself need not have any particular merit.

Regarding software, there is no definition of 'computer program' in the Copyright Ordinance. In *HKSAR v Chan Tak Tim*,[24] it was argued that while source code qualifies for protection, object code cannot be a literary work as it is unintelligible to humans. The authors' view is that this argument would not hold if properly tested; nevertheless object code should in any case be protected

20 Sections 58 and 59 of the Patents Ordinance.
21 Cap 528, Laws of Hong Kong.
22 Section 177(1) of the Copyright Ordinance.
23 Section 178(1) of the Copyright Ordinance.
24 [2004] 3 HKLRD 112 (CA).

as an adaptation[25] (arrangement or altered version, or a translation) of the source code, which attracts the same rights as the work itself.

Regarding databases, there is also no definition of 'compilation'. In practice it is likely that the Hong Kong courts would be guided by the approach of the European Court of Justice (ECJ). The ECJ found that fixture lists are entitled to copyright if the selection or arrangement of the contents constitutes the author's "intellectual creation", made with "free and creative choices" and not "dictated by technical considerations, rules or constraints".[26] Unstructured 'big data' would likely not qualify for copyright protection, but would when structured for efficient or useful processing.

5.2 Limitations on copyright protection

It is well known that copyright does not protect ideas, but only the expression or product of ideas. As there have been no cases in Hong Kong on this point, UK case law would be followed in practice.

In *Navitaire Inc v easyJet Airline Co and Bulletproof Technologies Inc*[27] and *Nova Productions Ltd v Mazooma Games Ltd & Ors & Bell Fruit Games Limited*[28] it was alleged that there was copying of the interfaces and business functionality of computer software. Both cases failed. In both cases, the source code was not copied.

More recently, in *SAS Institute Inc v World Programming Ltd*,[29] the UK High Court, following reference to the ECJ, held that neither the functionality of software, the programming language used to create it, nor the format of its data files were protected by copyright. Creating compatible or interoperable software is thus not an infringement of copyright.

5.3 Acts of infringement – what constitutes copying?

Infringement always requires copying, and restricted acts of copying incur civil liability. If the infringing copies are made for sale or hire, a criminal offence is also committed.

(a) Copying of a work

Copying of a work means reproduction of a whole or any substantial part of a copyright work either directly or indirectly[30] and in any material form.[31]

The burden is on the claimant to show that the defendant copied his or her

---

25 Section 29 of the Copyright Ordinance.
26 In C-604/10, on a reference from the Court of Appeal (England & Wales) in *Football Dataco Limited & Ors v Yahoo! UK Limited & Ors* [2010] EWHC 841.
27 [2004] EWHC 1725 (Ch).
28 [2007] EWCA (Civ) 219.
29 [2013] EWHC 69 (Ch).
30 Section 22(3) of the Copyright Ordinance.
31 Section 23(2) of the Copyright Ordinance.

work. The courts of Hong Kong have followed[32] the approach taken by the UK House of Lords in the Designer's Guild[33] case:

- identify the features of the defendant's design which the claimant alleges have been copied;
- compare the claimant's and defendant's designs, noting the similarities and differences (but the focus is on the similarities);
- decide whether the similarities the claimant relies on are sufficiently close, numerous or extensive as to be more likely to have resulted from copying than coincidence;
- disregard any commonplace or unoriginal similarities or similarities that consist only of general ideas; and
- if the claimant has then demonstrated sufficient similarity in the features it relies on (not the works as a whole), and establishes that the defendant had prior access to the copyright work, then the burden passes to the defendant.

The leading case on what constitutes a 'substantial part' is again *Designers Guild*: "This is a matter of impression, for whether the part taken is substantial must be determined by its quality rather than its quantity. It depends on its importance to the copyright work. It does not depend upon its importance to the defendants' work ..."

Copying can be inferred from the facts if disputed by a defendant, for example from the number and scope of alleged similarities, or the fame of the original work.

(b) *Copies made for sale or hire*

There are a number of criminal offences relating to copyright infringement which are set out in Section 118 of the Copyright Ordinance.

It is an offence for a person to make for sale, import, export, or sell an infringing copy of a computer program. Further, possession of infringing copies by a business with a view to them being used in the course of that business is also an offence.

5.4 Permitted acts

There are a number of exceptions to copyright infringement, which must be considered under the primary principle expressed by this question: "Does it conflict with normal exploitation of the work by the copyright owner, or unreasonably prejudice the copyright owner's legitimate interests?"[34]

---

32 *In Natuzzi SpA v De Coro Ltd* [2007] HKCFI 21.
33 *Designers Guild Ltd v Russell Williams (Textiles) Ltd* [2001] FSR 113.
34 Section 37 of the Copyright Ordinance.

The exceptions include fair dealing with a work for the purpose of research or private study,[35] criticism or review[36] and incidental inclusion of copyright material.[37]

5.5 Ownership and authorship

A work's first owner is the person who creates it, namely its programmer. Joint authorship is possible in cases where the contributions of different authors are not distinct.[38] In the case of a literary, dramatic, musical or artistic work which is computer-generated, the author is taken to be the person by whom the arrangements necessary for the creation of the work are undertaken; again, the programmer.[39]

Where a computer program is made by an employee in the course of employment, the employer will be the first owner of any copyright in the work subject to:

- any agreement to the contrary; and
- where such work is exploited by the employer or by someone else with the employer's permission in a way that could not reasonably have been contemplated by the employer and the employee at the time of making the work, the employer shall pay an award to the employee in respect of such exploitation at such amount as agreed between the employer and the employee or failing an agreement, as determined by the copyright tribunal.[40]

6. Registered design rights

The design of electronic products can also be protected by registration under the Registered Designs Ordinance if it is new and the appearance of the product is a material consideration when people buy or use that kind of product.[41] While designs are registered without examination, they will likely only be valid and enforceable if primarily aesthetic in nature, not functional. Registered designs can last up to 25 years, renewable every five years.

7. Confidential information

In Hong Kong, trade secrets and undisclosed commercial information are protected either by contract or by the common law duty of confidence. The duty of confidence can be imposed by way of contract. The three elements required to establish a breach of confidence were set out in *Coco v AN Clark (Engineers) Ltd*:[42]

---

35 Section 38 of the Copyright Ordinance.
36 Section 39 of the Copyright Ordinance.
37 Section 40 of the Copyright Ordinance.
38 Section 12 of the Copyright Ordinance.
39 Section 11 of the Copyright Ordinance.
40 Section 14 of the Copyright Ordinance.
41 Section 6 of the Registered Designs Ordinance, Cap 522, Laws of Hong Kong.
42 [1969] RPC 41.

- the information must be confidential in quality;
- that information must have been imparted in circumstances importing an obligation of confidence; and
- there must be an unauthorised use of the information to the detriment of the party communicating it.

There can be overlap between claims of copyright and confidential information in an action. Use which falls short of copyright infringement can nevertheless be misuse of confidential information.[43]

8. Other forms of protection

8.1 Trademarks and passing off

Trademarks used in software and electronic products and their related services to identify their origin can be protected under the Trademarks Ordinance.[44] Unregistered trademarks can also be protected through the common law principle of passing off.

8.2 Semiconductor topographies

The original layout-design or topography of an integrated circuit is protected[45] for 10 years from the end of the year of its initial commercial exploitation (or 15 years from its year of creation, if not exploited).

9. Counterfeiting

The legislation in Hong Kong provides for both civil remedies and criminal sanctions in relation to counterfeiting. The Hong Kong Customs and Excise Department is responsible for investigation and prosecution of criminal infringement cases relating to copyright piracy and trademark counterfeiting. In addition to pursuing counterfeiters, retailers and distributors can be pursued under the provisions of the Trade Descriptions Ordinance,[46] which states that is an offence to sell an infringing copy of software, and electronic products to which a forged trademark has been applied.

A particular problem in the south of China is the counterfeiting of integrated circuits – substandard, defective, used or simply fake chips with their part numbers scrubbed off, their surface finish corrected and branding and new part numbers applied. These integrated circuits are sometimes trans-shipped through Hong Kong.

---

43 *Dr Koo Chih Ling Linda v Dr Lam Tai Hing* [1992] 2 HKLR 314, 345.
44 Cap 559, Laws of Hong Kong.
45 Layout-design (Topography) of Integrated Circuits Ordinance, Cap 445, Laws of Hong Kong.
46 Cap 362, Laws of Hong Kong.

10. Enforcement and remedies

10.1 Obtaining information on the infringer and the infringement

Hong Kong court litigation includes discovery (disclosure), through which parties are obliged to produce documents relevant to the case. In practice, most IP cases in Hong Kong will include a mainland Chinese element, so it is vital to conduct thorough investigations first as disclosure is not available there, and to gather evidence in the proper manner (which in mainland China means via cooperation with a notary public). The availability of disclosure in Hong Kong proceedings can be a useful adjunct to parallel proceedings on the mainland.

10.2 Interim relief

The availability of interim remedies in Hong Kong is similar to that in the United Kingdom. *Anton Piller* (search and seizure) orders and *Mareva* (freezing) injunctions are available on the balance of convenience. While the applicant for such interim measures has a high burden to discharge in order to show that damages would not be an adequate remedy if the measures were not granted, nevertheless such measures are achievable in serious IP cases.

10.3 Threats

Relief is available against groundless threats of patent infringement proceedings,[47] but there is no corresponding provision in respect of copyright. The action by rights owners of merely notifying someone of their rights does not constitute a threat.

10.4 Remedies

Victims of patent infringement are entitled to compensatory damages or alternatively an account of profits, a declaration that the patent is valid and infringed, delivery-up and injunctions. Damages are not available against defendants who can prove that at the date of the infringement they were not aware, and had no reasonable grounds for supposing, that the patent existed. Simply stamping 'patented' on the product is not enough to disprove the defendant's state of knowledge – the actual patent number must be present.[48]

A full set of remedies is also available for relief in case of copyright infringement. These include compensatory damages (assessed on the basis of either the claimant's loss of profits, a reasonable licence fee as negotiated between willing licensor and licensee, or otherwise at the discretion of the judge) or alternatively an account of profits; additional damages, among other things for flagrancy of the defendant's infringement; delivery-up; and

47 Section 89 of the Patents Ordinance.
48 Section 81 of the Patents Ordinance.

injunctions.[49] Damages are not available where the defendant did not know or have reason to believe at the time of their infringement that copyright subsisted in the work.

11. Hot topic: Forthcoming implementation of Patents (Amendment) Ordinance 2016

The Patents (Amendment) Ordinance 2016 was enacted in June 2016 which will introduce major changes into Hong Kong patent law.

Major amendments include:

(a) introducing an Original Grant Patent system for grant of standard patents;
(b) refining the existing short-term patent system;
(c) implementing interim measures pending the development of a full-fledged regulatory regime for local patent practitioners in the long run; and
(d) other technical and miscellaneous amendments.

The amendments are not yet in force and will not come into force until a number of preparatory tasks have been completed.

The publisher would like to acknowledge Tom Grek for his contribution to this chapter in the previous edition.

49 Sections 107 and 108 of the Copyright Ordinance.

India

Prashant Phillips
Lakshmi Kumaran & Sridharan

India is globally synonymous with IT. The extensive IT community in India has become a source of IT/technology-driven innovation. This is evident from the increase in the number of patent filings, particularly in the field of computers and information, a predominant share of which are coming from Indian start-ups based in Bangalore.[1] As local innovations increase so has the demand for protection in India. Historically, cost and time have been two prohibitive factors which have generally kept the technical community away from patents. However, with the recent slew of amendments which provide start-ups with reduced fees and expedited examination, India is fast becoming a go-to patent destination for IT-based innovation.

1. National patent procedure

1.1 Examination system

Patent applications in India may either be a first filing (ie, for an application being first filed in India), or may be an application being filed either through the convention route or being filed as a national phase application pursuant to a Patent Cooperation Treaty (PCT) application. In the former case, an applicant has a time limit of 12 months during which they may enter India.[2] In the case of a national phase application, an applicant has 31 months from its priority date for entering the national phase in India.[3] Under the provisions of the Patents Act, a patent application may be made by either the first and true inventors or the assignee who has acquired the rights in the invention by way of an assignment.[4]

The Patent Rules 2003 provide a series of formal obligations which must be complied with. Among the various formal requirements, an applicant is obligated to appraise the Indian Patent Office of the status of all foreign

1 See www.forbes.com/siteds/sindhujabalaji/2017/10/16/indias-startups-are-filing-more-patents-than-ever-before-heres-why/#319cd20569f5.
2 Section 135(1) of the Patents Act.
3 Rule 20(4) of the Patent Rules 2003.
4 Section 6 of the Patents Act.

applications corresponding to the Indian application.[5] This obligation arises whenever a foreign application is filed and applies both to regular applications as well as where divisional applications or continuation applications are filed.

Indian patent applications are not automatically examined. Rather, examination has to be specifically requested.[6] Failure to file for a request for examination will result in an irrevocable loss of an application. A request for examination must be made within 48 months from the application's priority date. Only after the request for examination is made would the patent application be on course for examination. As a result, it is generally advisable that whenever possible the request for examination should be filed as early as possible. There is no provision of a preliminary examination under the Patents Act. All aspects of examination are conducted in one go, where the examiner and/or controller will examine the patent application on substantive grounds and also check whether all formalities as prescribed under the Patents Act are addressed.

Previously, patent applications were examined by the same branch of the patent office before which it was filed. Since the number of patent applications filed across the branches differed, the examination load was varied. The patent office, in an effort to arrest the mounting pendency, has done away with this constraint.[7] Irrespective of the branch before which the patent application is filed, it may now be examined by any patent office. This move is likely to balance the load between the various branches of the patent office and will also improve the overall pendency of the patent applications.

During substantive examination, the subject matter of a patent application being claimed is reviewed to ascertain whether the invention is novel, involves an inventive step and is capable of industrial application. The subject matter may be further examined to determine whether it falls within the scope of any of the excluded categories listed under Section 3 of the Patents Act along with any formal deficiencies that may have been noted during examination. The examination of the patent application leads to a first examination report (FER). The applicant is required to respond, failing which the application is deemed to be withdrawn.[8]

Based on the response submitted by the applicant, the controller may allow the patent application. In cases where the reply of the applicant was not satisfactory, the controller may provide the applicant with an opportunity for a hearing. During the hearing, the applicant may provide their oral and/or written submissions to address the outstanding objections, if any. If the submissions made are satisfactory, the controller may allow the application. If the submissions fail to address the objections of the controller, the controller

---

5 Section 8 of the Patents Act.

6 Section 11B of the Patents Act .

7 Rule 4(3) of the Patent Rules 2003.

8 Time period for putting the application in order for grant is six months from the date of the FER (Rule 24B(6) of the Patent Rules 2003). The time period for responding may be further extended by a period of three months.

may issue an order rejecting the patent application.[9] The order rejecting the patent application is not final and may be appealed before the Intellectual Property Appellate Board (IPAB).[10] It may be noted that the patent application (including the claims) may be amended during the course of examination, as permitted under the provisions[11] of the Patents Act.

Although the IPAB is established under the provisions of the Trademarks Act, it also acts as the appellate authority and administers various functions under the Patents Act. The IPAB is authorised to hear appeals arising out of the orders of the controller of patents. The order of the controller may not necessarily be issued only with respect to examination but also extends to orders issued during oppositions, compulsory licences, etc. The IPAB is composed of one judicial member and one technical member. For the purposes of the Patents Act, the technical member should have either been a patent controller or a registered patent agent having at least 10 years of practice.

1.2 Pre-grant opposition

The Patents Act provides for a mechanism whereby any person may object to the grant of patent for which an application is pending. The pre-grant opposition mechanism is a unique feature of the Patents Act through which an opportunity may be available for any member of the public to oppose any pending patent application. It may be considered as a supportive or complementary mechanism to the substantive examination conducted by the patent office.

A pending application may be opposed on a number of grounds, namely:[12]

- invention wrongly obtained by the applicant;
- claim is not novel with respect to a prior publication anywhere in the world, or with respect to a prior use in India only;
- claimed subject matter already claimed in a patent application filed before but published later than the application being opposed;
- claim is obvious or lacks an inventive step;
- claimed subject matter is not an invention under the act (ie, the subject matter falls within the scope of Section 3);
- complete specification does not clearly and sufficiently describe the invention;
- claimed subject matter is not enabled by the complete specification;
- applicant failed to keep the patent office appraised of applications filed outside India;
- convention application or national phase application not filed within the stipulated time;

---

9 Section 15 of the Patents Act.
10 Section 117A of the Patents Act.
11 Section 57 of the Patents Act.
12 Section 25(1) of the Patents Act.

- source of the biological material used in the invention is not disclosed or incorrectly mentioned in the specification; and
- invention anticipated by traditional or community knowledge.

In order to initiate pre-grant opposition, an opponent is required to file representations as per the prescribed rules.[13] The representations submitted by an opponent would, however, only be considered when a request for examination for the application being opposed has been filed. At this stage, the controller may provide an opportunity to the patent applicant to submit their written submission if the controller is of the opinion that the patent application in its present form cannot be allowed, or if he or she feels that certain amendments to the complete specification are required. Based on the representations made by the opponent and the written submissions presented by the patent applicant, the controller may either: reject the representations thereby terminating the pre-grant opposition; reject the application; or require that the application be suitably amended. Accordingly, the controller may assess whether the patent application is in a position to be allowed.

1.3 Post-grant opposition

As opposed to pre-grant oppositions, the post-grant opposition may be initiated only by a 'person interested'. The term 'person interested' had been analysed by the Supreme Court in the case of *Aloys Wobben and Anr v Yogesh Mehra*.[14] The Supreme Court while relying on the definition provided under Section 2(1)(t) of the Patents Act held that a person interested is one who "has a direct, present and tangible interest with a patent, and the grant of the patent, adversely affects his above rights and would include any individual who desires to make independent use of either the invention itself (which has been patented), or desires to exploit the process (which has been patented) in his individual production activity".[15] To such an extent, only persons (which may include juristic entities as well) either having tangible interests in the patent or plan to use the patented subject matter for their own commercial objectives, would be eligible for initiating a post-grant opposition against the patentee.

The grounds on which a post-grant action may be initiated against the patentee are largely similar to the grounds which may be relied on by an opponent while pursuing a pre-grant opposition. The post-grant opposition process is typically initiated through an opposition notice submitted before the controller of patents. The controller may subsequently notify the patentee of the same. At the same time, the controller further constitutes an Opposition Board which then

---

13 Rule 55 of the Patent Rules 2003.
14 (2014) 15 SCC 360.
15 (2014) 15 SCC 360, paragraph 21.

examines the opposition thus submitted and provides its recommendations to the controller. The controller, based on the recommendations of the Opposition Board, may provide a hearing where both the patentee and the opponent may provide their oral as well as written submissions. Based on the recommendations of the Opposition Board and the submissions made by the patentee and the opponent, the controller may either maintain the patent or allow the opposition thereby revoking the grant of a patent.[16]

A post-grant opposition may be filed only within one year from the date of grant of the patent. After that date an opponent would have to seek revocation of the patent under Section 64 of the Patents Act, before the IPAB.

1.4 Revocation

An interested person may also seek revocation of a granted patent. The revocation proceedings generally proceed before the IPAB pursuant to a petition based on any one or more grounds prescribed under Section 64 of the Patents Act. Section 64 provides for additional grounds when compared to sections pertaining to pre-grant and post-grant oppositions. For example, lack of enablement, prior secret use (which does not constitute a reasonable trial or experimentation), subject matter lacking utility, amendments to the complete specification were obtained through fraud, etc. It may also be noted that where a revocation is sought as a counter-claim in an infringement suit, the revocation is adjudicated by the appropriate High Court.[17]

A patent may also be revoked under the following circumstances:

- Revocation by the controller on direction by the central government where the patented subject matter relates to atomic energy. The patentee may be allowed an opportunity to amend their complete specification;
- Revocation by the central government where the patent is generally prejudicial to the public; and
- Revocation by the controller for non-working of a patent.

1.5 Timescales

Examination of the patent application only commences after an examination request is formally filed. Such a request may be made any time within 48 months from the priority date. Once the request for examination has been filed, it typically takes around three to five years for the application to get examined and perhaps another year before a final decision on an application (either an allowance or a rejection pursuant to a hearing) is made. Delaying filing a request for examination until the end of the 48-month period delays the commencement of the examination of the application. Therefore, if faster

---

16 Section 25 of the Patents Act.
17 Section 64(1) and Section 104 of the Patents Act.

prosecution is desired, it is preferable that the request for examination be filed as soon as possible

1.6 Costs

The official fees for some of the major activities are provided below. In 2014, the amendments included a separate fee structure for start-ups in addition to large and small entities. The applicable fee for small entities is roughly half of the fees specified for a large entity. The table below provides the indicative fees for some of the actions before the patent office.[18]

	Official fee for large entity (US$)	Official fee for start-ups (US$)
Patent application with 30 pages, 10 claims and single priority*	121	24
Requesting examination	303	61
Express request for examination	424	85
Request for early publication	189	38
Petition for condoning delay or other irregularity	121	24
Petition for one-month extension of time	121	24
Application for amendment – before grant	61	12
Application for amendment – after grant	121	24

* Fee may increase depending on the number of claims and number of extra pages.

Renewal fees are payable once the patent is granted. These range from $61 for a large entity in respect of the third year ($12 for start-ups) to $606 for the twentieth year of a patent ($121 for start-ups).

18 Table 1, First Schedule to the Patents Act.

2. Examination of software patents

Software inventions, or computer-related inventions as they are commonly referred to in India, are subject to the same standard of substantive examination as other types of inventions. Namely, the patent office while examining the patent application will broadly assess the following:

- whether the subject matter is novel, involves an inventive step and is capable of industrial examination (as per the definition of an invention provided under the Patents Act);
- whether the subject matter falls within the exclusions as prescribed under Section 3(k) or (m); and
- formal objections in addition to other requirements pertaining to claims, specification, drawings, etc.

2.1 Software-related inventions

In the case of almost all patent applications for computer-related inventions, an applicant will be confronted with a subject matter eligibility objection. The objection on the eligibility of CRIs derives from Section 3(k) of the Patents Act, reproduced below:

> *3. **What are not inventions** – the following are not inventions within the meaning of this Act –*
>
> *(k) a mathematical or business method or a computer program* per se *or algorithms.*

Section 3(k) imposes an absolute bar on mathematical methods, business methods and algorithms. Such an absolute bar does not extend to computer programs, in which case, only computer programs *per se* are excluded from patentability.

The wording of Section 3(k) is similar to the corresponding provisions in the UK Patents Act 1977 and the European Patent Convention (EPC). As with the UK Patents Act, claims relating to mathematical methods, methods of doing business and computer programs are not patentable.[19] This implies that claims directly claiming such subject matter would not be allowable. Similar provisions are also present in the EPC, wherein claims to the above-mentioned subject matter would also not be patentable.[20]

When being examined by the patent office, computer-implemented inventions may, in addition to being objected to as a computer program *per se*, be objected to as either claiming a business method, mathematical method or an algorithm, as discussed below.

(a) Mathematical methods

The Indian Patent Act does not specifically define what constitutes a

---

19 Sections 1(2)(a) and (c) of the UK Patents Act 1977.
20 Article 52(2) and (3) of the EPC.

mathematical method. However, an indication of what a mathematical method may be is found in the *Manual of Patent Office Practice and Procedure* (MPPP). Although the MPPP is not binding on the patent office, the MPPP does provide an indicative basis as to how any issues faced during the examination of the patent application are likely to be considered by the patent office, and to such an extent has a persuasive value.

In the case of mathematical methods, the MPPP provides that mathematical methods include acts involving mental faculty or mental skill. Examples of such processes may include methods of calculation, such as finding roots of numbers and formulating equations. In addition, the MPPP also stipulates that any computer program or software which implements such acts would also be excluded from patentability.[21] This is also consistent with the *Guidelines for Examination of Computer Related Inventions* (CRI Guidelines) released by the Patent Office in 2017.

The IPAB in its order in *Electronic Navigation Research Institute v Controller General of Patents & Design*,[22] while referring to its reasoning in the case of *Yahoo Inc v Controller of Patents*,[23] highlighted that the technical advance compared to the existing knowledge which forms the basis of inventive step, should itself not fall within the scope of Section 3(k). The subject matter of the application which related to principles of chaos theory was held by the IPAB to be subject matter pertaining to a mathematical method, and was therefore not patentable.

The fact that an invention should be considered for assessing subject matter eligibility is present in the CRI Guidelines. The CRI Guidelines at Section 4.5 prescribe that the substance of the claim taken together as whole has to be considered in order to assess whether the claimed subject matter pertains to mathematical methods, or any other categories enumerated in Section 3(k) of the Patents Act.

The Indian Patent Office is likely to raise objections based on this ground in cases where the subject matter involves mathematical functions. For example, subject matter involving digital signal processing may include one or more mathematical equations. In such cases, establishing that the claimed subject matter does not relate to processes which are performed on numerical value only, and also does not result in an output which is only numbers, may be sufficient to overcome the objections raised by the patent office.

Although the Patents Act excludes mathematical methods from patentability under Section 3(k), the CRI Guidelines also clarify that such an exclusion would not apply solely because the claims include mathematical formulae.

---

21 Paragraph 08.03.05.10(b) of the MPPP.
22 OA/26/2009/PT/DEL I Order 145 of 2013.
23 Order 222 of 2011.

(b) Business methods

The MPPP provides the following as business methods:[24]

- activities in a commercial or industrial enterprise;
- activities relating to transaction of goods or services; and
- B2B and B2C business transactions.

Inclusion of technical means may not be sufficient to traverse a business method-based objection. For example, if the subject matter is such that it incorporates a variety of technical means, such as computing networks, but inherently implements a business method, such an invention would not be patentable.

The scope of the business method exclusion has been discussed in the IPAB's decision in the case of *Yahoo Inc v Controller of Patents*.[25] In *Yahoo*, the IPAB on analysing the claimed subject matter concluded that the claimed invention and its technical advance were in fact related to business methods and therefore fell within the scope of excluded subject matter under Section 3(k). Reliance on technical advance to assess subject matter eligibility appears to be in line with the decision of the UK Court of Appeal in the case of *Symbian v Comptroller General of Patents*[26] which was also relied on by the IPAB in its order. While passing its order in the case of *Yahoo*, the IPAB refrained from defining a business method and held that an assessment has to be done on a case-by-case basis.

Objections contending that the claimed subject matter relates to a business method may be typically issued in cases where the invention is being implemented in a business environment, such as in the e-commerce space. Overcoming such an objection may involve establishing that the invention does not relate to business activity or transactions of goods and/or services. Furthermore, the applicant should also seek to establish before the patents office that the claimed subject matter involves a technical advance and that it seeks to address a problem which is technical.

(c) Algorithms

The Patents Act also precludes algorithms from patentability. The MPPP provides an exemplary indication as to what an algorithm may include, listing:[27]

- a set of rules or procedures;
- any sequence of steps;
- any method expressed by way of a finite list of defined instructions; and

---

24 Paragraph 08.03.05.10(c) of the MPPP.
25 Order 222 of 2011.
26 [2009] RPC 1.
27 Section 08.03.05.10 (d) of the MPPP.

- whether it solves a problem or otherwise, and whether it employs a logical, arithmetical or computational method, recursive or otherwise.

The exemplary list provided by the MPPP may appear to be broadly worded and therefore may not be a conclusive basis for determining the true scope of exclusion under the term 'algorithm'. It is also worth noting that the present term does not figure in either the UK Patents Act 1977 or the provisions of the EPC. However, it has been used in the context of mathematical algorithms. For example, the Technical Board of Appeals of the EPO in the case of *Vicom*[28] has treated algorithms as mathematical methods. In *Vicom*, while considering an algorithm to be no different from a mathematical method, the Technical Board held that an algorithm is an abstract concept prescribing how to operate using the numbers.[29] The Technical Board further stated that no direct technical result is produced as a result of an algorithm. The Technical Board also contrasted this by stating that where such concepts are utilised or implemented on a physical entity, the subject matter would be patentable.

Therefore, it may be gathered that the scope of the exclusion under algorithm is limited to inventions which are generally abstract, and which may have no technical or tangible bearing. Since the EPC is *pari materia* with the provisions of the act, it is quite possible that the ratio of the Technical Board in the case of *Vicom* is considered for its persuasive value by the Indian courts.

Claims may be objected to as relating to algorithms if the subject matter appears to be abstract or devoid of any structural or hardware features. Generally, the patent offices have dropped objections relating to algorithms on amending claims reciting such structural or hardware features.

(d) *Computer program* per se

Unlike the previously described categories which are absolutely barred from patentability, the bar on patentability of computer programs is not absolute. As with the wording of Section 3(k), only computer programs *per se* are excluded from patentability. The term *per se*, however, has not been defined either in the Patents Act or in the MPPP. It may be noted that even though claims may not claim a computer program as such, claims to a computing system or methods implemented through a computing program have been considered as claims for a computer program *per se*.

Owing to the similarity in the wording, the provisions of the Indian Patents Act and the EPC/UK Patents Act, are in *pari materia*. Article 52(2) and (3) of the European Patent Convention and Section 1(2) and (3) of the UK Patents Act 1977 contain exclusion for computer programs 'as such'. Both the European Patent

---

28 *Vicom v Computer related invention* – T 0208/84.
29 Paragraph 5 of *Vicom*.

Office (EPO) and the English courts have in their prior decisions concluded that claims which are directed to a computing system operable in accordance with a computer program constitute patent eligible matter provided they carry out or implement a technical process.[30] Under the act, the Indian courts are also likely to reach the same interpretation as the Indian provisions related to computer program *per se* and are in *pari materia* with that of European and UK provisions. In *Vicom*, the Technical Board while deciding on the patentability of inventions implemented using computer programs, stated that any subject matter cannot be rejected merely because it is implemented using a computer.[31] It has long been understood that the mere presence of technical elements would not be sufficient to render the subject matter patentable. Whether the subject matter involves something technical or not may be assessed on determining whether the subject matter under consideration solves a technical problem.

Similar to the above, the UK approach proposes to determine whether a technical contribution has occurred as a result of the invention.[32] In the case of *Symbian*, the Court of Appeal pointed out the importance of a technical contribution in assessing whether the subject matter falls within the non-patentable subject matter ambit. The Court further held that the logical basis for determining whether the claimed subject matter is patentable, is to determine whether it reveals a 'technical' contribution to the state of the art.[33]

Presence of a technical contribution or technical effect has also been relied on by the Delhi High Court in assessing patentability. While deciding on the issue of whether an interim injunction should be allowed, the Delhi High Court held that "any invention which has a technical contribution or has a technical effect and is not merely a computer program *per se* as alleged by the defendant and the same is patentable".[34]

The presence of a technical effect could be ascertained by any one or more of the following factors as had been outlined in the UK High Court in the case of *AT&T Knowledge Ventures/Cvon Innovations v Comptroller General of Patents*.[35] Some of these factors have been relied on – although informally and inconsistently – for determining the presence of technical effect:

- The technical effect is on a process which is carried on outside the computer;
- The claimed technical effect operates at the level of the architecture of the computer;

30 See T 208/84 *Vicom's Application; Symbian v Comptroller of Patents* [2009] RPC 1; *Aerotel Ltd v Telco Holdings Ltd* & Ors [2006] EWCA Civ 1371.
31 T 208/84 *Vicom's Application*, paragraphs 12 and 16.
32 See *Symbian v Comptroller of Patents* [2009] RPC 1.
33 *Ibid*, paragraph 48.
34 Paragraph 120 of *Telefonaktiebolaget LM Ericsson v Intex Technologies (India) Limited* in CS(OS) No.1045/2014.
35 [2009] EWHC 343 (Pat).

- The computer/system is being made to operate in a new way;
- The computer/system operates more efficiently and effectively as a computer; and
- The technical problem is overcome by the claimed invention not merely being circumvented.

Based on the above criteria, the claimed subject matter may be assessed to determine whether it falls within the scope of the exclusion provided under Section 3(k) of the Patents Act. Ascertaining whether the subject matter involves a technical effect would vary from case to case.

(e) ***Mere scheme or rule, performing mental act***

Section 3(m) of the Patents Act excludes activities which may be performed mentally. The MPPP provides that the subject matter which is the outcome of a mental process would be excluded from patentability. Some subject matter which has been exemplified by the MPPP include methods of teaching and learning.

Computer-implemented inventions may also be prone to objections under Section 3(m) of the Patents Act. In India, the patent office has been raising objections in cases where the claims fail to list any hardware or constructional features.

Amending the claims to include such features may be sufficient for overcoming the objections raised against the pending claims. It may be noted that overcoming objections based on this ground may be raised along with objections contending that the claimed subject matter pertains to a computer program *per se*.

2.2 Novelty and inventive step

For computer-related inventions, the criteria for assessing novelty and inventive step remain largely on a par with the UK Patents Act and the EPC.

3. Infringement

3.1 Acts of infringement

The rights that accrue from a patent are defined under Section 48 of the Patents Act. Under Section 48, the patentee is conferred with an exclusive right to prevent any third party from making, using, selling, offering to sell or importing the product embodying the invention. As per the Patents Act, the District Court is the court of first instance for patent infringement actions. If the defendant seeks to challenge the validity of the patent during an infringement action, the entire suit is to be transferred to the appropriate High Court.

3.2 Claim interpretation

During an infringement suit, the claim must be construed. In India, the patent claims have to be given a purposive construction as proposed in the House of Lords' judgment in *Catnic Components Ltd v Hill & Smith Ltd*[36] and also affirmed in its judgment in *Kirin-Amgen, Inc v Hoechst Marion Roussel Ltd.*[37] Reliance on purposive construction has been made by the High Court of Delhi in the case of the *Merck Sharp and Dohme Corporation and ors v Glenmark Pharmaceuticals.*[38]

4. Copyright protection

The Copyright Act 1957 provides the legislative basis for copyright law in India. The Copyright Act affords protection to literary and artistic works.

In Section 2(o) of the Copyright Act, literary works are defined as follows:

> *'Literary work' includes computer programmes, tables and compilations including computer databases.*

Although not evident from the above definition above, 'literary works' have been considered as any written or printed work. The definition notably includes computer program (or programmes as usually referred to in India). As per the Copyright Act, a computer program is defined as follows:[39]

> *'Computer programme' means a set of instructions expressed in words, codes, schemes or in any other form, including a machine readable medium, capable of causing a computer to perform a particular task or achieve a particular result.*

Owing to India being a signatory to the Agreement on Trade Related Aspects of Intellectual Property Rights (TRIPS) and the broad scope of the definition provided in the Copyright Act, reference to computer programs would include source code as well as the object code corresponding to such a computer program. As would be commonly understood, source code refers to the written code of the computer program with the source code being typically expressed using a high-level programming language. The compilation of the source code yields code which is in a machine-understandable form referred to as the object code.

4.1 Registering software programs for copyright protection

An application for registering a computer software in India can be made. Each application can be made in respect of one work only and has to be accompanied by the official fee. The application must be signed by the applicant, who may be the author or owner of the computer software. If the application is submitted by the owner of the copyright, a no objection certificate (NOC) issued by the author of the computer software in the owner's favour must be submitted.

---

36 [1982] RPC 183.
37 [2005] RPC 9.
38 2015 (63) PCT 257 (Del).
39 Section 2(ffc) of the Copyright Act.

Along with the application, the source code and the object code of the computer software are also required to be submitted with the application.

An application can be filed physically in the copyright office or can be sent via post.

An applicant has to notify every person who claims or has an interest in the copyright of the computer software or disputes the rights of the applicant and his or her intention to apply for the registration. If no objection is received from such notified individuals by the registrar of copyrights within 30 days of the receipt of the application, the registrar on being satisfied that the particulars provided are correct, enters the details of the application in the register of copyrights. If the registrar is not satisfied or an objection is raised by a third party in relation to the application, the registrar may further conduct an inquiry. Depending on the outcome of the inquiry, the registrar may register the application.

The registration process is considered to be complete only when a copy of the entries made in the register of copyrights is signed and issued by the registrar of copyrights or by the deputy registrar of copyrights in favour of the applicant.

4.2 Infringement of copyright

The Copyright Act authorises the copyright owner to create, translate and adapt his or her literary works.[40]

Translation has not been defined in the Copyright Act, but it is well established that any translated work has been held to be an original literary work.[41] In the context of computer programs, one may rely on the provisions of the Copyright, Designs and Patent Act 1988 which defines a translation, in the context of a computer program, as a version which is obtained from converting the computer language of the code into a different computer language or code.

Adaptations, on the other hand, are specifically defined in the Copyright Act[42] and cover works which are obtained by converting a work from one form to another. For example, works obtained through conversion of a dramatic work to a non-dramatic work, literary work to dramatic work, etc, would be adaptations and protected under Section 14(a)(vi) of the Copyright Act. It is

---

40 Section 14(a)(v) and (vi) of the Copyright Act.

41 For example, *Hafiz PH Abdul Gaffar Moulavi v Abdurahiman and Anr* [1999] 2 KLJ 686 at para 8, while referring to *Byrne v Statist Co* [1914] 1 KB 622 may be relied on.

42 Under Section 2(a) 'adaptation' means:
in relation to a dramatic work, the conversion of the work into a non-dramatic work;
in relation to a literary work or an artistic work, the conversion of the work into a dramatic work by way of performance in public or otherwise;
in relation to a literary or dramatic work, any abridgement of the work or any version of the work in which the story or action is conveyed wholly or mainly by means of pictures in a form suitable for reproduction in a book, or in a newspaper, magazine or similar periodical;
in relation to a musical work, any arrangement or transcription of the work;
in relation to any work, any use of such work involving its rearrangement or alteration.

worth noting that the definition also generally provides that any work which is obtained through an alteration or rearrangement of an existing work would also be an adaptation under the Copyright Act.

Instances involving a translation or an adaptation of a work, such as works of fiction or dramatic works, are easy to exemplify. However, assessing the same for a computer program may be challenging. For example, various software-based tools presently exist which may be used for converting a computer program from one language to another. It is pertinent to note that each programming language utilises different programming constructs for enabling certain core functionalities. During such a conversion, it is possible that the programming constructs within the parent computer program may undergo certain changes. Such an example would fall squarely within the definition as prescribed under the copyright legislation in the UK. However, such a similar provision is not present in the Copyright Act in India. Considering that any conversion from one programming language to another does in fact qualify as a translation, it is possible that the courts in India would consider such a 'translated' computer program is likely to be protected as a translation and would be an original literary work.

It may be argued that the changes made to the program by such a translation are not substantial and therefore copyright protection may not extend to such a translated computer program. In such cases, it is possible that the courts in India may construe such a translation to be a 'rearrangement' or an 'alteration' and consider such a translated/converted computer program as an adaptation as defined under the Copyright Act. It may be noted that this conclusion has yet to be confirmed through the courts.

Section 51 of the Copyright Act provides for infringement of copyright. While assessing whether the copyrighted work has been infringed, the courts may compare the works in their entirety. The criteria for ascertaining infringement has been laid down by the Supreme Court in *RG Anand v Delux Films and Ors.*[43] The Supreme Court in the present case held on the following aspects for the purposes of ascertaining infringement:

- In case of common sources, the courts are to determine whether the similarities are on substantial aspects of the mode of expression adopted in the creation of the work protected under copyright. To establish infringement, the allegedly infringing copy must be substantial. While considering the similarities, due consideration is to be given to the material differences between the two works. Where such material differences exist, no infringement would be possible.
- Literal imitation with few and/or minor modifications would be copyright infringement.

---

43 (1978) 4 SCC 118.

- The consumer of the work gets an unmistakable impression that the allegedly infringing work is, in his or her opinion, a copy of the prior work.

The above criteria would also be relied on while assessing whether any computer program infringes the copyright in a previously created computer program. It may be noted that infringement may be broadly categorised as follows:[44]

Textual copying: Since a computer program may be represented in a number of different forms (eg, source code, executable, etc), at any stage any one of the forms of computer program would be stored in the computer memory. The assessment in relation to infringement where the entire computer program is copied is straightforward. However, in cases where only a portion of the computer program has been copied, the situation is likely to be slightly different.

Since the copyright protection extends to portions which are a substantial part of the prior work, an assessment for determining whether copying a part of the program infringes the copyright of the prior work would involve in determining whether the copied portion constituted a substantial portion.[45]

Non-textual copying: Non-textual copying may involve copying of the idea behind the computer program, the structure employed by the computer program, and the function and/or language employed for the computer program.

In relation to copying of an idea, it has been held in numerous instances that no copyright exists in an idea, eg, in the case of *RG Anand*. This dichotomy would also extend to a computer program, as a result of which the idea or the functionality underlying the computer program under consideration is excluded from copyright. These principles have been extensively discussed and adjudicated upon by the European Court of Justice (ECJ) in the case of *SAS Institute Inc v World Programming Ltd*,[46] in which the ECJ held that in relation to computer programs "the intellectual creation in devising the form of expression of the program" is only liable to be protected under copyright law. This principle would be equally applicable when considering the functionality of the computer program. The same functionality (or function) being implemented would not necessarily imply that the expression (ie, the source code) of the two computer programs under consideration is overlapping.

---

44 *Copinger and Skone James on Copyright* 17th edition (Sweet & Maxwell, 2019), pp587–591, paras 7-84 to 7-88.

45 See Section 14 of the Copyright Act and the Supreme Court's decision in the *RG Anand* (1978) 4 SCC 118.

46 [2014] RPC 8 at paras 20–27.

4.3 Permitted acts

The Copyright Act also provides certain exceptions under Section 52. Such actions are not considered as infringing. The exemptions provided under the Copyright Act and the UK Copyright Act are similar. In relation to computer programs specifically, the following activities are considered as non-infringing and may be considered as providing the following:

- Backing up – Making copies of the computer program by the lawful possessor. The act of making such copies is permissible only if it is required for the purpose for which it is intended. Furthermore, the act of making back-up copies as a measure against loss or destruction is also permitted. The present exemption is available only to lawful possessors of the computer program.
- Interoperability – In certain circumstances, the programmer may have to study and assess the execution of a computer program so that compatible computer programs may be created. The assessment typically may involve a programmer studying various programmable interfaces so as to determine as to how the computer program (which may be in the process of being designed) is to communicate with other computer programs. The manner in which such assessment is to be performed may involve decompilation. As is generally understood, the process of decompilation generally involves generating a high-level code which may be then compiled later.[47]

It may be noted that the generation of a code through such a decompilation process under normal circumstances is likely to infringe the copyright of the original copyright holder. The Copyright Act provides specific exception to such activities. It may, however, be noted that the present exception is only available to lawful possessors of the computer program.[48]

- Observing, studying and testing of computer programs – The Copyright Act also exempts lawful possessors from utilising copyright-protected computer programs for the purposes of testing. The present exemption serves as an additional condition to the interoperability exemption discussed above.[49] As would be appreciated, the present and the above-mentioned exemption may be considered as generally aimed to encourage the development of an efficient and secure computer program.
- Copies or adaptation from a personally legally obtained copy – Under the present exemption, a person is permitted to either make copies or an

47 *Copinger and Skone James on Copyright* 17th edition (Sweet & Maxwell, 2019), p798, para 9-227.
48 Section 52(1)(ab) of the Copyright Act.
49 Section 52(1)(ac) of the Copyright Act.

adaptation of a computer program. However, this exemption can only be availed if: the computer program from which the copy is intended to be made is a personally legally obtained copy; and the copy or adaptation is only intended for non-commercial use.[50] The present requirement is applicable only in the case of a computer program in which personal use is considered as within the scope of fair dealing of other types of work which are not computer programs.[51]

4.4 Criminal sanctions

The Copyright Act also authorises the copyright owner to initiate criminal proceedings against an alleged infringer under Chapter XIII. These criminal remedies are distinct and independent of other remedies. The copyright owner may opt to simultaneously avail criminal remedies and civil remedies under the Copyright Act against an alleged infringer.

Section 63 of the Copyright Act deems deliberate infringement or abetment of infringement of a copyright in a work as an offence, which is punishable with imprisonment of between six months and three years. The Copyright Act also imposes a monetary penalty which may range between Rs 50,000 to 200,000. Second and subsequent convictions may invite an enhanced term of imprisonment which may extend from one year, and a minimum fine of Rs 100,000. It may be noted that the offence under Section 63 is a non-bailable offence.

For computer programs, the Copyright Act provides for a term of imprisonment of at least seven days to about three years. The courts may also impose a fine of Rs 50,000 up to Rs 300,000.

The Copyright Act also authorises a police officer of required rank to seize, without warrant, all copies of the work which may be allegedly infringing. The police officer may initiate such seizures if he or she is satisfied that an offence under Section 63 has been committed or is likely to be committed.

5. Ownership, inventorship and compensation

Under Indian law, the right to file a patent application lies with the inventors or their successors in title to whom the rights in an invention have been contractually transferred.

If a patent application is filed by anyone who is not the inventor, the applicant is required to establish, before the patent office, that title to the invention has been transferred to them prior to the filing date of the patent application. Rights in an invention must be specifically transferred or assigned. The mere presence of an employee-employer relationship is not sufficient for

---

50 Section 52(1)(ad) of the Copyright Act.
51 See Section 52(1)(a) which considers private or personal use as non-infringing.

the right in the invention to lie with the employer. In the case of an employee inventor, the rights in an invention would continue to subsist with an employee until a formal assignment is executed and the rights are specifically transferred.

The Patent Act is silent in terms of the quantum of monetary compensation that is payable to inventor employees.

Under Indian copyright law, normally the author of any work is the first owner of the copyright which subsists in such a work. However, in the case of works developed by an author during the course of employment under an employer, the ownership in copyright lies with the employer of the author, provided the author is employed under a contract of service or apprenticeship.[52] Section 17 further provides other circumstances in which the ownership may lie with an entity other than the author.

6. Confidential information

Article 39 of the TRIPS Agreement provides for various safeguards against disclosure of information which is secret. Under TRIPS, any proprietor is within his or her rights to prevent disclosure of any information which is lawfully within their control as long as:

- the information under consideration is secret;
- it has commercial value; and
- reasonable steps to keep such information secret have been undertaken.

Despite India being a signatory to TRIPS, no specific legislation on protection of sensitive information is currently in place.

Even though a specific legislation is absent, adequate contractual measures may be implemented for safeguarding confidential information. These can be implemented through non-disclosure agreements, which may be specifically enforced through the courts.

7. Other means of protection

7.1 Registered design protection

The Designs Act 2000 aims to protect the aesthetic features applied to any article. Under the Designs Act, a design is defined as any features of shape, configuration, pattern, ornament or composition of lines or colours. To qualify as a design, such features are to be applied to an article, which in the finished article result in the aesthetic appeal or appeals in the eye of the observer.[53]

As per the definition, a design does not include various mechanisms of

---

52 Section 17(c) of the Copyright Act.
53 Section 2(d) of the Designs Act.

constructions or articles which are in substance mere mechanical devices. A design is not permitted to include a trademark.

Many computer programs are accessible through a graphical user interface (GUI). The registrability of GUIs under the Designs Act in India is still an open issue. It is, however, pertinent to note that nothing in the Designs Act imposes any bar on the registration of GUIs. If a GUI is considered to be a feature under Section 2(d) of the Designs Act, then applying such a feature to a display screen of a computing device may result in the GUI being treated as a design under the Designs Act. It may also be noted that the List of Classes accompanying the Design Rules 2001 also include a classification pertaining to screen displays and icons.[54]

Although there is no jurisprudence that supports the above, the registrability of GUIs is also supported by various UK case law. The registrability of GUIs was discussed by the UK Registered Designs Appeal Tribunal in its order in *Apple Computer Incorporated v Design Registry*.[55] The present order was passed under the Registered Designs Act 1949 which is similarly worded to the provisions of the Designs Act. In its order, the Tribunal held that icons can be considered as patterns or ornaments applied to an article by an industrial process and hence liable to be registered under the Registered Designs Act 1949. While passing the order, the Tribunal had also held that such icons are in fact patterns or ornaments applied to an article (ie, a display screen), and therefore are designs which can be registered. Considering that the UK and the Indian provisions are *pari materia*, it is very likely that the courts in India will find in favour of registrability of GUIs.

7.2 Semiconductor topographies

The Semiconductor Integrated Circuits Layout Design Act 2000 (SICLD Act) came into effect on 1 May 2011. The present legislation specifically aims to protect a layout design of transistors and other circuitry elements, which may be expressed in a specific manner in semiconductor integrated circuits. Under the SICLD Act, the registered proprietor of a registered layout-design possesses the exclusive right to the use of the layout-design and to obtain relief in case of infringement.

The period of protection granted under the SICLD Act is 10 years from the date of filing an application for registration. It may be noted that up until 2019, two certificates for layouts have been issued by the SICLD Registry.

8. Enforcement

Intellectual property may be enforced either before a court of law, or by way of border enforcement mechanisms.[56] It may be pertinent to note that the border

54 Class 14-04: Screen Displays and Icons.
55 [2002] FSR 38.

enforcement route is no longer available for patent infringement matters owing to an amendment to the Border Enforcement Rules which specifically excluded Patents and the Patents Act 1970 from its purview.

The Commercial Courts, Commercial Division and Commercial Appellate Division of High Courts Act 2015 (Commercial Courts Act) took effect on 23 October 2015. The Commercial Courts Act provides for dedicated courts for adjudicating commercial disputes. Under this legislation, commercial disputes include any disputes that may arise out of IP rights, such as patents.[57] Therefore, all patent disputes would be adjudicated before the Commercial Courts. The Commercial Courts Act was further amended in May 2018. As per the Commercial Courts Act, the states were empowered to set up Commercial Courts at district level. Furthermore, the amendments also reduced the specified value of the commercial dispute to INR 300,000. So far, the majority of patent infringement actions have been instituted before the High Court of Delhi, followed by the High Court of Bombay. The Commercial Courts Act is likely to have a positive impact on the time taken for adjudication patent infringement disputes. This is likely to result in the judges becoming more familiar with patent law and its concept.

8.1 Obtaining information on the infringer and the infringement

Appointment of a local commissioner is prescribed within the Civil Procedure Code 1908 by virtue of which a patent right holder may request the same, while seeking an injunction. The appointment of the local commissioner is through a court order, which allows the local commissioner to inspect and confiscate infringing goods.

8.2 Interim relief

In cases of patent or copyright infringement, the patentee or copyright owner may seek to obtain an interim injunction in order to secure a business advantage over the alleged infringer. The courts in certain cases have granted interim injunctions in a patent infringement suit, for example, based on consideration of the following objective criteria:

- whether the plaintiff has a *prima facie* case;
- whether the balance of convenience is in favour of the plaintiff;
- whether the plaintiff would suffer irreparable injury if a temporary injunction is refused. Mere monetary loss has been held not to be an irreparable injury; and
- public interest considerations in certain circumstances.[58]

---

56 Intellectual Property Rights (Imported Goods) Enforcement Rules 2007 under the Indian Customs Act 1962.

57 Section 2(1)(c)(xvii) of the Commercial Courts Act.

58 *Merck Sharp and Dohme Corporation and ors v Glenmark Pharmaceuticals* 2015 (63) PCT 257 (Del).

8.3 Remedies

Under the Patents Act, the patentee asserting infringement may seek reliefs such as damages, accounts of profit, permanent injunction, destruction of goods etc. In addition, urgent reliefs such as interim injunction, seizure of goods, etc, may also be paid for. It may be pertinent to note that even though the patent may have been granted at a later date, claims for damages may be made from the date of publication.[59]

Similarly, the Copyright Act provides for civil remedies to the copyright owner in the event that their copyright work is infringed besides permanent injunction. Section 55 of the Copyright Act provides that a copyright owner is entitled to remedial action in the form of an injunction, damages and accounts in the event of infringement of their copyright.

9. Hot topics

The Indian patent regime has undergone a series of administrative changes in recent years. On the administrative front, the patent office has inducted a fresh set of examiners and appointed new controllers to handle the pending backlog, particularly for applications in the field of computer science and electronics. The Patent Rules have also been amended recently, with additional amendments in the process of discussion. The recent proposed amendments seek to address the issues that have been raised by various stakeholders in relation to working statement (Form 27). There is also a proposal to introduce a Pilot PPH programme between the Indian Patent Office and the Japan Patent Office. This is likely to ease the ever-increasing backlog of pending applications.

---

59 Section 11A of the Patents Act.

Japan

Nobuhito Ishihara
Nobuyuki Kashiwa
Takashi Katagiri
Keisuke Soeda
Shoji Yoshida
Kyowa Patent and Law Office

There are numerous electronics manufacturers and software companies in Japan and they file a large number of patent applications, not only in Japan, but also in foreign countries. Foreign enterprises may be blocked from entering the Japanese market by the number of pending patent applications and existing patents.

In order to establish Japanese patent portfolios and have patents to negotiate cross-licensing with other Japanese companies, it is very important to obtain patents in Japan. Nowadays, a large number of patent applications on electronics technology and software are needed to cover and protect a single electronics product.

1. National patent procedure

1.1 Examination system

The Japanese patent system is a deferred examination system. Substantive examination of a patent application starts after submission of an examination request. A request for examination must be filed within three years of the filing date of a patent application in order to have the patent application undergo a substantive examination.[1] In the case of a Japanese patent application derived from a Patent Cooperation Treaty (PCT) patent application, an examination request must be filed within three years of the international filing date.

If the examiner rejects a patent application, the applicant can file a demand for trial with the Japan Patent Office (JPO). The examination of the demand for trial is conducted by a panel consisting of three trial examiners who will issue a trial decision of rejection or a trial decision of grant. If necessary, an applicant can file a juridical action to appeal a trial decision of rejection to the Tokyo High Court.

1 Patent Law, Article 48-3.

1.2 Timescales

Currently, the JPO normally issues notifications of reasons for rejection, allowance, etc around one year after the filing of the request for examination. Prosecution timescales can be accelerated filing a request for accelerated examination.

The JPO also participates in the Patent Prosecution Highway (PPH) system, an internationally recognised system in which grant and/or prosecution in one jurisdiction can be used to accelerate prosecution in another.

If either acceleration is requested or a request to use the PPH system is filed, a response to the request (examination report or an indication of allowance) is normally issued by the JPO within about three months of any such request being filed.

Once an examination request has been filed with the JPO, a notice of rejection or a grant of patent is normally received approximately one year later. Normally grant of patent occurs one or two years after filing an examination request. The speed with which examination occurs is the same regardless of whether an examination request is filed when the application is filed or if the examination request is delayed. Therefore, the time between filing a patent application and grant varies between approximately one to two years (if examination is requested when the application is filed) to approximately four to five years (assuming a request for examination is delayed for the full three years).

1.3 Costs

The costs for filing a new Japanese patent application are normally between ¥250,000 and ¥350,000 including official fees, attorney's fees and translation fees. The costs for filing an examination request are normally around ¥150,000 including official fees and attorney's fees.

In order to register a patent after receiving a notice of grant, the first to third annuities must be paid. The fourth and subsequent annuities must be paid before the expiration of the effective term for patent.

The annuity fees increase as the years go by. For example, the first to third annuity fees are 2,100 + the number of claims × ¥200, whereas the 10th to 25th annuity fees are 55,400 + the number of claims × ¥4,300.

2. Examination of software patents

The Japanese Patent Act, defines an 'invention' as a highly advanced technical idea created by using the laws of nature.[2]

The Japanese Patent Act categorises inventions into 'product inventions', 'process inventions', and 'inventions of processes for producing products'.[3] Prior

---

2 Patent Law, Article 2, paragraph 1.
3 Patent Law, Article 2, paragraph 3.

to the Patent Act revision of 2002, computer programs were not protected under the Japanese Patent Act. Since then computer programs have been included under the 'product invention' category. Electric and electronic products, semiconductor devices, and electric circuits are of course protected as product inventions, while processes for their control are protected as process inventions. The use and the transfer etc of a product manufactured by means of an invention of a 'process for producing the product' are regarded as carrying out the invention, and are also protected by the patent right.

2.1 Software-related inventions

According to the Examination Guidelines for Patent and Utility Model in Japan issued by the JPO,[4] a 'software-related invention' means an invention utilising computer software (software), which is a program or any other item that may be considered to be an equivalent of a program. The term 'program' is defined in the Japanese Patent Act as "a set of instructions given to a computer, which work to produce a specific result".[5] An "item that may be considered to be an equivalent of a program" if it has characteristics similar to those of a program since the item defines computer processing, cannot be called a program since the item does not directly give instructions to a computer. For example, a data structure that is expressed by correlations between data elements qualifies as an "item that may be considered to be an equivalent of a program".

The scope of software-related inventions include information processing apparatuses that operate in cooperation with software, methods for operating such apparatuses, and computer-readable recording media that store software. Inventions relating to business methods realised by means of information and communication technology (ICT) are also included in the scope of software-related inventions.

2.2 Patentable subject matter

In order to meet the patent eligibility requirements, all inventions, including software-related inventions, must be "a highly advanced creation of technical ideas utilizing the laws of nature".[6]

The Japanese Patent Act does not consider the following to be inventions: natural laws, a mere discovery that is not creation, anything that is against natural laws, anything that does not utilise natural laws, anything that is not a technical idea, and anything that cannot apparently solve a given problem.

Thus, for example, a claimed invention is not regarded as an invention under the Japanese Patent Act if it is a mere presentation of information (such

4 Examination Guidelines for Patent and Utility Model in Japan (Provisional Translation), www.jpo.go.jp/e/system/laws/rule/guideline/patent/tukujitu_kijun/index.html.

5 Patent Law, Article 2, paragraph 4.

6 Patent Law, Article 2, paragraph 1.

as image data taken with a digital camera), an artificial convention (such as the rules of chess or a method of doing business), or a mathematical formula. While an invention relating to a method of encoding video signals is considered to meet patent eligibility requirements, a bit stream generated by the method or a recording medium that records the bit stream are not since they are regarded as mere presentations of information. On the other hand, test charts for television receivers, which themselves have technical features, are not considered to be mere presentations of information.

Software inventions which either: concretely control an apparatus or perform a specific process relating to its control; or inventions that perform a concrete processing operation based on technical characteristics of a subject, are automatically considered to be eligible for patent protection in Japan.

Examples of the first type of invention (inventions relating to control of an apparatus or process) would include, for example, control software for a rice cooker, a washing mashing, an engine, a hard disk drive, a chemical reaction apparatus, a nucleic acid amplification apparatus, and so on.

Technical characteristics which would cause software to fall into the second category of invention would include physical characteristics, chemical characteristics, biological characteristics, electric characteristics, and so on. Specific examples of technical characteristics include engine rotation speed, rolling temperature, relation between gene sequences and expression of a trait in a living body, and physical or chemical bonding relationships of substances. An example of a software-related invention belonging to this second type of invention would be an invention of a specific method for compressing image data to be displayed on a display screen.

If a software-related invention does not correspond to either of the above categories, it may still be considered patentable if either it concerns concrete processing of information using hardware resources (ie, physical devices or physical elements used for processing, operation, or implementation of a function such as a CPU, a memory device, an input device, and an output device, and a physical device connected to the computer) or an information-processing device which acts in cooperation with such processing.

For example, a program for determining a railway route from a departure station to an arrival station may correspond to this type of invention and meets the patent eligibility requirements since the program cooperates with hardware resources such as input means for inputting search conditions, a memory storing railway network information, and display means for displaying the route found in the search to constitute a specific information-processing device for the purpose of searching a railway route.

Whether a software-related invention relating to a method of doing business is an 'invention' as stipulated in the Japanese Patent Act is determined based on whether the information processing performed by the software used in the invention is specifically realised by using hardware resources, not on whether the method has a specific feature.

As with other types of invention, commercial success or equivalent facts may help positively indicate the presence of inventive step in a software-related invention relating to a method of doing business. It is required, however, that such facts be based on technical features of the claimed invention, and not caused by other reasons such as sales techniques and advertising.

(a) *Novelty and inventive step*

The novelty and inventive step of a software-related invention are determined in the same manner as all other types of inventions. Unlike in European patent practice, where a non-technical feature that does not contribute to a technical feature is not taken into consideration when examining whether an invention has inventive step, the Japanese Examination Guidelines instruct patent examiners to take into consideration all the features recited in the claims of a software-related invention.

Similar to inventions in other categories, advantageous effects of the invention and the presence of factors that prevent combining cited techniques are elements in support of inventive step.

(b) *Description requirements*

Sufficiency: The examination with respect to the enablement requirement of a software-related invention is performed in principle in the same manner as the enablement requirement examination of inventions in other categories. A claim reciting a technical procedure or function of a software-related invention may be objected to as failing to meet the sufficiency requirement if the detailed description of the invention in the specification does not describe how the technical procedure or function is carried out by hardware or software, and those skilled in the art would not be able to understand how the technical procedure or function is carried out based on the technical common knowledge at the time of the filing of the patent application for the software-related invention.

Clarity: A claim will be objected to as being unclear if the recitation of the claim is unclear, no technical relationship can be found between features of the claimed invention, or the category of the claimed invention is unclear.

For example, an invention of "an information transmission medium transmitting a specific computer program" would be determined as failing to satisfy the clarity requirement since the technical relationship between the information transmission medium and the computer program is not defined and thus no technical relation can be found between features.

Similarly, an invention of "a program signal array for causing a computer to execute a step A, a step B, a step C ..." would be determined to fail to satisfy the clarity requirement since whether the claimed invention is a product invention or a process invention is unclear because the preamble of the claim is 'a program signal array'.

Terms other than 'program' (eg, 'module', 'library', 'neural network', 'support vector machine', or 'model') may be used if it is clear that the claimed invention is a 'program' from the descriptions of the specification and the drawings and the technical common knowledge at the time of filing of the application.

2.3 Artificial intelligence (AI)-related inventions

As of March 2019, the JPO has not provided specific examination guidelines for AI-related inventions. AI-related inventions are therefore examined based on the examination guidelines for software-related inventions mentioned above. However, in Annexes A and B of the Examination Handbook for Patent and Utility Model in Japan,[7] the JPO has provided examples of AI-related inventions for the benefit of applicants of patent applications for AI-related inventions which will be discussed in greater detail below.

(a) 'Trained models' may be granted patent

Case Study 2-14 in Chapter 1 of Annex B shows that a 'trained model' may correspond to a program invention and may be granted a patent.

Claim 1 of Case Study 2-14 reads as follows:

A trained model ... the model causes the computer to perform a calculation based on the said trained weights in the said first and second neural networks in response to appearance frequency of specific words obtained from the text data of accommodations reviews input to the input layer of the said first neural network and to output the quantified values of accommodation reviews from the output layer of the said second neural network.

The JPO's position is that a model invention that is clearly a program, like the invention defined by Claim 1, may be regarded as a technical idea created by using natural laws and considered to be an invention as stipulated in the Japanese Patent Act.

As explained in the previous section, a program invention may be granted a patent in Japan. Providing a patented program through telecommunication lines is regarded as the use of the patent.[8] Therefore, if a third party installs a server for distributing a patented trained model in Japan and distributes the

---

7 Examination Handbook for Patent and Utility Model in Japan, www.jpo.go.jp/e/system/laws/rule/guideline/patent/handbook_shinsa/index.html.

8 Patent Law, Article 2, paragraph 3, sub-paragraph 1.

trained model from the server through telecommunication lines, the patentee may claim that distribution be ceased, the trained model be erased, etc.[9]

(b) AI-related inventions may be rejected if data items of training data lack correlation

Case Study 49 of the 'Inappropriate Descriptions' section in Annex A describes a body weight estimation model. This model is generated through machine learning using a known machine learning algorithm such as a neural network, in which training data include face-outline angles obtained by analysing face images of people and actual measured height and weight values.

The specification defines 'face-outline angle' as an angle between a cheek line and a jaw line, and states that a statistically significant correlation is found between a cosine of the face-outline angle and the BMI (defined as the weight divided by the square of the height) of a person. A figure shows that data plotted in a coordinate system with the horizontal axis representing the BMI and the vertical axis representing the cosine of the face-outline angle are approximated to a linear function. The specification also discloses that, in addition to the face-outline angle, an arbitrary characteristic value, which is obtained from a face image and represents a face shape, may be used as a training data item, but does not disclose a specific correlation between such a characteristic value and physical information such as height, weight and BMI of a person.

Claim 1 of this case study recites an estimation model that estimates the body weight of a person based on a characteristic value representing the face shape and the body height of the person, using training data containing the characteristic values representing face images of people and the actual measured height and weight values. Claim 2 recites that the characteristic value representing a face shape is a face-outline angle.

The JPO admits that the invention defined by Claim 2 is disclosed in the specification in a clear and sufficient manner for those skilled in the art to carry out the invention.

However, the JPO determines that the recitation of Claim 1 is insufficient and that those skilled in the art would not have understood that a person's weight may be estimated based on the arbitrary characteristic value representing the face shape and on the person's height for the following reasons:

- the specification does not disclose a specific correlation between the arbitrary characteristic value and the physical information that is based on the height and the weight of the person;
- the technical common knowledge at the time of the filing of the corresponding patent application does not suggest that there is a specific correlation between those; and

9 Patent Law, Article 100, paragraphs 1 and 2.

- the specification does not disclose any performance evaluation results of the estimation model actually generated by using the arbitrary characteristic value.

The JPO thus concludes that Claim 1 does not meet the support requirement set forth in Article 36, paragraph 6, sub-paragraph 1 of the Japanese Patent Act, stating that the scope of the disclosure of the specification cannot be broadened or generalised to cover the scope of the invention defined by Claim 1.

The JPO also determines that it cannot be said that a desired body weight estimation system may be made by generating an estimation model using, as training data, an arbitrary characteristic value representing a feature of the face shape other than a face-outline angle, and actual measured values. As a result, the JPO objects to Claim 1 as not meeting the enablement requirement set forth in Article 36, paragraph 4, sub-paragraph 1 of the Japanese Patent Act.

It should be noted, therefore, that a generalised recitation like the recitation of Claim 1 cannot be accepted unless supporting content can be found in the specification if no correlation between a plurality of training data items, such as input data and output data of an estimation model, can be found even if the technical common knowledge at the time of the filing of the patent application is considered.

(c) ***Differences and effects of training data may provide novelty and inventive step***

Case Study 34 in the 'Inventive Step' section in Annex A shows a hydroelectric power generation capacity estimation system that estimates the capacity to generate hydroelectric power using a neural network.

The hydroelectric power generation capacity estimation system defined by Claim 1 of Case Study 34 includes a neural network, a machine learning unit, and an estimation unit. The neural network includes an input layer and an output layer. Input data that are input to the input layer contain a precipitation amount of upstream of a river, a water flow rate of the upstream of the river, and a water flow rate into a dam in the downstream of the river during a predetermined period between a reference time and a predetermined time before the reference time. The data output from the output layer contain the hydroelectric power generation capacity after the reference time. The machine learning unit trains the neural network using actual values of the input and output data as training data. The estimation unit inputs the input data into the neural network trained by the machine learning unit, setting a current time as the reference time, and calculates an estimated value of a future hydroelectric power generation capacity based on the output data, of which the current time is the reference time.

Claim 2 recites that the input data further contain a temperature of the upstream of the river during the predetermined period between the reference time and the predetermined time before the reference time.

An invention cited against this invention estimates hydroelectric power generation capacity by means of a regression equation model. The JPO's position is that although the cited invention uses a regression equation model instead of a neural network, those skilled in the art would have easily used a neural network instead of the regression equation model. The JPO therefore denies inventive step of the invention defined by Claim 1.

On the other hand, the JPO admits that the cited invention does not have the feature recited in Claim 2 that the input data contain a temperature of the upstream of the river during the predetermined period, and that the invention defined by Claim 2 would be able to perform a highly accurate hydroelectric power generation capacity estimation that may cope with increased flow rate caused by melted snow in spring since the input data contain the temperature of the upstream of the river. The JPO considers this effect advantageous and not expected from the cited invention. The JPO therefore admits that the invention defined by Claim 2 involves an inventive step.

As described above, the inventive step of an invention may be obtained if training data of the invention differs from training data of the cited invention and the difference results in an advantageous effect that cannot be expected from the cited invention. Obviously, even if training data used in an invention differ from that of a cited reference, the invention lacks inventive step if the scope of the training data is determined to be within the scope of training data those skilled in the art would have generally tried.

(d) *Preprocessing of training data may lead to inventive step*

Case Study 36 of the 'Inventive Step' section in Annex A shows an apparatus for assisting dementia stage diagnosis by using a trained neural network to estimate a dementia stage. The specification discloses that a question topic of a questioner and the transcript (character strings) of a response of a subject extracted from a speech segment are put into the trained neural network. Claim 1 recites: "the neural network is trained through machine learning using training data so as to output an estimated dementia stage, in response to an input of the character string of the speech segment by the respondent corresponding to the question topic of the questioner".

A dementia stage estimation apparatus in a cited reference uses a neural network trained through machine learning using training data so as to output an estimated dementia stage in response to an input of a character string obtained by speech recognition processing of speech information.

The JPO finds that there is a difference between the neural network defined by Claim 1 and the neural network disclosed in the cited invention in that, in the former, machine learning is used to output an estimated dementia stage by inputting the question topic of the questioner and the character string of the speech segment of the respondent in response to the corresponding question topic, while in the latter machine learning is used to output an estimated dementia stage by inputting the character string obtained by speech recognition processing without separating the speech segments of the questioner and the respondent.

The JPO considers that it is a common technical matter for those skilled in the art to perform preprocessing so as to modify a form of training data to be input before training a neural network using the training data in order to improve accuracy of estimation performed by the neural network, but that no prior art reference is found, which discloses a specific method for evaluating a dementia stage by specifying a question topic of a questioner in character strings obtained from speech information relating to dialogues between the questioner and the respondent, and using the question topic in association with a response from the respondent. The JPO determines that those skilled in the art would not have easily arrived at this technique, and that a simple design modification or an adaptation of a design matter to improve the estimation accuracy of the invention disclosed in the cited invention would not arrive at this technique, either.

The JPO further finds that the invention defined by Claim 1 brings about the significant effect of providing a highly accurate estimation of dementia stage since it enables a neural network to effectively learn from the training data the insights of a well-experienced doctor as a result of associating a specific question topic of a questioner and a response (character string) of a respondent. Based on the foregoing, the JPO admits the inventive step of the invention defined by Claim 1.

Thus, preprocessing of training data may provide inventive step to an invention.

2.4 Internet of Things (IoT)-related Inventions

(a) *Patent eligibility, novelty, and inventive step*

Advances in AI technology have enabled the use of big data (ie, very large or complex data sets), and advances in semiconductor technology have led to performant, miniaturised sensors and communication devices. As a result, IoT-related technology has also considerably progressed in recent years. In Japanese patent practice, IoT-related inventions are examined basically using the same criteria as for software-related inventions. The AI technology portions of IoT-related inventions using AI technology are examined in the same manner as AI-related inventions.

The patent eligibility of an AI-related invention is first determined by assessing whether the invention specifically performs (information) processing by means of the above-mentioned criteria for examining software-related inventions. Whether the AI-related invention is an invention as stipulated in the Japanese Patent Act is determined by whether specific information processing that is suitable for the intended use is performed as a result of the cooperation of software and hardware resources. Thus, patent eligibility of IoT-related inventions using software is determined in the same manner as that of software-related inventions. The specification of a Japanese patent application for an IoT-related invention therefore needs to describe at least specific means or specific processes for achieving the invention. The following are specific examples of such cases.

According to Case Study 2-9 in Chapter 1 of Annex B of the Examination Handbook for Patent and Utility Model in Japan, the following claimed invention corresponds to an invention as stipulated in the Japanese Patent Act since each unit includes specific means for causing software and a hardware resource to operate in cooperation:

A system for allocating an unmanned autonomous vehicle, the system including a server, a portable terminal, and an unmanned autonomous vehicle, the portable terminal including a transmission unit for transmitting a user ID to the server, the server including a storing unit for storing a face image, an acquiring unit for acquiring the face image corresponding to the user ID ... and the autonomous vehicle including an autonomous driving unit for performing autonomous driving of the autonomous vehicle up to a vehicle allocation position, a processing unit for performing face authentication processing with respect to surrounding people, and a judging unit for judging a person having a face matching the face image acquired by the acquiring unit.

However, according to Case Study 2-10 in the same chapter, the following claimed invention may not be an invention since information processing is not recited:

A system for allocating an unmanned autonomous vehicle, the system including a server, a portable terminal and an unmanned autonomous vehicle, wherein when receiving a request for an unmanned autonomous vehicle from a user who specifies a vehicle allocation position, the server allocates the unmanned autonomous vehicle to the user.

In this case, the claim may be amended to provide patent eligibility to the claimed invention if the specification specifically supports information processing. If it does not, such an amendment is not allowed.

The examination of novelty and inventive step of IoT inventions using software is also conducted basically in the same manner as that of software-related inventions. Since the examination of novelty and inventive step of AI-related portions of an IoT invention using AI technology is similar to the examination of their patent eligibility, please refer to the section on patentable subject matter on software-related inventions, above.

(b) Sub-combination inventions

A 'sub-combination' invention is each of the product or process inventions that are combined to form a combination product invention (which includes two or more devices) or a combination process invention (which includes two or more processes).

An IoT-related invention generally includes a system with a plurality of devices and terminals that are connected by a network. Patent protection is sometimes sought for a part of such a system, which corresponds to a sub-combination invention.

According to the Examination Guidelines for Patent and Utility Model in Japan, the following matters are taken into consideration when examining a sub-combination invention:

- matters relating to other sub-combinations recited in the claims shall also be examined;
- the claimed sub-combination invention shall be examined by determining the meanings or roles that such matters have in connection to the claimed sub-combination invention; and
- the specification, the drawings, and the technical common knowledge at the time of the filing of the patent application shall also be considered in determining the meanings or roles of the matters.

If the examiner considers that the matters relating to other sub-combinations specify the structure, the function, etc of the claimed sub-combination invention, the examiner shall determine that the claimed sub-combination invention has such a structure or function. If there are differences between the claimed sub-combination invention and a cited invention, the examiner shall determine that the claimed sub-combination invention has novelty. If the matters relating to other sub-combinations only define such other sub-combinations without defining the structure, the function, etc of the claimed sub-combination invention, the examiner shall determine that the matters relating to other sub-combinations do not have a meaning (role) in defining the claimed sub-combination invention. In this case, even if the wording of the matters relating to other sub-combinations and that of the features of the cited invention differ, unless there are other differences the structure and the function of the sub-combination invention shall be found to be similar to those of the cited invention, and the claimed sub-combination invention shall be found to lack novelty.

For example, Case Study 38 of the 'Novelty' section in Annex A of the Examination Handbook for Patent and Utility Model in Japan discusses the following invention:

A drone monitoring system including a terminal including means for acquiring a current position as first information and sending the first information, a server including means for selecting a drone that is the closest to a target based on the first information and means for sending the first information to the selected drone, and a plurality of drones each having means for acquiring a current position of the drone as second information, means for receiving the first information from the server, and means for performing flight control of the drone based on the second information and the first information.

This invention is determined to be an invention as stipulated in the Japanese Patent Act and to have novelty over the following cited invention: "A drone monitoring system for specifying a done corresponding to received terminal identification information."

Case Study 38 also discusses the following invention:

A drone connected to a server, the drone including means for acquiring a current position of the drone as second information, means for receiving position information of a terminal as first information from the server, and means for performing flight control based on the second information and the first information, wherein the server includes means for selecting a drone that is the closest to a target terminal to be monitored based on the first information received from the target terminal to be monitored, and means for sending the first information to the selected drone.

While this invention is directed to a drone device, the technical features of the invention are only that of the server, which is an element of just another sub-combination, but are not that of the drone device. The other features are the same as those of the cited invention. In this case, the criteria used by the server to select the drone do not have any influence on the structure and the function of the claimed drone. Therefore, the features of the other sub-combination do not define the structure or the function of the drone. Thus, although the wording relating to the other sub-combination is different, there is little difference between the structure and the function of the drone of this invention and those of the cited invention. Since there is no other difference, this invention is determined not to have novelty.

In conclusion, the system defined by Claim 1, "A system including a server (including a novel structure, function, or the like) and a terminal (not including a novel structure, function, or the like)", is determined to have novelty. The server defined by Claim 2, "A server used in the system according to Claim 1", is also determined to have novelty since the server has novelty in its structure, function, or the like. However, the terminal defined by Claim 3, "A terminal used in the system according to Claim 1", is determined to lack novelty since the terminal does not have a novel structure, function, or the like.

2.5 Data

While data that is not subjected to specific processing are not protected under the Japanese Patent Act, a program, structured data, and a data structure may be protected as product inventions. A specific example of such an invention relating to 3D printing technology is given below (Case Study 3-3, Patent Eligibility, Annex A):

3D printing data read by a control unit of a 3D printer, including a three-dimensional shape and a color of a doll to be made.

In this invention, the means or method by which the data are read by the control unit of the 3D printer does not have a characteristic technical feature. The content of the data, "a three-dimensional shape and a color of a doll to be made", is a characteristic technical feature. Therefore, a main purpose of this invention is mere presentation of information, which is not an invention as stipulated in Article 29, paragraph 1 of the Japanese Patent Act. The claimed data therefore is not an invention.

On the other hand, the following invention is presented as Case Study 2-15 in Chapter 1 of Annex B of the Examination Handbook for Patent and Utility Model in Japan:

3D printing data used in a 3D printer which laminates a model material that finally constitutes a 3D-modeled object and a supporting material that supports the model material during modeling, the 3D printing data having a structure including first data indicating quantity and position of discharge of the model material, a first pointer that points to data used for modeling based on the first data, second data indicating quantity and position of discharge of the supporting material, a second pointer that points to data used for modeling based on the second data ... herein a control unit of the 3D printer is used in a process for acquiring the first data or the second data from a memory unit in accordance with the first pointer or the second pointer after the modeling based on the first data or the second data is performed.

This invention is determined to meet the patent eligibility requirement since the invention can be said to be data having a structure to enable information processing performed by the control unit to acquire data used for the next step of the modelling, after the modelling based on the first data or the second data is performed. Thanks to this structure, the data for modelling defines the information processing performed by the control unit, thus acquiring characteristics similar to those of a program. Therefore, the data is determined to be equivalent to a program (due to having a structure that is equivalent to a program) as stipulated in Article 2, paragraph 4 of the Japanese Patent Act, and therefore regarded as software. Furthermore, a specific method for controlling a 3D printer or performing processes accompanying a 3D printer control corresponds to "a highly advanced creation of technical ideas using the laws of nature". Therefore, the invention of Case 2-15, which is software for causing a control unit to carry out such a method, corresponds to the invention as stipulated in the Japanese Patent Act.

Although a data structure that does not carry out a direct processing operation, such as a database, is not protected under the Japanese Patent Act, it may be protected under the Copyright Act. Valuable data may be protected under the Unfair Competition Prevention Act to prevent improper acquisition and improper use of such data. Valuable data, which may also be useful as intellectual property, can be easily copied and distributed, and damage caused by its unfair distribution may rapidly and broadly extend. From the recent use of big data, it appears that the scope of the above-mentioned laws may need to extend to the protection of big data.

3. Infringement

3.1 Claim interpretation

The technical scope of a patented invention is determined based on the recitations of the claims.[10] The meaning of the terms used in the claims is interpreted in consideration of the descriptions in the specification and the drawings,[11] excluding the Abstract.[12]

Japanese patent law does not exclude the possibility of using the doctrine of equivalents, which is often used as a means to expand the scope of patent right and becomes a point of dispute in patent infringement cases brought to court. Nevertheless, as the conditions for the applicability of the doctrine of equivalents are strict, it is important that, in the patent application proceedings, the claims are prepared in consideration of the literal infringement.

3.2 Acts of infringement

If an unauthorised third party carries out a patented invention as a business activity, the patent right is infringed. The term 'carries out' means producing, using, transferring etc, exporting, importing, or offering of transfer etc of the products if the patented invention is a product invention.[13] As explained above, a program invention is a product invention. The scope of transferring etc includes transferring and lending, and also supplying through telecommunication lines if the product is a program.

In the case of a process invention, the term 'carries out' means the use of the process.[14] In the case of an invention of a process of producing a product, the term 'carries out' means using, transferring etc, exporting, importing, or offering of transfer etc of the product produced by the process.[15]

---

10 Patent Law, Article 70, paragraph 1.
11 Patent Law, Article 70, paragraph 2.
12 Patent Law, Article 70, paragraph 3.
13 Patent Law, Article 2, paragraph 3, sub-paragraph 1.
14 Patent Law, Article 2, paragraph 3, sub-paragraph 2.
15 Patent Law, Article 2, paragraph 3, sub-paragraph 3.

3.3 Contributory infringement

Though not constituting a direct infringement (literal infringement), but being considered preparatory or aiding to infringement, certain actions which are highly possible to induce direct infringement are regarded as indirect infringement of patent rights.[16]

3.4 Network inventions

Whether or not a network-related invention in which multiple elements (eg, a server and one or more client devices) are connected through a network such as the internet is infringed, is judged based on the 'All Element Rule' in the same way as the other types of inventions. Therefore, a patent on a network-related invention is judged to be infringed if the system of the defendant includes the server and the client device recited in the claim. When considering such inventions, the Tokyo District Court has ruled that, identifying who implements a patent, is determined by identifying who dominantly manages the allegedly infringing system.[17]

4. Copyright protection

4.1 Copyright protection for software programs

Computer programs are given as an example of 'works' protected under the Copyright Law of Japan.[18] However, works are further defined as products in which thoughts or feelings are expressed in a creative way and which falls within the literary, scientific, artistic or musical domain.[19] As a result, computer programs are not protected if they do not express thoughts or feelings in a creative way.

For the above reasons, program copyright infringement cases in Japan depend upon whether the programs at issue are such creative works.

Precedent cases indicate that, for a program to be a creative work, the following conditions must be satisfied:

- The program consists of a wide range of command representations, combinations of such command representations, and various sequences of representation;
- The representations are not common and showcase the creator's personality and creativity.

Even if a program is determined to be a work, a later program will only infringe the copyright in that work if a later work embodies the relevant creative work. Determining whether a later program infringes copyright is performed by comparing parts of these programs where recognised creative work exists and determining:

---

16 Patent Law, Article 101.
17 Eyeglass lens supply system case H16 (Wa) 25576, 14 December 2007.
18 Copyright Law, Article 10, paragraph 1, sub-paragraph 9.
19 Copyright Law, Article 2, paragraph 1, sub-paragraph 1.

- whether the creative expressions are identical; and
- whether creative characteristics of expression can be directly perceived.

In view of this approach, it is often difficult to establish that the copyright infringement of a program has occurred.

In Japan, programs may be registered with the Software Information Center (SOFTIC). Though such registration makes it easy to produce evidence in a trial as to the content of the program or the time of its creation, it does not necessarily demonstrate that the program is a work.

4.2 Anti-circumvention measures

Revision of the Copyright Law in 2012 expanded the definition of 'technological protection measures' provided in Article 2(1)(xx) to include devices for conversion or decryption of works, performances, phonograms or broadcasts, or wire-broadcasted sounds or images that are recorded on a recording medium after specific conversions (Cryptographies, such as CSS used for DVD and AACS used for blu-ray). The revision also means that the restoration of works from the converted information also falls within the definition of 'circumvention'.

Article 30(1)(ii) of the Copyright Law provides that in the case where reproduction has become possible by the circumvention of technological protection measures or as a result of such circumvention, the results of acts deterred by such technological protection measures have ceased to be obstructed, and reproduction is made with the knowledge of the facts described above, such reproduction is not permissible even if the work is for his or her personal or family use or other equivalent uses.

Further, Article 120-2 of the Copyright Law provides that a person who either: transfers to the public the ownership of, or rents to the public, manufactures, imports or possesses for transfer of ownership or rental to the public, or offers for use by the public, a device the sole function of which is to circumvent technological protection measures (including a set of parts of such a device capable of being easily assembled) or reproductions of a computer program the sole function of which is to circumvent technological protection measures, transmits to the public, or makes transmittable, the aforementioned computer program, or as a business, circumvents technological protection measures in response to a request from the public, shall be punishable by imprisonment with work for a term not more than three years or by a fine of not more than ¥3 million, or by both.

5. Ownership, inventorship and compensation

In Japan, ordinarily an inventor originally acquires the right to obtain a patent from the time when such an invention is produced.

However, an employer can acquire the right to obtain a patent for an employee invention from the time when such invention is conceived and it is normal practice for Japanese companies to include provisions transferring the right to apply for a patent to the company in their standard contracts for employment as is permitted by Article 35, paragraphs 2 and 3 of the Patent Law.

An employer in Japan is statutorily required to provide 'reasonable benefits' to an employee in return for the right to obtain a patent for any employee inventions.

To meet the criterion of 'reasonable benefit' provided in Article 34(4) of the Patent Law, it must include not only the amount of consideration but also the process to determine the amount considered. In other words, even if the amount of consideration is relatively low, it is considered that the amount meets the criterion of 'reasonable benefit' as long as the regulation was established after proper procedures. To meet the requirement of the 'proper procedures', negotiation between the employees and their employer as to the proper amount of the consideration or the method for the calculation, disclosure of the regulation to the employees to whom the regulation applies in proper method, and hearing opinion conducted for the purpose of determining the details of the reasonable benefit, are essential.

When such reasonable benefits are monetary, the method for calculating the amount of reasonable benefits and the timing of payment thereof, should be included in the employee's employment contract.

Revision of the Patent Law allowed the reasonable benefits to be non-monetary. According to the Guidelines issued by the Minister of the Economy, Trade and Industry, the following are listed as the specific examples of non-monetary reasonable benefits:

- opportunity to study abroad at the expense of the employer;
- stock options;
- career advancement or promotion resulting in improvement of monetary aspect;
- paid leave in excess of annual paid leave prescribed by the laws and regulations or employment regulations.

Before the revision in 2004, the amount of the remuneration had to be strictly calculated to be equivalent to the benefit which the employer obtained due to the contribution of the inventor. Therefore, there were many lawsuits including the blue LED case, where the Tokyo District Court ordered the payment of ¥20 trillion. However, since the change in the law in 2004, as long as the amount of benefit is determined by due processes as mentioned above and found to be reasonable benefits, an employer does not have to pay additional consideration even if it is still lower than the benefit which the employer obtained due to the contribution of the inventor.

In principle, the ownership of copyright in a work is originally attributed to the author of such a work. However, ownership of a work created by an employee in the course of the performance of his or her duties in connection with the employer's business on the initiative of an employer, made public by the employer as a work under the employer's own name, is attributed to the employer, unless otherwise stipulated by contract, work regulations or the like at the time of the making of the work.

6. Confidential information

The Unfair Competition Prevention Act provides provisions as to protection of 'trade secret' in Article 2(1)(iv)–(ix), and Article 2(6) provides that "the term 'trade secret' as used in the Act means a production method, sales method, or any other technical or operational information useful for business activities that is controlled as a secret and is not publicly known".

As shown in the definition, the trade secret has to be useful for business, controlled as a secret, and not publicly known. Therefore, the possessor of the trade secret has to control the information strictly as a secret in order for it not to be publicly known.

7. Other means of protection

7.1 Registered design protection

In Japan, it is possible to register an industrial design by filing a design application with the JPO or an international application with the WIPO.

After filing, the JPO formally and substantially examines the registrability of designs (ie, novelty, industrial reproducibility, etc) in line with Japanese design examination guidelines.[20]

Registration of an industrial design may be maintained, by paying an annuity, for a maximum of 20 years counted from the registration date in Japan.

The official fee for filing an application for a design is ¥16,000 and for registering a design for the initial three years is ¥25,500.

Under Japanese design practice, graphical user interfaces for computers (GUI), including icons, are registrable. If such protection is desired, the device for which the GUI is used (eg, portable information terminal device, video recorders, 3D printers) has to be shown in the drawings. It is acceptable to disclaim the shape of the device using the GUI by depicting it in broken lines.

7.2 Unfair competition law

The Unfair Competition Prevention Act defines various kinds of unfair competition in Article 2(1)(i)–(xvi), including the acts of providing any device or

20 Available at: www.jpo.go.jp/e/system/laws/rule/guideline/design/shinsa_kijun/index.html.

program having a function of circumventing technological restriction measures (Article 2(1)(xi) and (xii)). A person whose business interests have been infringed by the unfair competition can seek an injunction suspending or preventing the infringement against the person who sells such kinds of devices or programs, and the infringers are liable to compensate for damages resulting therefrom.

7.3 Semiconductor topographies

The Act on the Circuit Layout of a Semiconductor Integrated Circuits ensures the proper exploitation of the layout design of semiconductor integrated circuits. When an application for a registration of establishment is filed, the registration of establishment is made except in some cases. One of the cases is where a creator had transferred the semiconductor integrated circuits manufactured utilising the layout design in the course of trade, in relation to a layout design pertaining to the application more than two years prior to the date on which said application is filed. The duration of a layout design exploitation right shall end 10 years after the date on which a registration of establishment is made. A holder of the right shall have the exclusive right to registered layout design in the course of trade. However, there are some limitations on the right. In particular, the effect of the right shall not extend to any exploitation of layout design which is created by other persons, such as copyright, to any acts for the purpose of analysis or evaluation, nor to any integrated circuits (including articles incorporating the integrated circuits as a part thereof) manufactured by the holder. And no holder of the right may exploit the registered layout design in the course of trade, when such exploitation of the registered layout design falls under the working of another person's patented invention or registered utility model.

The holder has a right to seek injunction. The holder may demand that a person who infringes or is likely to infringe his or her own layout design exploitation right or exclusive exploitation right should discontinue or prevent such infringement. And the holder may further demand that the measures necessary for the prevention of such infringement should be taken. Claims for damages are based on the Civil Code. Presumption of amount of damages is provided in Act on the Circuit Layout of a Semiconductor Integrated Circuits.

8. Enforcement

8.1 Obtaining information on the infringer and the infringement

The Code of Civil Procedure provides for an "Order to Submit Documents" in Article 223, wherein the court, upon petition, can issue a ruling ordering the person in possession of the document to submit the document. The Code also provides "Preservation of Evidence" in Article 234, wherein the court, upon petition, can conduct an examination of the evidence in advance. However, those provisions are rarely used in patent infringement cases since the requirements are very strictly handled. In this regard, very recently, the Patent Law has introduced the Visa System, wherein the court, upon petition, can issue a ruling ordering the officer of the court to enter the factory of the defendant and to collect evidence of the infringement. While the system may be similar to the system of 'discovery' in United States or 'disclosure' in United Kinghdom, it remains to be seen how the system will be handled.

8.2 Interim relief

As to interim injunctions of patent infringement litigation cases, the Civil Provisional Remedies Act provides the 'order of provisional disposition'. This order which determines a provisional status, where the infringer cannot sell the infringing products, may be issued when such status is necessary in order to avoid any substantial detriment or imminent danger that would occur to the patentee with regard to the relationship of rights in dispute. The court may not reach a ruling on a filing of an objection to a provisional remedy without holding oral arguments or holding a hearing date at which both parties may be present, and it will take four to six months on average to obtain the order of provisional disposition.

8.3 Threats

In Japan, theoretically a patentee can be sentenced with criminal intimidation or found to be liable for tort, in the case where the patentee threatens a competitor with a patent infringement suit. However, such cases are almost vanishingly rare.

In contrast, if a patentee threatens business customers of the competitor and the court finds non-infringement later, this can be found to constitute unfair competition (announcement or dissemination of a falsehood that is injurious to the business reputation of a competitor) provided in Article 2(1)(xv) of the Unfair Competition Prevention Act and there are many cases where a court will find against a patentee if such threats have occurred.

8.4 Remedies

If a patentee finds an infringement of its patent, the patentee may take such steps as warning the infringer and negotiating a licence of use of the patent. The patentee may also demand, as civil remedies, the injunction of the infringement,[21] compensation for damages caused by the infringement,[22] the restitution of unjust enrichment,[23] and measures for restoring the business credibility of the patentee, which was harmed by the infringement.[24]

---

21 Patent Law, Article 100.
22 Article 709 of the Civil Code.
23 Articles 703 and 704 of the Civil Code.
24 Patent Law, Article 106.

9. Hot topics

9.1 Advisory opinion for Standard Essential Patent (SEP)

Thanks to advances in IoT, various products are being connected to communication networks. The necessity of complying with standards in technical fields such as ICT is increasing as a result of various industries offering such products and services. Under the circumstances, licensing negotiations, which have conventionally involved companies in telecom industries such as telecommunications carriers and telecommunications equipment manufacturers, are now including companies in other industries such as automobile manufacturers and service providers. The shift in parties sitting at the negotiation table has made it difficult to solve problems by conventional ways, such as cross-licensing within the same industry. A system which facilitates licensing negotiations among companies from different industries is demanded.

Responding to such demands, on 1 April 2018 the JPO started a new operation of the advisory opinion system defined in the Article 71 of the Japanese Patent Act. According to the new operation, the JPO, as an official organisation, can provide an advisory opinion as to whether a patented invention is essential to a standard when there is a dispute over the essentiality of the patented invention between the concerned parties.

An advisory opinion under the conventional operation of the advisory opinion system (conventional advisory opinion) determines whether an object A, which is a product or a method to be manufactured or carried out, is within the technical scope of a patented invention.

An advisory opinion under the new operation of the advisory opinion system (advisory opinion on SEP) determines whether a patented invention is essential to a specific standard. Specifically, if the virtual object A is within the technical scope of the patented invention, the panel concludes that the patented invention is essential to the standard. Here, the virtual object A is a hypothetical object which is specified only by indispensable features of standard documents.

The following matters are common to the conventional operation and the new operation:

- An advisory opinion may be requested not only by the patentee, but also a party who plans to manufacture or carry out the object A;
- A panel of three trial examiners prepares the advisory opinion. The examination is in principle conducted based on documents. After the examination is finished, the panel issues an advisory opinion, which is sent to the concerned parties. The advisory opinion is also made public in the form of a publication;
- The advisory opinion is merely an expert opinion, and does not have

any legal binding force. However, advisory opinions provided by the JPO are considered to be highly respectable and authoritative;

- No petition can be filed against the conclusion of advisory opinion;
- One request for advisory opinion can target only one object A (or one virtual object A);
- The official fee for an advisory opinion is ¥40,000.

A request for advisory opinion on SEP submitted to the JPO shall indicate items such as 'purport of request', 'demandee (respondent)', and 'reasons for request'.

The following is an example of information in the 'purport of request' section:

For the purpose of essentiality check, we request an advisory opinion as to whether a virtual object product A, which complies with the standard ... falls within the technical scope of the patented invention of Japanese Patent No. ...

The particulars, including the title and the version of the standard, should be specified.

The other party to a dispute over the patented invention essentiality is indicated in the 'demandee' section. In the absence of the demandee, it is considered that a dispute over the essentiality of a patented invention between the concerned parties has not been established, and no benefit may be obtained from the request for advisory opinion as to the essentiality of the patented invention. The dispute herein needs to relate to the essentiality of a specific patented invention. A licensing negotiation that does not address the essentiality of a specific patented invention is thus regarded as a dispute that cannot require an advisory opinion on SEP.

A request for advisory opinion that is not considered to bring about a benefit shall be dismissed. Then, a decision on request is issued and published with the name of the demandant, the name of the demandee, the name of the agent, the conclusion, and the reason for dismissal.

The 'reasons for request' section includes:

- a statement as to the necessity of the request;
- a statement explaining a virtual object A;
- a statement of technical comparison between a patented invention and the visual object A; and
- a statement for explaining the reason why the virtual object A is within the technical scope of the patented invention, and the patented invention is essential to a standard.

The statement as to the necessity of the request for advisory opinion on SEP shall state the existence of a dispute over the essentiality of a specific patented invention in a licensing negotiation between concerned parties.

The virtual object A must be specified only by features that are determined to be essential in standard documents compiled by standards setting organisations (SSOs) or the like. The standard documents are limited to those prepared by standard setting organisations and being acceptable as evidence. The virtual object A shall be specified in accordance with the descriptions of the standard documents, without substantially modifying the features thereof and without broadening or narrowing the scope of the features.

A feature that, according to the standard documents, cannot be determined as being essential or selective (optional), may not be specified as a feature of a virtual object A.

If no feature of a virtual object A can be found in the standard documents, the request for advisory opinion on SEP may be dismissed. It should be noted that more than one virtual object A could be specified from a single standard.

The reasons why a virtual object A is within the technical scope of a patented invention, and why the patented invention is essential in a related technical field is explained by comparing the features of the specified virtual object A and the respective features of the patented invention. If the interpretation of a term used in the standard documents is an issue, the demandant's interpretation should be presented in the request with the grounds (evidence) thereof.

The flow of the new operation is substantially the same as that of a conventional operation, and therefore relatively simple. When a request for advisory opinion is filed by a demandant, the JPO sends a copy of the request to the demandee and requests that it submits a written reply. A panel of trial examiners compares the patented invention and the virtual object A and determines whether the virtual object A is within the technical scope of the patented invention. If the virtual object A is within the technical scope of the patented invention, the panel concludes that the patented invention is essential to the standard.

Since whether a patented invention is essential to a standard is determined by the JPO as a fair and neutral party, the new operation for an advisory opinion on SEP is expected to greatly contribute to facilitating licensing negotiations and resolving disputes between concerned parties. For more of the advisory opinion system on SEP, see the English handbook issued by the JPO.[25]

9.2 Case study: software and business model inventions

Finally, we briefly introduce a decision given by the Intellectual Property High Court, which was widely disputed in the Japanese intellectual circles.[26] In this case:

---

25 Implementation of the New Operation of Advisory Opinion System to Determine Standard Essentiality and Publication of the "Manual of 'Hantei' (Advisory Opinion) for Essentiality Check", www.jpo.go.jp/e/system/trial_appeal/hantei_hyojun.html.

26 Intellectual Property High Court 2017(Gyo-Ke)10232, www.ip.courts.go.jp/app/files/hanrei_en/327/002327.pdf.

- an invention was examined and granted a patent by the JPO;
- as a result of an opposition filed against the granted patent, the registration of the patent was cancelled by the JPO;
- the JPO's decision was cancelled in an appeal filed with the Intellectual Property High Court; and
- the patent registration was restored.

Here is Claim 1 of the final claim set registered:
A steak-providing system for implementing a method of providing steaks, the method comprising: guiding a customer to a table of a standing type restaurant; asking the customer for a request as to a weight of steak; cutting meat having the weight requested by the customer out of a meat block; grilling the meat; and conveying the grilled meat to the table of the customer, the system comprising:

- *a label on which a table number of the table of the customer is indicated;*
- *a measuring device for measuring the meat that is cut in accordance with the request of the customer; and*
- *an indication item for distinguishing the meat cut in accordance with the request of the customer,*
- *wherein a sticker on which a weight of the meat measured by the measuring device and the table number of the table indicated on the label are shown is outputted, and*
- *wherein the indication item is the sticker.*

The only point of issue in the lawsuit seeking the cancellation of the decision on the granted patent was 'patent eligibility', and therefore novelty and inventive step of the patent were not discussed. The Intellectual Property High Court provided the following statement:

It can be understood that Patent Invention 1 includes the element of a procedure performed by a person from the reception of an order to the service of food at a steak restaurant (the present steak providing method), but goes beyond that and satisfies requirements unavoidably generated when the present steak providing method is implemented and solves the problem of Patent Invention 1 that 'the preferred amount of steak is inexpensively provided to the customer' by employing the configuration related to the present measuring device and the like including the specific articles or instruments (devices) such as the label, the measuring device, and the sticker (indication item) so as to prevent confusion with another customer's meat.

In view of the technical problem of Patent Invention 1 and the configuration of the technical means for solving the problem and the technical significance such as the effects derived from the configuration, and the like, Patent Invention 1 has the specific articles or devices (the present measuring device and the like) such as the label, the measuring device, and the sticker (indication item) as the technical means for solving the problem of Patent Invention 1 by preventing confusion with another customer's meat and can be considered to correspond to 'creation of the technical ideas utilizing laws of nature.' Therefore, it can be considered that Patent Invention 1 corresponds to the 'invention' as stipulated in Article 2, paragraph 1 of the Patent Act.

Opinions concerning the final result of this case vary widely.

The publisher acknowledges Takeshi Sekine for his kind assistance with the preparation of this chapter.

South Korea

Kenneth K Cho
Wiggin and Dana LLP
Gregory B Kang
Nath, Goldberg & Meyer
Hyejung Lee
Yoon Sang Lee
Chang Hwan Shin
Kim & Chang

South Korea continues to be an important player in key global technology sectors with many Korean companies enjoying great success in the global marketplace. In the electrical and telecommunications technology sectors, Korean companies are well-known brands for consumer electronics, semiconductors, white goods and mobile phones. As these markets continue to grow, IP protection and enforcement will equally remain an important focus in Korea, for Korean companies and for foreign companies doing business in Korea that compete with Korean companies in other jurisdictions. Reflecting those trends, South Korea has consistently ranked fourth in the world since 2008 when measuring volume of patent and utility model applications.[1] The Korean Intellectual Property Office (KIPO) is a member of the IP5 which are the top five patent offices in the world based on annual filings.

1. National patent procedure

1.1 Examination system

The Korean patent system is a full examination system where a patent right is conferred only after full substantive examination of novelty and inventive step by KIPO.

An applicant may defer examination for up to five years from its filing date and KIPO will not examine a patent application until an applicant files a request for examination. That request may be filed when the application is first filed at KIPO or at any time up to three years (five years for patent applications filed before 1 March 2017) from the application filing date. Failing to file an examination request before the deadline results in the patent application being deemed irrevocably withdrawn.[2]

1 *WIPO IP Facts and Figures* 2012 edition.
2 Article 59, Paragraphs 1–5 of the Patents Act.

1.2 Timescales

The time taken for prosecuting a patent application varies greatly from application to application, depending on when the applicant files a request for examination, the number of office actions issued by the KIPO examiner, and backlogs at KIPO. Substantive examination is only initiated once a request for examination is filed. Ordinarily, a first office action (a preliminary rejection or an allowance) is issued about 12 to 16 months from the date when the request for examination is filed.

Since 1 October 2008, KIPO has instituted a three-track examination system to provide applicants with some measure of control over the duration of examination. Under this system, applicants can select one of three examination tracks: normal, expedited or deferred.

(a) Normal examination

This is the default examination track. Other than filing a routine request for examination, no separate request is required.

(b) Expedited examination

For certain qualified patent applications that fall under Article 61 of the Korean Patent Act and Article 9 of the Enforcement Decree of the Patent Act, applicants may submit a request for expedited examination after filing the routine request for examination. Examples of such qualified patent applications include, among others, applications:

- directed to a claimed invention put into commercial use by a person who is not the applicant after the application is opened ('third party practice');
- directed to a claimed invention practised by or under preparation to be practised by the applicant ('self-practice');
- to be examined under the Patent Prosecution Highway; and
- for which the applicant paid a prescribed fee and requested a prior art search from one of the expert search authorities provided under Article 58, Paragraph 1 of the Patent Act.

(c) Deferred examination

An applicant may submit a request to defer examination by designating a deferred examination commencement date:

- when requesting examination; or
- within nine months from the filing of the request for examination.

The examiner will then commence examination of the application within about three months from the deferred examination commencement date as designated by the applicant.

1.3 Costs

The official fees[3] for filing an application with KIPO are as follows with approximate indicative equivalent costs in US dollars. Actual costs of course fluctuate according to the prevailing exchange rate:

Application filing fee (electronic filing)	₩46,000 ($41.21)
Claim for foreign priority (per priority electronic filing)	₩18,000 ($16.12)
Application filing fee (paper filing)	₩66,000 ($59.13) plus ₩1,000 ($0.90) for each page exceeding the 20-page limit[4]
Claim for foreign priority (per priority paper filing)	₩20,000 ($17.92) (paper filing)

Official fees for filing a request for examination are proportional to the number of pending claims at the time of filing a request for examination, as follows (fluctuating according to the exchange rate):

Basic fee for the first claim	₩187,000 ($167.56)
Fee for each additional claim	₩44,000 ($39.42)

Annuity fees may be categorised into:

- a registration fee which should be paid at the time of patent registration and covers the first three years; and
- annuities which should be paid every year after registration.

Fees for registration (issuance) include:

Basic fee for three years	₩45,000 ($40.32)
Additional fee for three years for each claim	₩39,000 ($34.94)

Similarly, annuity fees for each year after the first three years comprise a basic fee and an additional fee for each claim in the application. These range from a basic fee and a per-claim fee of ₩40,000 ($35.84) and ₩22,000 ($19.71) in respect of the fourth year progressively rising to ₩360,000 ($322.58) and ₩55,000 ($49.28) for the final year.

3 Comprehensive information on the official fees can be found on the KIPO website at www.kipo.go.kr/kpo/user.tdf?a=user.english.html.HtmlApp&c=92004&catmenu=ek03_04_01.

4 An additional fee is incurred for each additional page when the total number of pages of the application (including the detailed description, the drawings, and the abstract) exceeds 20 pages.

2. Examination of software patents

2.1 Patentable subject matter

The Patent Examination Guidelines (the guidelines), published in August 2018, are the prevailing standard for KIPO examiners in examining software-related inventions. Specifically, the guidelines provide a chapter (Section 9, Chapter 10) dedicated to the examination standards for computer-related inventions.

Unlike the United States and some other jurisdictions, the Korean Patent Act explicitly provides a definition of what constitutes an 'invention'. Under Article 2 of the Patent Act, a patentable invention is defined as an advanced creation of a technical idea utilising the laws of nature.

According to the guidelines, for a computer-related invention to qualify as a creation of a technical idea utilising the laws of nature, information processing by software must be concretely realised by utilising hardware. In general, this is achieved by:

- an information-processing device (machine) cooperatively operating with software;
- the operating method of the computer;
- a computer-readable storage medium storing recorded software or a specific data structure; and
- software recorded in a computer-readable storage medium (applicable to patent applications filed on or after 1 July 2017).

KIPO accepts all of these as constituting patentable subject matter. However, a computer program list *per se*, and data structure *per se* etc are not patentable subject matter. Further, it should be noted that, while a computer program recorded in a computer-readable medium is patentable under the guidelines, a computer program itself (eg, a computer program that is not tied to a computer-readable medium) is not. Thus, while the act of practising a computer program in relation to a computer-readable medium may be protected (eg, the act of manufacturing, using, assigning, or leasing a computer program recorded in a computer-readable storage medium), the act of practising a computer program in a manner not tied to a computer-readable storage medium (eg, the act of transmitting a computer program over an information transmission medium) may not be protected.

2.2 Description requirements

Korean patent law requires that "a specification should be described in a manner such that those skilled in the art can easily practise the invention".[5] This requirement applies to patent applications in general including patent applications for software inventions.

5 Article 42, Paragraph 3(1) of the Patent Act.

According to the guidelines,[6] it is understood that, while a specification need not provide details of well-known hardware components, a specification should describe hardware and/or software components needed in realising functional features recited in claims, if such hardware and/or software is not well known and would not be easily implemented by those skilled in the art.

Claims should be directed to a method, a tangible object (eg, an information-processing device or a computer-readable storage medium), or a computer program recorded in a computer-readable storage medium. A claim directed to a computer program not recorded in a computer-readable storage medium, program signal, program article, or program product does not fall under one of allowable categories of patentable subject matter (because it is deemed to be indefinite).

In the case of method claims it should be clear from a claim which component performs each step of the claimed method. Thus, for example, in the case of a method claim for a software invention, the component undertaking each step should be clearly identified; for example, a method ... comprising calculating ... by a proxy server etc.

2.3 Novelty and inventiveness requirements

Even if a software-related invention qualifies as patentable statutory subject matter and meets the description requirements, it must still pass further tests as to novelty and inventiveness. The novelty and inventiveness of a computer-related invention is considered in light of both known technologies and common general knowledge in the field of computers.

A patent will not be granted for an invention which could have been easily conceived by a person having ordinary skill in the pertinent art from technologies publicly known, practised or described in a publication distributed in or outside of Korea before the priority date.

In general, inventive step is determined by comparing the objective, constitution (the structural elements of the invention), and effect of the invention with the prior art. In cases where the difference in constitution between the invention and the prior art would make the claimed invention particularly unique or difficult to conceive from the prior art, then the invention is deemed to be inventive. However, if the difference is not unique or difficult, it is usually necessary to prove that there are unexpected or superior effects, and/or a unique objective (problems to be solved) in the invention.

Furthermore, where two or more prior art references are cited, the difficulty in combining the references to reach a claimed invention based on certain factors may also be considered for assessing the inventive step of the invention.

---

6 Section 1.2 of the Examination Guidelines for Computer Related Inventions.

The guidelines[7] identify the following as examples of inventions which lack inventive step:

- the simple implementation of a combination of conventional methods and/or means in a computer system, unless there is technical difficulty in combining/implementing them in the computer system and/or an unexpected advantageous effect arises from such combination/implementation; and
- simple application of a method or means used in a specific computer-related field to another computer-related field.

2.4 Court's approach to software patents

There are no Korean court rulings explicitly stipulating a comprehensive standard regarding the subject matter requirements of software-related inventions. However, there are a number of court rulings on business model or business method inventions, discussed below, stating that information processing by software on a computer must be concretely realised utilising hardware (the same wording as appears in the guidelines in relation to the software-related inventions) in order to be a patentable invention.

Supreme Court Decision Case No 1997 Hu 2507 was an example of an invention which was found to fulfil the requirements for patentable subject matter. In that case the invention related to a method for controlling a numerical controller device. The invention involved executing a sub-word supplemental processing program (ie, software) using a numerical control input format which caused a numerical controller device (ie, hardware) to identify, control and operate a machine in a particular manner. The court held that the invention caused physical transformations outside the hardware, the practical use of which is disclosed in the specification. The invention was therefore found not to be based solely on pure human mental activities that did not utilise the laws of nature and hence the invention was considered to relate to patentable subject matter. The court held that the invention caused physical transformations outside the hardware, whose practical use was disclosed in the specification, and found that the invention was considered to relate to patentable subject matter.

Supreme Court Decision Case No 2001 Hu 3149 was a case which related to a business model invention regarding the implementation of a new business system utilising information technology. There the court held that the invention was allowable to the extent that the information processing by software on a computer was concretely realised utilising hardware.

---

7 Section 2.3.4 of the Examination Guidelines for Computer Related Inventions.

3. Infringement

3.1 Claim interpretation

The scope of protection for a patented invention is determined by what is written in the claims.[8] However, when the claim language is unclear and the technical constitution of the patent cannot be identified, the claim may be interpreted with reference to the detailed description and drawings of the specification.[9] The Supreme Court also recognises the doctrine of equivalents when there is no literal infringement and has set out the following requirements for infringement:[10]

- The problem to be solved by both inventions is the same;
- Even if a required claim element is substituted, the same object of the patented invention is accomplished, providing substantially the same effect;
- This substitution is so clear that those with ordinary knowledge in the technical field of the invention could easily conceive of it.
- The invention either:
  - is not one of ordinary knowledge in the art that could easily be invented from the prior art available at the time of filing; or
  - was not known at the time of filing the patent application; and
- The substituted claim element of the invention was not intentionally disclaimed in the prosecution history of the patented invention.

3.2 Acts of infringement

Acts of infringement include the following acts carried out for business purposes without consent of the patentee or exclusive licensee:

- in relation to product inventions, acts of manufacturing, using, assigning, leasing, importing or offering for assignment or lease the invented product (including advertising it for assignment or lease);
- for process inventions, acts practising the process; and
- for inventions of processes for manufacturing products, acts of using, assigning, leasing, importing or offering the product manufactured by the process for assignment or lease.[11]

3.3 Contributory infringement

Under Article 127 of the Patent Act, the following acts are considered to constitute 'indirect infringement':

---

8 Article 97 of the Patents Act.

9 Supreme Court Decision 90 Hu 1499 of 26 November 1991; Supreme Court Decision 96 Hu 1040 of 10 April 1998.

10 Supreme Court Decision 97 Hu 2200 of 28 July 2000.

11 Article 2(iii) and Article 94 of the Patents Act.

- A patent is related to an invention of a product, an act of producing, transferring, leasing, importing or offering for sale or lease (including displaying for sale or lease), as a business, articles used exclusively for the production of the patented product; or
- A patent is related to an invention of a process, an act of producing, transferring, leasing, importing or offering for sale or lease (including displaying for sale or lease), as a business, articles used exclusively to practise the patented process.

It is important to note that 'indirect infringement' in Korea under the Patent Act does not necessarily involve knowledge on the part of the infringer that a particular component was especially made or adapted for use in the patent infringement as in other jurisdictions such as the United States.

3.4 Cross-border issues

There is no Korean precedent addressing the issue of infringement of a patent, such as a network patent, where part of the infringing activity takes place in Korea and part in another jurisdiction. It is unlikely that direct infringement would be found in such a situation under the 'all elements' rule, and there would be no indirect infringement under Article 127 of the Patent Act unless the infringing activity involved articles used exclusively for the patented invention.

4. Ownership, inventorship and compensation

4.1 Initial ownership, assignment of ownership, and licensing

In Korea, the right to a patent belongs to the inventor or his or her successor in title.[12] The right to a patent may be transferred by assignment in whole or in part. On issuance of a patent, the holder of the patent right may grant an exclusive or non-exclusive licence to a third party to use or exploit the patented invention.

4.2 Inventorship

An inventor has the right to be identified in a patent application.[13] Conversely, the applicant for a patent application need not be the inventor, as the right to a patent may be transferred before the filing of the application, for example by contract, inheritance, etc. Further, wrongly attributed inventorship is not a ground for rejecting a patent application, nor a ground for invalidating a patent.[14]

---

12 Article 33, Paragraph 1 of the Patents Act.
13 Article 42, Paragraph 1(5) of the Patents Act.
14 Article 62 and Article 133, Paragraph 1 of the Patents Act, which set out a list of grounds for rejection and invalidation, do not recite incorrect inventorship as a ground for rejection or invalidation.

4.3 Employee inventions and compensation

The normal rules of entitlement (ie, the right to apply for a patent belonging to the inventor or his or her successor in title) also applies to inventions created in the course of employment. Thus, absent an assignment of the right to an invention/patent from employees to their employer (or any employment agreement requiring assignment of employees' rights to all their work-for-hire inventions to the employer), the same rule should apply. That is, the employee who is the inventor retains rights to the invention including patents.

If a patented invention is assigned to the employer and commercialised for profit, then reasonable compensation will be calculated by considering the employer's profits from the invention under the Invention Promotion Act. The act's main provisions are summarised below.

(a) *Definition of work-for-hire inventions and resulting rights*

The Invention Promotion Act defines a work-for-hire invention as an invention that results from:

- present or past duties of the employee; and
- being within the scope of the employer's business.

Unless otherwise stipulated, the ownership of a work-for-hire invention (which can be registered as a patent, utility model or design) belongs to the inventor-employee. However, small or medium-sized entity employers (SMEs) as defined in the Small or Medium Sized Entity Framework Act are automatically entitled to a royalty-free non-exclusive licence to use the invention.[15]

No other agreement assigning the employee's invention rights to the employer is valid – only an agreement in respect of work-for-hire inventions.

(b) *Employee's duty to report and the employer's duty to notify*

An employee must immediately report the completion of a work-for-hire invention to the employer in writing. If two or more employee–inventors contributed to the invention, they must make a joint report.

An employer receiving a report must notify the employee-inventor in writing within four months as to whether it will acquire ownership of the work-for-hire invention. Such acquisition of ownership becomes legally effective upon the employer's notification. However, absent a prior employer-employee agreement or work regulation, the employer cannot acquire the work-for-hire invention against the employee's will.

---

15 In order for an employer who is not an SME to receive a non-exclusive licence, the employer must have a pre-invention contract or employment rule that requires the employee to transfer all their in-service inventions to the employer. Under the Small or Medium Sized Entity Framework Act, if a foreign parent company owns 30% or more shares of its Korean subsidiary and the gross amount of assets of the parent company is W500 billion or more, the Korean subsidiary is not classified as an SME.

If an employer fails to issue its notice within the four-month period, it is deemed to have abandoned its claim and is thus not entitled to a royalty-free non-exclusive licence without the employee's consent.

(c) Assessing reasonable compensation

The employee is entitled to 'reasonable compensation' for any work-for-hire invention acquired by the employer.

Further, the Invention Promotion Act provides several factors to be taken into account when making such an assessment. If the compensation is provided under an employment agreement or work regulations, the following factors (and any other relevant considerations) will be relevant:

- consultations between the employer and employees determining a policy for deciding on the type and amount of compensation;
- communication of the policy to employees (eg, by announcement, mail, etc);
- employee input in determining the type and amount of compensation.

In other words, the Invention Promotion Act requires a company's internal compensation policy to provide reasonable compensation taking into account the above factors.

Further, the act stipulates that if a company has no internal policy to determine compensation, or if the compensation is found to be unreasonable taking into account the above criteria, whether compensation is reasonable will be determined looking at the employer's expected profits and the degree of the contribution made to the invention by the employer versus the employee (eg, within the context of a lawsuit filed by the employee).

The Invention Promotion Act also stipulates that even if a patent application is not filed for a work-for-hire invention acquired by an employer, the employee is still entitled to reasonable compensation.

5. Copyright protection

5.1 Copyright protection for software programs

The term 'computer program work' is defined in the Copyright Act to mean "creative works expressed as a series of instructions or commands which are used directly or indirectly to obtain a specific result in a computer or other device having information processing capability".[16] Therefore, both source code and object code, as well as any other form in which a program can be embodied, are protected; derivative programs, made by adapting a prior program, are also protected. However, as used in creating programs, programing languages, program rules and algorithms are not protected.

---

16 Copyright Act, Article 2(16).

However, several exceptions to this apply to computer programs. End-users may make a back-up copy of software as long as they legally control the initial copy.[17] Also, software may be reproduced or used in the following cases:[18]

- reproduction for individual, non-profit purposes in limited circles (eg, at home);
- reproduction and use for the purpose of analysing, studying, and determining specific elements of a program;
- reproduction and use for educational purposes in elementary and secondary educational institutions, as long as such use does not seriously damage the interests of rights holders;
- reproduction in textbooks for educational purposes in such institutions, subject to payment of a licence fee to be determined by the Minister of Culture, Sports and Tourism;[19]
- reproduction and use in entrance examinations to such institutions or other tests of knowledge or skills in non-profit organisations; and
- where required in legal proceedings.

5.2 Copyright protection for consumer electronics

In principle, copyright protection can extend to drawings and even to 3D reproductions of drawings. However, if an idea can only be expressed in a limited number of ways, then it may be difficult to show infringement. For example, in one case, the plaintiff, Samsung SDS, drew up a design layout for a video-transmission system in a bid for a subway project. This layout was secretly acquired by the defendant, another bidder on the project, who simply presented the layout as its own. The plaintiff brought a civil action for violations of trade secret laws and copyright laws. However, the Supreme Court held that the design layout was not protected by copyright law because it was a so-called 'functional work'. Its expression was limited by its functionality and thus did not show individual creativity.[20] Thus, 2D drawings and 3D reproductions must not be limited by functionality, but must show individual creativity.

Likewise, icons and screen layouts can be copyright-protected; but if deemed 'functional works', that protection may be limited to certain special cases.

6. Confidential information

Confidential information is protected under the Unfair Competition Prevention and Trade Secret Protection Act and the Act on the Prevention of Divulgence and Protection of Industrial Technology. Where information is considered to be

17 Copyright Act, Article 101-5.
18 Copyright Act, Article 101-3.
19 Copyright Act, Article 101-3(2).
20 *State v LG Industrial Systems Co Ltd*, Case No 2002Do965 (Supreme Court, 27 January 2005).

a 'trade secret' under the former act and 'industrial technology' under the latter, such information may be protected under both laws.

Under the Unfair Competition Prevention and Trade Secret Protection Act, a trade secret means technical or business information that can be used in business activities that is unknown to the public and possesses independent economic value, and where reasonable effort is being made by its owner to keep it secret. Under the act, the following activities are defined as infringing trade secrets:[21]

- acquiring a trade secret through larceny, embezzlement or coercion or other improper methods, or subsequently using or revealing the acquired trade secret (including disclosing it to certain people while generally maintaining the confidentiality of the trade secret);
- acquiring a trade secret knowing (or being grossly negligent in not knowing) that it was the subject of illegal acquisition, or subsequently using or revealing a trade secret so acquired;
- using or revealing a trade secret having learned (or having been grossly negligent in not learning) subsequent to its acquisition that its acquisition was illegal;
- using or revealing a trade secret in breach of a contractual or other obligation to maintain that trade secret for the purposes of acquiring an improper benefit or harming the owner of the trade secret;
- acquiring a trade secret knowing (or being grossly negligent in not knowing) that the trade secret had been revealed in the manner set out in the fourth bullet above, or that the trade secret had otherwise been the subject of such disclosure, or subsequently using or revealing a trade secret so acquired; and
- using or revealing a trade secret having learned (or been grossly negligent in not learning) subsequent to the acquisition of the trade secret that it had been revealed in the manner set out in the fourth bullet above, or that the trade secret had otherwise been the subject of such revelation.

The Unfair Competition Prevention and Trade Secret Protection Act provides the following remedies in case of trade secret infringement:

- injunctive relief;
- damages;[22] and
- restoration of goodwill or business reputation of the trade secret holder.

---

21 Article 2(3) of the Unfair Competition Prevention and Trade Secret Protection Act.

22 Under new Article 14-2(6) and 14-2(7) of the Unfair Competition Prevention and Trade Secret Protection Act, which at the time of writing was due to come into effect on 9 July 2019, courts may award treble damages for intentional or wilful acts of trade secret infringement as a punitive measure.

Also, the acquisition, usage or disclosure of trade secrets for the purpose of making an illegal profit or causing damage to the trade secret holder will attract criminal sanctions, as will acts of attempt, preparation or conspiracy to illegally acquire usage or disclosure of a trade secret.

The maximum penalty for trade secret infringement is imprisonment for up to five years or a fine of ₩50 million or less or at least twice (not to exceed 10 times) the pecuniary amount gained from the infringement if the 10 times the pecuniary amount gained from the infringement exceeds ₩50 million.[23]

7. Other means of protection

7.1 Registered design protection

Under Article 2(1) of the Design Protection Act,[24] the appearance of an article (or a part of an article) and typefaces can be protected as registered designs. In principle, in order to be registered as a design under the act, a design should have:

- novelty;
- creativity; and
- industrial applicability.

Depending on the kind of article, designs may be registered without any substantive examination: so the act classifies articles which are classified into the Locarno Classes 2, 5 and 19 as sensitive to trends, having short lifecycles, and therefore registrable without substantive examination.

A registered design is protected for 20 years from the application date upon registration, and the owner must pay annual maintenance fees. The design right confers an exclusive right to the owner to commercially produce, use, assign, lease, import or export articles associated with the registered design and similar designs.[25] If this exclusive right is infringed, the owner of the design right may file a civil action and/or criminal complaint.

7.2 Unfair competition law

The Unfair Competition Prevention and Trade Secret Protection Act prohibits the following activities:[26]

- to cause confusion of identity of the source of goods by using marks identical or similar to a famous mark;
- to cause confusion of business facilities by using a sign identical or similar to a famous business sign;

---

23 This clause was amended on 30 July 2013 and became effective on 31 January 2014.

24 Article 2(1) of the Design Protection Act available at www.kipo.go.kr/upload/en/download/DESIGN%20PROTECTION%20ACT_201703.pdf.

25 Article 92 of the Design Protection Act.

26 Article 2(1) of the Unfair Competition Prevention and Trade Secret Protection Act.

- to dilute distinctiveness of a famous mark;
- to falsely indicate the origin of goods;
- to misleadingly represent the goods' production, manufacture, or processing;
- to cause other misleading representations of the source of goods or false advertising;
- for a local agent or representative to misuse a foreign mark;
- to register a domain name incorporating a famous name, trademark or trade name of another person, without legitimate or proper rights to the domain name;
- to transfer, lend, import, or export exact copies of products; and
- any other acts of infringement, for one's own business without permission, on other persons' economic interests by using the outcomes achieved through substantial investment or efforts in a manner contrary to fair commercial practice or competition order.[27]

The following remedies are available under the act:
- injunction (preliminary and permanent);
- compensation for damage (in cases of 'dilution', only damage caused by intentional acts);
- restoration of damaged goodwill; and
- criminal penalties (imprisonment for up to three years or ₩30 million).[28]

7.3 Semiconductor topographies

Semiconductor topographies (layout designs) can be protected in Korea by the Act on the Layout Designs of Semiconductor Integrated Circuits by filing an application for registration within two years of the date of first commercial use of the topographies. The owner of a registered semiconductor topography will be granted an exclusive right to use the topography for profit for 10 years from the registration date (but not exceeding 15 years from the creation of the topography).

8. Counterfeiting

Counterfeit products made to closely imitate the appearance of another product may constitute various types of infringement of IP rights, such as patent, design patent, copyright, and trademark rights as well as violate the Unfair Competition Prevention and Trade Secret Protection Act.

One may contact a law firm or private investigator; or directly contact the

---

27 This clause was added by the 30 July 2013 Amendment.

28 Criminal penalties are not available for anti-cybersquatting claims and any other acts of infringement defined above.

police, prosecutor's office, or KIPO for assistance in dealing with counterfeit products in Korea. In the case of imports or exports of counterfeit products, one may also contact the Korean Customs Service.

9. Enforcement

9.1 Court structure

In Korea, patent infringement actions are heard by one of the five designated district courts,[29] and patent validity is determined by the Intellectual Property Tribunal established within KIPO.

If a patent infringement action is initiated a defendant can challenge the validity of a patent by instituting a separate invalidation trial before the tribunal. However, the district court, which is the court of original jurisdiction for patent infringement actions, proceeds under the presumption that a patent is valid until a decision is issued stating otherwise. Having said that, the Supreme Court has held that the court can refuse to enforce a patent and dismiss an infringement action without having to await a formal finding from the IPT that the patent is invalid, if lack of novelty or inventive step can be shown.[30]

9.2 Obtaining information on the infringer and the infringement

Parties can obtain information either in the form of documents or testimony submitted as evidence. There is no formal 'pre-trial' discovery under the Korean Code of Civil Procedure, and the general rule is that a party must produce its own evidence. It would be unusual for a court to order production of evidence, and parties are not obliged to disclose the existence of documents in their possession relating to a case. All evidence submitted by one party is available to the other.

In some instances, a party may file a motion for preservation of evidence requesting the court to preserve evidence prior to the filing of the initial complaint, for example in cases where there is a danger that the evidence may not be available for presentation at the hearing; and parties can petition the court to order another party or a third party to submit a known document that relates to the merits of the case. However, attorney confidentiality may be asserted as a defence to such a court order.

---

29 The Korean Civil Procedure Act and Court Organization Act vests exclusive jurisdiction of IP (patents, utility models, trademarks, designs, and plant variety rights) infringement cases with the Seoul Central, Daejeon, Daegu, Busan, and Gwangju district courts. Appeals of IP infringement cases are heard exclusively by the Patent Court.

30 For example, in a case where the plaintiff patent holder claimed damages against the defendant infringer for patent infringement, the Supreme Court ruled in favour of the defendant infringer, finding the patent at issue to be invalid for lack of inventive step. The Supreme Court made it clear that even when there was no final IPT decision on the validity of a patent, the court could deny the plaintiff's patent infringement claim if it was obvious that the patent would be invalidated for lack of inventive step (see Supreme Court Decision 2010 Da 95390, 19 January 2012).

It is also possible for courts to issue protective orders for trade secrets produced during an infringement action. Pursuant to the Korea–US Free Trade Agreement, the Unfair Competition Prevention and Trade Secret Protection Act was amended to introduce a protection order for confidential information in infringement litigations.

9.3 Interim relief

Preliminary injunctions and pre-judgment attachments are available in court actions.

Courts may issue a preliminary injunction upon a *prima facie* showing of infringement and of the necessity for provisional relief. In deciding the latter, the courts will:

- balance the irreparable harm posed to the plaintiff from ongoing infringement against economic harm to the defendant if the relief is granted;
- consider the adequacy of damages to redress the injury to the plaintiff from infringement; and
- assess the likelihood of invalidation of the patent, although the validity of a patent will be presumed until the patent is found to be invalid in an invalidation action.

Courts may, at their discretion, require the patentee to post a security bond before issuing a preliminary injunction for the purpose of indemnifying the other party for possible damages. When a preliminary injunction order issues, the court may:

- prohibit the respondent from continuing the manufacturing or sale of the goods which infringe the petitioner's right;
- direct that the infringing articles or other articles used for the infringement be transferred to the custody of a court bailiff; and
- instruct the court bailiff to post an appropriate public notice of the order on the premises of the respondent.

Pre-judgment attachments (preliminary attachments) are broadly accepted in Korea in order to secure the assets of the defendant for satisfaction of future judgment. However, the courts normally require the claimant to deposit between one-third and one-eighth of the amount of the claim as security should the underlying claim be ultimately determined to be groundless. When parties apply for pre-judgment attachment, they must identify the assets to be attached.

9.4 Threats

Before commencing an infringement lawsuit in a civil court, IP owners usually send a cease-and-desist letter to infringers demanding that they stop the infringing activity. Demand letters must be based on supportable allegations and, if not, the alleged infringer may seek relief by commending a civil action against the IP owner for tortious interference with business. Also, under Article 117 of the Code of Civil Procedure, if an infringement lawsuit is filed and the court finds the allegations to be baseless, the court may order the plaintiff to post a security deposit with the court to cover the defendant's legal and other fees, with or without the defendant's request. If the plaintiff does not post the deposit despite the court order, the court may dismiss the lawsuit.[31] The deposit amount is determined by the court at its discretion, and when the case is finalised, it may be returned to the plaintiff after deduction by the court of the legal costs.

9.5 Remedies

Both civil and criminal remedies are available for infringement of IP rights.

The following three types of civil remedies are available in Korea:

- injunctive relief (preliminary and permanent);
- damages;[32] and
- restoration of injured business goodwill or reputation.

In determining the amount of damages for patent infringement, courts presume the amount to be the number of infringing goods multiplied by the estimated profit that the patentee could have obtained by selling the products if there had been no infringing activity, but that amount can vary at the sole discretion of the court taking into account all the evidence.

A person infringing an IP right or exclusive licence may face criminal charges as well as any civil action. To initiate a criminal prosecution, the rights owner must file a criminal complaint. The maximum penalty for patent infringement is imprisonment for up to seven years or a fine of up to ₩100 million.

Permanent injunctions may be sought in court actions together with other civil remedies. Under Article 126 of the Patent Act,[33] a patentee or exclusive licensee may demand:

- discontinuation of the infringement;
- the destruction of infringing products (including products made by an infringing manufacturing process);
- the removal of the equipment to commit the infringement; and
- other measures necessary to prevent the infringement.

---

31 Article 124 of the Code of Civil Procedure.

32 Under new Article 128(8) and (9) of the Korean Patent Act, which at the time of writing was due to come into effect on 9 July 2019, courts may award treble damages for intentional or willful acts of patent or utility model infringement as a punitive measure.

33 Available at www.kipo.go.kr/upload/en/download/PATENT_ACT.pdf.

10. Hot topics

10.1 Korea–US Free Trade Agreement and Patent Act amendments

The Patent Act was amended to implement the Korea–US Free Trade Agreement, such that:

- the term of a patent may be adjusted to compensate for delays in examination of the patent caused by KIPO;
- the grace period for preserving the novelty of an invention notwithstanding public disclosure of the invention by the applicant is extended from six months to 12 months;
- the patent revocation provision for patents that have not been practised in Korea for more than two consecutive years from the grant of a compulsory licence is abolished; and
- protective orders may be granted for trade secrets acquired during a patent infringement action.

These changes took effect on 15 March 2012.

The most recent amendments to the Korean Patent Act became effective on 1 March 2017 and included the following:

- implementation of a post-grant opposition system referred to as a cancellation proceeding; and
- reducing the period to request examination from five to three years.

10.2 The Intersection between Antitrust and Intellectual Property Laws

The primary antitrust and competition law in Korea is the Monopoly Regulation and Fair Trade Law (FTL) which was enacted in 1980. The Korea Fair Trade Commission (KFTC) is the government agency responsible for interpreting and enforcing the FTL. During the last decade, the KFTC has become increasingly wary about the anti-competitive application of IP licensing and enforcement practices in light of some of the more public global and domestic IPR licensing and litigations that are reported in the press and marketplace.

The KFTC issued a set of Guidelines on the Unfair Exercise of Intellectual Property Rights (the Guidelines). The Guidelines were amended four times between 2010 and 2017 and provide definitions, recommended clauses, examples and explanations for determining fair licensing and enforcement practices. Early versions of the Guidelines focused on the following five competition law violations in patent licensing and explain how they have the effect of restricting competition and how we might reduce such anti-competitive effects:

- coordinated action and restriction of competition;
- unilateral action and restriction of competition;

- unfair agreement by abusing superior position;
- use of unfair competition methods; and
- concentration of economic power by supporting affiliates.

The December 2014 amendments to the Guidelines which took effect on 24 December 2014 added new sections on the enforcement of standard essential patents (SEPs) and non-practising entities (NPEs). More specifically, these amendments provide guidance on SEP holders who have made FRAND commitments and the ability to seek injunctive relief. The new section on NPEs illustrated possible abusive acts such as imposing an unusually high royalty rate, denying FRAND commitments that were made by a previous owner of the IPR, forming a consortia to unfairly refuse to license or license on a discriminatory basis, or sending a warning letter or initiating patent lawsuits where important information is omitted or the defence is made more difficult.

The 2016 amendments to the Guidelines which took effect on 23 March 2016 changed the definition of SEPs to delete the reference to *de facto* SEPs, removed the reference to "choice of law" provisions unfavourable to a party as a factor in evaluating fairness, and provided additional guidance on a refusal to license factors that would help evaluate the fairness thereof. On 7 December 2016, the KFTC created a new Knowledge Industry Anti-Monopoly Division consisting of five KFTC officials charged with enforcing fair trade and competition laws in knowledge-based industries such as the pharmaceutical, information and communication technology, and biotechnology industries.

In the summer of 2017, the KFTC issued a new IP survey questionnaire to about 70 pharmaceutical companies (both foreign and domestic) seeking detailed information regarding products, patents and litigation disputes in IPR in Korea. These surveys were similar to the 2010–2011 surveys that targeted potential abuse of IP rights in the pharmaceutical, electronics, chemical and mechanical industries.

This continuing interest by the KFTC in the anti-competitive effects of IPR licences and enforcement programs, and the recent formation of the Knowledge Industry Anti-Monopoly Division continues to highlight the fact that the KFTC remains one of the most active competition authorities in the world on the intersection between competition law and IP rights.

United States

Christopher Johns
Timothy May
Finnegan LLP

The United States is one of the most dynamic countries in the world for intellectual property. Close to 350,000 patents were issued in 2017,[1] and nearly three million patents are in force.[2] The United States is also a major outsourcer of intellectual property. In 2018, the United States was the originating country for about 22% of the quarter-of-a-million Patent Cooperation Treaties (PCTs) (international) applications filed.[3] The United States is also unique in certain aspects. Unlike many major jurisdictions, more than three-quarters of all cases are decided by juries, and punitive damages for 'wilful' infringement can be awarded.[4,5] Litigating in the United States can also be expensive – average costs for a single case range from $1 to 6 million.[6] Understanding the quirks of the US system can be very important for manufacturers, licensors and attorneys alike.

1. Outline of national patent procedure

1.1 Filing

Prior to 15 March 2013, the US patent system was relatively unique in that it operated under a first-to-invent regime.[7] Applications filed after that date, however, are examined on a first-to-file basis, as in much of the rest of the world.[8]

To obtain a filing date for a new US non-provisional application, the application must contain a specification – no claims are required at filing.[9] Applicants are recommended to include at least one claim because, among

---

1 2018 Patent Litigation Study, PricewaterhouseCoopers, p2, www.ipwatchdog.com/wp-content/uploads/2018/09/2018-pwc-patent-litigation-study.pdf.

2 Statistical Country Profiles: United States of America, 2019, www.wipo.int/ipstats/en/statistics/ country_profile/profile.jsp?code=US.

3 Patent Cooperation Treaty Yearly Review 2019, World Intellectual Property Organization, 2019, p11, www.wipo.int/edocs/pubdocs/en/wipo_pub_901_2019.pdf.

4 PwC 2018 Patent Litigation Study, p3.

5 An overview of patent litigation systems across jurisdictions, World Intellectual Property Organization, 2018, p13, www.wipo.int/edocs/pubdocs/en/wipo_pub_941_2018-chapter1.pdf.

6 *Id.*

7 35 USC Section 102(A) (pre-AIA).

8 Leahy-Smith America Invents Act, Pub L No 112-29, 125 Stat 285–293 (codified as amended at 35 USC Sections 100, 102, 103, 134, 135, 146, 291).

9 37 CFR Section 1.53(b).

other reasons, the claims form part of the application's original written disclosure.[10]

1.2 Examination system

(a) *Classification of applications and examination*

Applications are classified based on the technology involved in the invention.[11] Each classification is associated with a different art unit, and each art unit is part of a Technology Centre.

Applications are examined on a first-in-first-out basis, based on workloads in each art unit. Earlier-filed applications – or those having an earlier priority date (eg, from a provisional or foreign application) – tend to be examined more quickly.

(b) *Examination process*

Substantive examination generally begins with an office action 'on the merits'. Office actions on the merits cover procedural issues (including requirements to correct figures, specification language, claim language form, and priority issues), as well as substantive issues (prior art rejections under Sections 102 and 103, patent eligibility under Section 101, and written description, enablement, and indefiniteness rejections under Section 112).[12] In general, the mailing of an office action sets a three-month period for reply, extendable by up to three months with payment of fees.[13]

The first action issued by the US Patent and Trademark Office (USPTO) is referred to as a 'non-final' office action.[14] This means that prosecution is still open and that amendments to the claims may be made as a matter of right.[15] Any second or subsequent actions may be made 'final', except where the examiner introduces a new ground of rejection that was not necessitated by either the applicant's amendment of the claims, or information submitted in an information disclosure statement filed after examination has begun.[16] After the mailing of a final office action, the applicant's options for reply are limited.

A final office action limits an applicant's ability to amend the claims or continue the examination of the claims.[17]

When responding to a final office action:

---

10 *Manual of Patent Examining Procedure* (hereafter MPEP) Section 608. The MPEP is a reference work published by the USPTO which contains instructions for examiners, current procedures, case law, statutes, and rules. While it does not have the force of law, it is frequently cited as evidence of patent practice. The MPEP is available at https://mpep.uspto.gov/RDMS/MPEP/current.

11 MPEP Sections 902–903, 905.

12 37 CFR Section 1.104; MPEP Section 707.

13 MPEP Section 710.02(b).

14 37 CFR Section 1.111(a).

15 37 CFR Section 1.111(c).

16 MPEP Section 706.07(a); 37 CFR Section 1.97(c).

17 37 CFR Section 1.113.

- An applicant may cancel all rejected claims to obtain an allowance.[18] An applicant may do so by filing an after-final response, such as an after-final amendment that incorporates dependent claims into independent claims.

  An exception to this restrictive rule is provided by a pilot programme called the After Final Consideration Pilot 2.0 (AFCP 2.0) which enables limited after-final practice.[19] The AFCP 2.0 programme gives applicants and examiners the opportunity to come to an agreement about allowable subject matter. Applicants must submit a minor narrowing amendment to at least one independent claim, and examiners are given additional time to search for relevant prior art.
- The applicant may appeal the case to the Patent Trial and Appeal Board (PTAB).[20] This involves filing a notice of appeal before the expiration of the six-month period for reply, and filing an appeal brief within two months thereafter (which is extendible by up to five additional months).[21] After filing the brief, the examiner mails an 'examiner's answer' in which the examiner may withdraw rejections in the office action or may introduce new grounds of rejection.[22] The applicant then has two months (non-extendible) to file a reply brief, pay the appeal forwarding fee, and optionally request oral argument.[23] Generally, the only evidence that will be considered in an appeal includes evidence that was introduced during prosecution of the application.[24]

  Another pilot programme called the Pre-Appeal Brief Conference Pilot Program offers applicants the opportunity to plead their cases in front of a three-examiner panel prior to proceeding to appeal. It is intended to spare the time and expense required for the appeal process. To participate in the programme, an applicant submits a concise set of arguments in five or fewer pages.[25]
- An applicant may commence a civil suit against the USPTO by filing a complaint in the US District Court for the Eastern District of Virginia (located one block from the USPTO's headquarters in Alexandria, VA).[26] Unlike the PTAB appeal process, the plaintiff may introduce new evidence that was not previously introduced during prosecution, but may not present new issues (eg, legal theories) for consideration.[27] Such lawsuits are relatively rare.

---

18 37 CFR Section 1.113(c).

19 AFCP 2.0, 78 Fed Reg 29117 (17 May 2013); see also www.uspto.gov/patent/initiatives/after-final-consideration-pilot-20 (providing most up-to-date information).

20 37 CFR Section 41.37; MPEP Section 1205.

21 37 CFR Section 41.37(e); MPEP Section 1205.01.

22 37 CFR Section 41.39(a); MPEP Section 1207.

23 37 CFR Sections 41.41, 41.45, 41.47; MPEP Sections 1208–1209.

24 37 CFR Section 41.30; MPEP Section 1204.04.

25 "New Pre-Appeal Brief Conference Pilot Program", Off Gaz Pat & Trademark Office (12 July 2005).

26 MPEP Section 1216.02.

27 *Id*; see also *Hyatt v Kappos*, 625 F3d 1320 (Fed Cir 2010) (*en banc*), aff'd, 132 S Ct 1690 (2012).

- An applicant may file a request for continued examination (RCE) and pay the associated fees. This continues examination and places the case back on the examiner's docket for review. In general, these tend to get picked up more quickly than new, unexamined applications.

1.3 Timing

As of May 2019, first office actions are mailed, on average, within 16.5 months of the application's original US filing date. The average pendency of an application (ie, the amount of time before issuing as a patent or being abandoned) was 30.2 months from filing. In the business methods art units (in Technology Centre 3600), these time periods are slightly longer. First office actions were mailed (on average) 19 months from filing, and total average pendency was 30.7 months.

It is possible to vary these timescales either by asking for prioritised examination or requesting deferred examination.

Track one – or prioritised examination – is a form of accelerated examination at the USPTO. This programme, instituted in late 2011, guarantees a final disposition within 12 months.[28] To qualify, the application must have no more than four independent claims, 30 total claims, and no multiple dependent claims.[29] A track one request may be filed with a new application or with a request for continued examination. The USPTO defines final disposition broadly. Track one processing ends when the USPTO mails a final office action or a notice of allowance, or the applicant files a petition for extension of time, an RCE, a notice of appeal, a request for suspension of the application, or abandons the application, among other things.

Deferred examination is also available. An applicant may request deferral of examination of an original application by up to three years from the earliest filing date, so long as the application has not already been examined and the applicant has not requested non-publication.[30]

1.4 Costs

Each step of examination has associated fees with it. The basic application fees permit utility applications to have 20 claims, three of which may be independent. There are fees that can be paid for additional claims, additional independent claims, or to include multiple dependent claims (ie, a claim that depends on more than one other broader claim).[31] The table below summarises the relevant fees for a new US non-provisional application and the additional fee for a track one application as of May 2019.

---

28 "Changes to Implement the Prioritized Examination Track (Track I) of the Enhanced Examination Timing Control Procedures Under the Leahy-Smith America Invents Act", 76 Fed Reg 59050 (23 September 2011).

29 *Id.*

30 MPEP Section 709; 37 CFR Section 1.103.

31 All fees below are in US dollars (US$), as of 17 May 2019.

Filing fee	$300
Search fee	$660
Examination fee	$760
Each claim in excess of 20	$100
Each independent claim in excess of three	$460
Each multiple dependent claim	$820
Track one fees	$4,140

Unlike some jurisdictions, there are no annuity fees to maintain pendency of the application. Instead, the USPTO requires an issue fee, followed by payment of the following maintenance fees due at set periods after the patent issues.

Issue fee	$1,000
3.5 year maintenance fee	$1,600
7.5 year maintenance fee	$3,600
11.5 year maintenance fee	$7,400

Each of the above fees are reduced by 50% for small entities and by 75% for micro entities.

A party, including an individual inventor,[32] may qualify as a small entity if they do not (and are under no obligation to) assign, grant, convey, or license the invention to any other entity that would not qualify as a small entity. A small business concern may qualify[33] if it satisfies additional requirements (including having fewer than 500 employees).[34] A nonprofit organisation may qualify if it is an institution of higher education, is exempted under the tax

32 37 CFR Section 1.27(a)(1).
33 37 CFR Section 1.27(a)(2).
34 See 13 CFR Sections 121.801–121.805.

laws, is a US nonprofit scientific or educational organisation qualified under a nonprofit statute, or is a foreign nonprofit organisation that would otherwise qualify if it were located in the United States.[35]

A party may qualify as a micro entity if it:

- qualifies as a small entity;
- has not been named as an inventor on more than four previous filed patent applications;
- did not have a gross income exceeding three times the median household income for the preceding calendar year;[36] and
- has not assigned, granted, or conveyed a licence or other ownership interest in the application (or under an obligation to do so) to an entity that has a gross income higher than exceeding three times the median household income for the preceding calendar year.[37]

A party may also qualify as a micro entity if the applicant's employer is an institution of higher education, and the applicant has assigned, granted, conveyed, or is under an obligation by contract or law, to assign, grant, or convey, a licence or other ownership interest in the application.[38]

2. Examination of software patents

2.1 Patentable subject matter

The patent law covers eligibility for patenting:

> *Whoever invents or discovers any new and useful process, machine, manufacture, or composition of matter, or any new and useful improvement thereof, may obtain a patent therefor, subject to the conditions and requirements of this title.*[39]

Examples of claims that fall outside of 35 USC Section 101 include claims directed to structure lacking a physical or tangible form (eg, data *per se* or ideas), transitory signals (such as a carrier wave), and human activity.[40]

2.2 Supreme Court's approach to software patents

The Supreme Court has only directly addressed software patents or business methods patents a few times in recent years.

One of the first times it addressed software was the *Benson* case in 1972.[41] In that case, the computer program in question converted binary-coded decimal

---

35 37 CFR Section 1.27(a)(3).
36 In 2017, the median household income was $61,372.
37 35 USC Section 123(a)(1)–(4).
38 35 USC Section 123(d)(1)–(2).
39 35 USC Section 101.
40 Leahy-Smith America Invents Act (AIA), Pub L No 112-29, Section 33, 125 Stat 284 (2011); see also MPEP Section 2106.03.
41 *Gottschalk v Benson*, 409 US 63 (1972).

numbers into pure binary numerals. The court found that the process claim was "so abstract and sweeping as to cover both known and unknown uses of the [binary-coded decimal] to pure binary conversion".[42] The court's problem with the claims was that they attempted to protect the basic *idea*, rather than a specific practical application of the idea.[43]

In 1980, the Supreme Court took up a case (*Diehr*) covering a program that implemented a formula for curing synthetic rubber.[44] Unlike *Benson*, the claim did not attempt to cover all implementations of the formula. Instead, it implemented the formula as a repetitive process operated on a computer that would automatically open a tyre press when the formula equalled a specific amount.[45] The *Diehr* court thus found the claim patent eligible.

In 2010, the Supreme Court decided *Bilski v Kappos*, which considered a patent relating to hedging investment risk on energy markets.[46] While the Supreme Court's decision invalidated the patent, it explicitly found that there was no statutory bar against business method patents or software patents.[47]

In 2014, the Supreme Court decided *Alice v CLS Bank*, a case relating to patents for mitigating settlement risk in a financial exchange.[48] The Court laid down a test that it had previously applied to patents related to disease diagnosis,[49] finding that a claim would be ineligible for patenting if it:

- is directed to an abstract idea (or other judicial exception such as a law of nature or a natural phenomenon); and
- did not recite significantly more than the identified exception.

The court's decision did not lay out the precise contours of either part of the test.

2.3 Federal Circuit's approach to software patents

After *Alice*, the Federal Circuit attempted to define the contours of patent-eligible subject matter with mixed results on seemingly similar patents.

For example, in the *Ultramercial III* case, the Federal Circuit considered a claim concerning an 11-step method for distributing copyrighted media over the internet to consumers in exchange for them viewing an advertisement.[50] The Federal Circuit reasoned that the claim described "only the abstract idea of showing an advertisement before delivering free content",[51] and that the claim

42 *Benson*, 409 US at 68.
43 *Benson*, 409 US at 71.
44 *Diamond v Diehr*, 450 US 175.
45 *Diehr*, 450 US at 179 n 5.
46 *Bilski v Kappos*, 561 US 593 (2010).
47 *Bilski*, 561 US 606–609.
48 *Alice Corp Pty v CLS Bank Int'l*, 573 US 208 (2014).
49 *Mayo Collaborative Servs v Prometheus Labs, Inc*, 566 US 66, 78 (2012).
50 *Ultramercial, Inc v Hulu, LLC and Wildtangent, Inc*, 772 F3d 709 (Fed Cir 2014).
51 *Id*, at 715.

otherwise had only routine additional steps of updating an activity log, requiring a request from the consumer to view the advertisement, and restrictions on public access. According to the court, these remaining steps were "conventional", "specified at a high level of generality", and did not supply the required "inventive concept".[52]

In contrast, roughly three weeks later, the Federal Circuit decided DDR involving a claim that created a composite web page that combined the content of one web page with the look and feel of another web page.[53] The Federal Circuit stated that "identifying the precise nature of the abstract idea" was not as straightforward as in other cases. Regardless of the abstract idea, the Federal Circuit found that the claimed solution was "necessarily rooted in computer technology in order to overcome a problem specifically arising in the realm of computer networks".[54]

Examples of software inventions analysed by the Federal Circuit are:

Eligible

- Merging data in a network-based filtering and aggregating platform, and enhancing network accounting data records;[55]
- Computerised user interfaces for trading on fast-moving markets;[56]
- Interfaces on mobile devices enabling deep-linked functionality in an application without starting the application;[57]
- An internet content filtering system (eg, for parental controls) located on a remote ISP server that flexibly associates network accounts with filtering schemes and filtering elements;[58]
- A self-referential table for a computer database that stores information and adds new columns or rows according to an algorithm;[59] and
- Creating animations of facial gestures based on rules that relate subsequences of phonemes, timings, and morph weight sets.[60]

Ineligible

- Requesting and receiving at a cellular telephone, located outside the range of a regional broadcaster, network-based content from the broadcaster;[61]
- Providing a first menu (eg, restaurant menu) that has categories and items, and generating a second menu from that first menu by allowing categories and items to be selected;[62]

---

52 *Id*, at 716
53 *DDR Holdings, LLC v Hotels.com, LP*, 773 F3d 1245 (Fed Cir 2014).
54 *Id*, *at* 1257.
55 *Amdocs (Israel) Ltd v Openet Telecom, Inc*, 841 F3d 1288 (Fed Cir 2016).
56 *Trading Techs Int'l, Inc v CQG, Inc*, 675 FApp'x 1001 (Fed Cir 2017).
57 *Core Wireless Licensing SARL v LG Elecs, Inc*, 880 F3d 1356 (Fed Cir 2018).
58 *BASCOM Glob Internet Svcs, Inc v AT&T Mobility LLC*, 827 F3d 1341 (Fed Cir 2016).
59 *Enfish LLC v Microsoft Corp*, 822 F3d 1327 (Fed Cir 2016).
60 *McRO, Inc v Bandai Namco Games Am Inc*, 837 F3d 1299 (Fed Cir 2016).
61 *Affinity Labs of Texas, LLC v DirecTV, LLC*, 838 F3d 1253 (Fed Cir 2016).
62 *Apple, Inc v Ameranth, Inc*, 842 F3d 1229 (Fed Cir 2016).

- Detecting events on an electric power grid in real time over a wide area and automatically analysing the events;[63]
- Detecting improper access to a patient's protected health information in a computer environment based on received data and application of rules;[64]
- Classifying and storing digital images in an organised manner;[65] and
- Analysing/displaying investment data generated using statistical formulas.[66]

The Federal Circuit and lower courts have made clear that the Section 101 inquiry is a very fact-specific inquiry.[67]

2.4 USPTO Guidance

The USPTO has issued several memoranda and other publications since 2014 that provide guidance for patent examiners and patent applicants.

Step 1 – Claims within statutory category

Step 1 of the Section 101 text involves determining whether the claims fall within the four categories of statutory subject matter mentioned above: process, machine, manufacture, or composition of matter.[68]

Step '2A' – Claims directed to judicial exception

After determining that the claim is directed to statutory subject matter (process, machine, manufacture, or composition of matter), the eligibility analysis moves to determine whether the claim is directed to a judicial exception, such as an abstract idea.[69] This test does not "simply ask whether the claims involve a patenteligible concept";[70] rather, examiners are directed to distinguish claims that recite an exception (such as a claim explicitly reciting a physics equation) and claims that they merely involve an exception (such as a claim to a child's toy that involves natural principles, such as physics).[71]

While the Supreme Court did not go into detail in *Alice* as to what constitutes an abstract idea,[72] the courts and USPTO have provided some guidance on the boundaries of that concept.

Claims directed to improvements in computer functionality or other

---

63 *Electric Power Group, LLC v Alstom SA*, 830 F3d 1350 (Fed Cir 2016).
64 *Fairwarning IP, LLC v Iatric Sys, Inc*, 839 F3d 1089 (Fed Cir 2016).
65 *Digitech Image Techs, LLC v Elecs. for Imaging, Inc*, 758 F3d 1344 (Fed Cir 2014).
66 *SAP Am, Inc v InvestPic, LLC*, 898 F3d 1161 (Fed Cir 2018).
67 See *Berkheimer v HP Inc*, 881 F3d 1360 (Fed Cir 2018).
68 35 USC Section 101.
69 MPEP Section 2106.04(II).
70 *Enfish, LLC v Microsoft Corp*, 822 F3d 1327, 1335 (Fed Cir 2016).
71 MPEP Section 2106.04(II).
72 *Alice Corp Pty v CLS Bank Int'l*, 573 US 208 (2014).

technology, for example, are generally not abstract.[73] Also, a claim that is based on or that involves an abstract idea, but that does not recite the abstract idea, is generally not abstract.[74] And if a claim does recite an abstract idea, but does not seek to tie up the abstract idea, then the claim is generally not abstract.[75]

In contrast, the courts have found claims abstract[76] when they are directed to:

- concepts relating to fundamental economic practices like financial transactions[77] or risk mitigation;[78]
- methods of organising human activity like transaction or relationship management;[79]
- mitigating settlement risk;[80]
- organising information;[81]
- an idea of itself (an uninstantiated concept, plan or scheme);[82] and
- mathematical relationships or formulas.[83]

Step '2B' – Significantly more than an abstract concept

While abstract ideas, natural phenomena, and laws of nature are not eligible for patenting, claims that integrate these exceptions into something significantly more *are* eligible.[84] Thus, the second part of the *Alice/Mayo* test is often referred to as a search for an inventive concept.[85]

Some examples[86] of claims that have been found to qualify as significantly more than an abstract concept include:

- Generating a composite web page that combines the look and feel of one web page with the content of another web page;[87]
- Operating a computer program that implements a mathematical formula to operate a tyre press;[88] and
- A content filtering system on a remote server that flexibly associates network accounts with filtering schemes and filtering elements.[89]

---

73 See, eg, *Enfish, LLC v Microsoft Corp*, 822 F3d 1327 (Fed Cir 2016); *McRO, Inc v Bandai Namco Games America Inc*, 837 F3d 1299 (Fed Cir 2016); *Visual Memory LLC v Nvidia Corp*, 867 F3d 1253 (Fed Cir 2017); see also MPEP Section 2106.04(a)(I).

74 See, eg, *DDR Holdings, LLC v Hotels.com, LP*, 773 F3d 1245 (Fed Cir 2014); see also MPEP Section 2106.04(a)(1)(I)–(II).

75 *Id.*

76 A larger list of abstract claims can be found at MPEP Section 2106.04(a)(2).

77 *Credit Acceptance Corp v Westlake Svcs*, 859 F3d 1044 (Fed Cir 2017).

78 *Bilski v Kappos*, 561 US 593 (2010).

79 *buySAFE, Inc v Google, Inc*, 765 F3d 1350 (Fed Cir 2014).

80 See footnote 72 above.

81 *Bascom Glob Internet Servs, Inc v AT&T Mobility LLC*, 827 F3d 1341 (Fed Cir 2016); *In re TLI Communications LLC Patent Litigation*, 823 F3d 607 (Fed Cir 2016).

82 *Versata Software, Inc v SAP America, Inc*, 793 F3d 1306 (Fed Cir 2015).

83 *Diamond v Diehr*, 450 US 175 (1981); *Gottschalk v Benson*, 409 US 63 (1972).

84 *Alice Corp Pty Ltd v CLS Bank Int'l*, 573 US 208, 217 (2014) (citing *Mayo Collaborative Servs v Prometheus Labs, Inc*, 566 US 66, 71–72 (2012)).

85 *Id.*

86 MPEP Section 2106.05(I)(A) contains a longer list.

87 *DDR Holdings, LLC v Hotels.com, LP*, 773 F3d 1245 (Fed Cir 2014).

88 *Diamond v Diehr*, 450 US 175 (1981).

89 *BASCOM Glob Internet Servs, Inc v AT&T Mobility LLC*, 827 F3d 1341 (Fed Cir 2016).

In 2018, the Federal Circuit reversed a lower court's determination of ineligibility under Step 2B, finding that there was a genuine issue of material fact as to whether the claims contained only well-understood, routine, and conventional elements, or recited something significantly more.[90] The Federal Circuit explained that while eligibility is a question of law, there are factual inquiries underpinning it. Step 2B, therefore, requires a factual determination of "whether something is well-understood, routine, and conventional to a skilled artisan at the time of the patent".[91]

It should be noted that the search for an inventive concept is different from obviousness analysis under 35 USC Section 103. Specifically, lack of novelty or obviousness of a claimed invention does not necessarily indicate that additional elements are well-understood, routine, conventional elements[92] – nor does a finding of all elements in the prior art mean that the claim is ineligible.[93]

2.5 Anticipation

Whether a claim is anticipated in the United States (known as 'novelty' in other countries) is determined based on the effective filing date of the claimed invention. The effective filing date of a claimed invention is the actual filing date or the filing date of the earliest application for which the patent/application is entitled.[94]

Under US patent law,[95] applications may claim priority from a pending earlier application in a few ways:

- A continuation application claims priority to an earlier application.[96] The effective filing date for the purpose of applying the prior art is the date of the earlier application.[97]
- A divisional application claims priority to an earlier application and claims an invention that is independent and distinct from the earlier application.[98] Such applications result from USPTO determinations known as 'restriction requirements'.[99] Like continuation applications, the effective filing date for the purpose of applying the prior art is the date of the earlier application.[100]

---

90 *Berkheimer v HP Inc*, 881 F3d 1360 (Fed Cir 2018).

91 881 F3d at 1369.

92 See, eg, *Intellectual Ventures I LLC v Symantec Corp*, 838 F3d 1307, 1315 (Fed Cir 2016) ("the 'novelty' of any element or steps in a process, or even the process itself, is of *no relevance* in determining [eligibility]").

93 See, eg, BASCOM, 827 F3d at 1350 (while each element was known in the art, the "non-conventional and non-generic arrangement of known, conventional pieces" was sufficient to find the patent eligible).

94 35 USC Section 100(i).

95 The US Patent Act is codified in 35 USC Sections 1–376. In some instances, this chapter refers to only the section number of 35 USC.

96 35 USC Section 120.

97 *Id.*

98 35 USC Section 121.

99 See generally MPEP Sections 800 *et seq.*

100 35 USC Section 121.

- A continuation-in-part (CIP) application claims priority to an earlier application (eg, as a continuation) but adds additional subject matter that was not disclosed in the earlier application.[101] Claims in such CIP applications that are not supported by the parent application have a later effective filing date;[102] however, the earlier application does not qualify as prior art, unless it was published more than a year before the effective filing date.

Another manner of priority claim is using a US provisional application. To obtain a filing date for a provisional application, only a specification is required.[103] These applications are much less formal than non-provisional applications in that they are not examined and cannot be amended.[104] There is a 12-month period in which a provisional application must be converted to a nonprovisional application or in which a new non-provisional application must be filed claiming priority to the provisional application.[105]

Under Section 102(a), a claimed invention is entitled to a patent unless:

- the claimed invention was patented, described in a printed publication, or in public use, on sale, or otherwise available to the public before the effective filing date of the claimed invention; or
- the claimed invention was described in a patent issued under Section 151, or in an application for patent published or deemed published under Section 122(b), in which the patent or application, as the case may be, names another inventor and was effectively filed before the effective filing date of the claimed invention.

There are a few exceptions to this first-to-file system.

An earlier reference, disclosed one year or less before the effective filing date of the claimed invention, does not qualify as prior art under Section 102(a)(1) if:

- it was made by the inventor or joint inventor or by another who obtained the subject matter disclosed directly or indirectly from the inventor or a joint inventor; or
- the subject matter had been publicly disclosed by the inventor or a joint inventor or another who obtained it directly or indirectly from the inventor or a joint inventor.[106]

An earlier reference also does not qualify under Section 102(a)(2) if:

---

101 MPEP Section 201.08; 37 CFR Section 1.78.
102 MPEP Section 706.02(VI).
103 37 CFR Section 1.53(c).
104 *Id.*
105 37 CFR Section 1.53(i).
106 35 USC Section 102(b)(1).

- the subject matter was obtained directly or indirectly from the inventor or a joint inventor;
- the subject matter had, before being effectively filed under Section 102(a)(2), been publicly disclosed by the inventor or a joint inventor or another who obtained the subject matter disclosed directly or indirectly from the inventor or a joint inventor; or
- the subject matter and the claimed invention, not later than the effective filing date of the claimed invention, were owned by the same person or subject to an obligation of assignment to the same person or entity.

Earlier references can also be disqualified if the claimed invention and the earlier reference were developed as part of a joint research agreement between parties.[107]

2.6 Obviousness

Whether a claim is obvious in the United States (known as 'inventive step' in other countries) is determined based on what would have been obvious to a person of ordinary skill at the time of filing for the patent.[108] This determination requires a three-factor test known as the *Graham* factors:

- determining the scope and content of the prior art;
- ascertaining the differences between the claimed invention and the prior art; and
- resolving the level of ordinary skill in the pertinent art.[109]

After resolving the *Graham* factors, examiners must determine whether the claimed invention would have been obvious to one of ordinary skill in the art at the time of filing for the patent, by articulating one or more reasons why the claimed invention would have been obvious. Reasons include:

- combining prior art elements, according to known methods, yielding predictable results;
- simple substitution of a known element for another, yielding predictable results;
- using a known technique to improve similar devices, methods, or products, in the same way;
- applying a known technique to a known device, method, or product, ready for improvement, yielding predictable results;
- choosing from a finite number of identified, predictable solutions, with a reasonable expectation of success;

---

107 35 USC Section 102(c).
108 35 USC Section 103 *and* 35 USC Section 103(a) (pre-AIA).
109 *Graham v John Deere Co of Kansas City*, 383 US 1, 17 (1966).

- work in one field of endeavour prompting variations for use in the same or a different field based on design incentives or other market forces, if the variations would have been predictable; and
- a teaching, suggestion, or motivation in the prior art that would have led one of ordinary skill to modify or combine prior art to arrive at the claimed invention.[110]

There are a number of avenues for rebutting an examiner's finding of obviousness:

- If a modification of prior art would render the prior art unsatisfactory for its intended purpose, there would be no suggestion or motivation to perform that modification.[111] For example, the Federal Circuit reversed a finding that it would have been obvious to turn a prior art device upside down to meet the claimed invention, because the prior art device would no longer operate in that orientation.[112]
- If a modification of prior art would change the principle of operation of that prior art, there would similarly be no suggestion or motivation to perform that modification.[113] For example, it would not be obvious to modify a prior art reference requiring a stiffened sealing to use a resiliently deformable seal.[114]
- Secondary considerations are relevant as well. Evidence of commercial success, long-felt but unsolved needs, failure of others, and unexpected results[115] can be used to rebut an examiner's finding of obviousness.[116] However, for evidence of secondary considerations to be accorded weight, the party offering the evidence must establish a nexus between it and the merits of the claimed invention.[117]

2.7 Written description and enablement

There are two separate requirements under Section 112(a): the specification must contain a written description of the claimed invention, and the "manner and process of making and using" the claimed invention in terms that "enable any person skilled in the art ... to make and use the same".[118] These are widely referred to as the 'written description requirement' and the 'enablement requirement', respectively.[119]

---

110 MPEP Section 2143; see also *KSR Int'l Co v Teleflex Inc*, 550 US 398 (2007).
111 MPEP Section 2143.01(V).
112 *In re Gordon*, 733 F2d 900, 902 (Fed Cir 1984).
113 MPEP Section 2143.01(VI).
114 *In re Ratti*, 270 F2d 810, 813 (CCPA 1959).
115 *Graham v John Deere Co of Kansas City*, 383 US 1, 17–18 (1966).
116 MPEP Section 2141(V).
117 *In re Huai-Hung Kao*, 639 F3d 1057, 1068 (Fed Cir 2011).
118 35 USC Section 112(a) *and* 35 USC Section 112, para.1 (pre-AIA).
119 *Ariad Pharm, Inc v Eli Lilly & Co*, 598 F3d 1336, 1344 (Fed Cir 2010).

Satisfying the written description requirement means that a specification must describe a claimed invention in such detail that one of ordinary skill could conclude that the inventor had possession of the claimed invention as of the filing date.[120] This does not mean that the specification must describe the claimed subject matter in exactly the same terms as used by the claims.[121] Claims that are original to the application are presumed to have adequate written description.[122] Written description can be met using any part of the application – including the words of the application, illustrated structures, figures, diagrams, etc.[123]

The enablement requirement means that one skilled in the art must be able to make and use the claimed invention without undue experimentation. Determining whether or not the specification discloses enough to avoid undue experimentation requires considering:

- the quantity of experimentation necessary;
- the amount of direction or guidance presented;
- the presence or absence of working examples;
- the nature of the invention;
- the state of the prior art;
- the relative skill of those in the art;
- the predictability or unpredictability of the art; and
- the breadth of the claims.[124]

(a) Definiteness – Section 112(b)

Section 112(b) requires that the claims particularly point out and distinctly claim the subject matter which the inventor(s) regard as the invention. This is commonly referred to as the 'definiteness requirement'.

While claims are under examination, they must be given their "broadest reasonable interpretation" consistent with the specification as understood by one of ordinary skill in the art.[125] Under this interpretation, each word in a claim must be given its plain meaning unless that meaning would be inconsistent with the specification.

Whether a claim is indefinite under 35 USC Section 112(b) (or 35 USC Section 112, second paragraph (pre-AIA)) requires determining whether one skilled in the art would understand what is claimed when the claim is read in light of the specification.[126] (Note that this standard is slightly different for

120 *Vasudevan Software, Inc v MicroStrategy, Inc*, 782 F3d 671, 682 (Fed Cir 2015).
121 *Eiselstein v Frank*, 52 F3d 1035, 1038 (Fed Cir 1995).
122 *Application of Wertheim*, 541 F2d 257, 264 (CCPA 1976).
123 *Lockwood v Am Airlines, Inc*, 107 F3d 1565, 1572 (Fed Cir 1997).
124 *In re Wands*, 858 F2d 731, 737 (Fed Cir 1988).
125 *In re Packard*, 751 F3d 1307, 1312 (Fed Cir 2014); *In re Morris*, 127 F3d 1048, 1054 (Fed Cir 1997); *Phillips v AWH Corp*, 415 F3d 1303, 1316 (Fed Cir 2005); see also MPEP Section 2111.
126 *Power-One, Inc v Artesyn Techs, Inc*, 599 F3d 1343, 1350 (Fed Cir 2010).

claims appearing in issued patents, as those claims are definite unless, when interpreted in light of the specification and the prosecution history, they fail to "inform those skilled in the art about the scope of the invention with reasonable certainty".[127])

(b) *Means-plus-function – Section 112(f)*

Claims are also examined to determine if they recite means-plus-function elements under 35 USC Section 112(f). This section reads as follows:

> *An element in a claim for a combination may be expressed as a means or step for performing a specified function without the recital of structure, material, or acts in support thereof, and such claim shall be construed to cover the corresponding structure, material, or acts described in the specification and equivalents thereof.*[128]

The broadest reasonable interpretation of such means-plus-functions claims is "the *structure, material* or *act* described in the specification as performing the entire claimed function and equivalents to the disclosed structure, material or act".[129]

Determining whether or not to apply Section 112(f) is a three-step process in front of the USPTO:

- Determine whether the claim recites any limitations having the term 'means', 'step', or a generic placeholder term used as a substitute for performing a claimed function (like '*module* for *processing*').
- If so, determine if the term is modified by functional language, typically, but not always linked by the transition word 'for' (eg, means for) or another linking word or phrase (like '*processing* means for sending a communication').
- If so, determine if the 'means', 'step', or generic placeholder term is not modified by sufficient structure, material, or acts for performing the claimed function (like '*connector assembly*').[130]

If the claim is found to recite a means-type term and associated functional language, and does not also recite sufficient structure, material or acts for performing the claimed function, the specification must have adequate disclosure showing the structure corresponding to the Section 112(f) language. Importantly, if the claim term relates to software, the corresponding structure is not a computer or processor, but the algorithm run by the computer or processor to implement the term's claimed function. Software applications should thus include a disclosure of the algorithm (eg, using flow charts) used by the computer or processor to implement the claimed functions. Without such

127 *Nautilus, Inc v Biosig Instruments, Inc*, 572 US 898, 910 (2014).
128 35 USC Section 112(f).
129 MPEP Section 2181 (emphasis added); see also *In re Donaldson Co*, 16 F3d 1189, 1194 (Fed Cir 1994).
130 MPEP Section 2181(I).

disclosure, the claim reciting the Section 112(f) language must be rejected under Section 112(b) as failing to particularly point out and distinctly claim the invention.[131] For example, if a specification recites a means-plus-function term, such as "database editing means for editing a row of the database", but lacks disclosure of the algorithm used to program the computer to perform the claimed editing, the claim may be doomed as indefinite.[132]

3. Infringement

3.1 Acts of infringement

An infringer is one who makes, uses, offers to sell, or sells any patented invention in the United States, or imports into the United States any patented invention, during the term of the patent.[133]

3.2 Claim interpretation

The words in a claim are the most important aspect of the patent. As the Chief Judge of the Federal Circuit put it in 1990, "the name of the game is the claim".[134] During litigation, US courts will conduct one or more *Markman*[135] hearings during which a judge will determine constructions for one or more claim terms. In support of various claim constructions, each party offers intrinsic evidence (eg, the specification/figures/claims and prosecution history) and/or extrinsic evidence (eg, dictionary definitions or technical literature).

The procedure and law for construing patent claims in the United States often play critical roles in effective strategies for drafting new US patent applications.

One type of intrinsic evidence – the prosecution history – is especially important as it can affect the applicability of a patent claim to an allegedly infringing item or process. For example, in the 2002 *Festo* case,[136] the Supreme Court explained that statements and claim amendments made during prosecution could prevent the patentee from arguing that a patent's claims are infringed under the doctrine of equivalents (a doctrine by which claims may be read to cover infringing acts or items with insubstantial differences from the claim language).[137]

131 *In re Donaldson Co, Inc*, 16 F3d 1189, 1195 (Fed Cir 1994).
132 *Finisar Corp v DirecTV Grp, Inc*, 523 F3d 1323, 1340 (Fed Cir 2008).
133 35 USC Section 271(a).
134 Giles S Rich, "The Extent of the Protection and Interpretation of Claims – American Perspectives" (1990) 21 Int'l Rev Indus Prop & Copyright L 497, 499.
135 Named for *Markman v Westview Instruments, Inc*, 517 US 370 (1996).
136 *Festo Corp v Shoketsu Kinzoku Kogyo Kabushiki Co*, 535 US 722 (2002).
137 *Festo*, 535 US at 733.

3.3 Direct infringement

A direct infringer is one who performs one of the proscribed acts in Section 271(a): making, using, offering to sell, or selling a patented invention in the United States, or importing into the United States any patented invention.[138]

Selling a product that is preprogrammed to perform a patented method cannot, without more, constitute direct infringement.[139] For example, a mobile phone manufacturer was found not liable for direct infringement for phones that were programmed to infringe the claims.[140]

3.4 Indirect infringement

(a) *Induced infringement*

Induced infringement[141] occurs when one induces another to infringe a patent, and the inducing party is liable as an infringer.[142]

Induced infringement requires showing that an accused infringer actively encouraged infringement, knowing that the acts they induced constituted patent infringement, and that the encouraging acts *actually resulted* in direct patent infringement. Although inducement may be proven by circumstantial evidence (eg, advertisements and user manuals), the inducement must be found to have actually occurred.[143]

(b) *Network claims (divided infringement/joint infringement)*

In 2015, the Supreme Court's decision in *Limelight v Akamai* had an impact on enforcement of patents directed to distributed systems involving electronic networks.[144] In Limelight, the patents concerned a method of delivering electronic data using a content delivery network.[145] The court held that if there is no direct infringement of a claim (eg, by a single party running the overall distributed system), there can be no induced infringement of a claim directed to that overall system.[146]

One takeaway from *Limelight* is to draft multiple sets of claims, one from the viewpoint of each single party to a multi-party system. For example, in a system involving a mobile device communicating with a banking system and to effect purchases at a point-of-sale device, it could be advisable to draft claims from the viewpoint of the mobile device (sending and receiving data over a wireless network, transmitting information locally to the point-of-sale device), the

---

138 35 USC Section 271(a).
139 *Adaptix, Inc v Apple, Inc*, 78 F Supp 3d 952 (ND Cal 2015).
140 *Adaptix*, 78 FSupp 3d at 956.
141 35 USC Section 271(b).
142 *Id.*
143 *Power Integrations, Inc v Fairchild Semiconductor Int'l, Inc*, 843 F3d 1315 (Fed Cir 2016).
144 *Limelight Networks, Inc v Akamai Techs, Inc*, 572 US 915 (2014).
145 *Id.*
146 572 US 915 at 922.

point-of-sale device (receiving information locally from the mobile device, processing sales and calculations), and any financial network (receiving and sending communications to the mobile device and the point-of-sale device).

(c) *Contributory infringement*

Contributory infringement occurs when an accused party offers to sell or sells a component of a patented item or used in a patented process. If the accused party knows that the component is especially made or adapted for infringement of the patent, and the component is not a "staple article or commodity of commerce suitable for substantial non-infringing use", the party may be found liable as a contributory infringer.[147]

There are also provisions in Section 271(f) regarding infringement of exported items. Infringement may be found when an entity supplies "all or a *substantial portion*" of the components of a patented invention in or from the United States, uncombined, in such a manner to actively induce combination of the components *outside* of the United States.[148] In short, where the accused infringer supplies some or all of the component parts of an accused product from the United States to an overseas location for final assembly, Section 271(f)(1) gives the patentee an opportunity to recover damages from the infringer based on the export of only a part of the patented invention. However, recent case law has curtailed the reach of this theory. The Supreme Court ruled in 2017 that a single component cannot ever constitute a substantial portion of a patented invention so as to trigger liability under Section 271(f)(1).[149]

In contrast, another section of the law (Section 271(f)(2)) differs in that even a single component of a patented invention that is *especially made or adapted for use in the invention* (and not a staple article or commodity of commerce suitable for substantial non-infringing use) may lead to liability for infringement when similarly exported.[150]

With respect to software, the Supreme Court considered software that originated in the United States and was sent abroad for duplication (either using a golden master disk or via a network) in *Microsoft v AT&T* in 2007.[151] Because the master disk or electronic transmission itself is "never installed on any foreignmade computers", and copies are made which are used for installation, the copies that are actually installed are not exported and thus there was no infringement under Section 271(f).

147 35 USC Section 271(c).
148 35 USC Section 271(f)(1) (emphases added).
149 *Life Techs Corp v Promega Corp*, 137 S Ct 734, 741 (2017).
150 35 USC Section 271(f)(2); *Life Techs*, 134 S Ct 741–742.
151 *Microsoft Corp v AT&T Corp*, 550 US 437 (2007).

4. Inventorship and associated issues

4.1 Oath/declaration

Inventorship is determined based on the claims.[152] Patent applications must include, or be amended to include, the name of the inventor for *any* invention claimed in the application. If a person is the one who first reduced to practice any element of any claim, that person is an inventor. If multiple people contributed to the invention, they must all be named – regardless of the amount of contribution from each inventor or the number of claims that the inventor contributed to.[153] The oath/declaration must be filed no later than the date on which the issue fee is paid.[154]

To be named as an inventor, the inventor must sign an oath or declaration.[155] Before the America Invents Act came into effect on 16 March 2013, the requirements for such a document (beyond a signature) were that the inventor "believes himself to be the original and first inventor of the process, machine, manufacture, or composition of matter, or improvement thereof, for which he solicits a patent; and ... state of what country he is a citizen".[156]

Since 16 March 2013, the requirements are slightly different.[157] There is no longer a requirement to name each inventor's citizenship, but the oath/declaration now requires affirmation that the application was made at the inventor's direction or by the inventor. These small differences are important for applications with an effective filing date before 16 March 2013, because any continuing applications (continuation/divisional/continuation-in-part) that are filed after this date must comply with the new guidelines.

As discussed above, an oath/declaration must be executed for all named inventors in an application. A substitute statement may be filed in lieu of an oath/declaration, if an inventor is deceased, legally incapacitated, cannot be found after diligent effort, or refuses to sign despite being under obligation to assign the invention to another.[158]

4.2 Assignment

Ownership grants the patent owner the right to exclude others from making, using, offering for sale, selling, or importing into the United States the claimed invention.[159] The applicant of a patent application is dependent upon the ownership of an application. For AIA applications (those filed on or after 16

---

152 35 USC Section 115(a).
153 35 USC Section 116(a).
154 35 USC Section 115(f).
155 See generally 35 USC Section 115.
156 35 USC Section 115 (pre-AIA).
157 35 USC Section 115(b).
158 35 USC Section 115(d)(2).
159 35 USC Section 154(a)(1).

September 2012), the original applicant is presumed to be the initial owner of an application for an original patent.[160] In contrast, for pre-AIA applications, the ownership of the patent (or patent application) initially vests in the named inventors of the invention of the patent.[161]

Federal patent law does not place many requirements on actual assignment of the patent right. For example, unlike in some countries, the amount of compensation from an assignee to an inventor is not required to be a large amount. In one case from 2014, the Federal Circuit found that an accused infringer was a part owner of a patent, owing to a clause in the assignment that assigned ownership in the patent to the accused infringer for the consideration of one US dollar.[162]

However, employment agreements – commonly relied upon to establish ownership of all inventions developed by inventors while employed by a company – must be carefully written to ensure ownership is clear. In a 2018 case, a clause in an employment agreement stating that the inventor "will assign" rights in the invention provided an obligation to assign, but did not create an immediate assignment.[163]

Care should be taken if a US patent or patent application has co-owners or joint owners. Unless there is some agreement otherwise, each co-owner owns an equal and undivided interest in the entire patent.[164] As such, each co-owner of a patent is free to make, use, offer to sell, or sell the patented invention within the United States, or import the patented invention into the United States *without* the consent of the other co-owner. As a result, one co-owner could sell and/or license its rights in the patent to a third party without the consent of the other co-owner.[165] The co-owner would not even be required to give an accounting or share with the other co-owner any compensation for such a sale or license. For these reasons, it is often strategically necessary for co-owners of a US patent to have a written agreement that addresses licensing, enforcement and other issues.

5. Copyright protection

5.1 Requirements

The US Copyright Act provides protection for "original works of authorship" once they are "fixed in any tangible medium of expression".[166] This includes computer programs as well as more traditional forms of copyrighted goods, like

160 See 37 CFR Section 3.73(a).
161 *Beech Aircraft Corp v EDO Corp*, 990 F2d 1237, 1248 (Fed Cir 1993).
162 *Memorylink Corp v Motorola Solutions, Inc*, 773 F3d 1266 (Fed Cir 2014).
163 *Advanced Video Techs LLC v HTC Corp*, 879 F3d 1314 (Fed Cir 2018).
164 35 USC Section 262.
165 See, eg, *Schering Corp v Roussel-UCLAF SA*, 104 F3d 341, 344 (Fed Cir 1997).
166 17 USC Section 102(a).

literary works, musical works, dramatic works, pantomimes and choreographic works, pictorial, graphic, and sculptural works, motion pictures, sound recordings, and architectural works. Copyright does not extend to mere concepts (eg, ideas, principles, concepts, discoveries) or to processes, systems, or methods of operation.[167]

Copyright is automatic: once an original work of authorship is fixed into a tangible medium of expression, copyright attaches and registration with the US Copyright Office is not required to obtain protection.[168] However, except under limited circumstances, a copyright owner may not institute a suit alleging copyright infringement until registration has been made.[169]

To obtain registration, two complete copies of published works (and one complete copy of unpublished works) must be deposited at the Copyright Office.[170]

(a) Computer programs

A computer program is defined as a "set of statements or instructions to be used directly or indirectly in a computer in order to bring about a certain result".[171] Just as with other forms of copyrightable subject matter, computer programs can only be protected if they are fixed. Fixing programs into a tangible medium of expression requires making it "sufficiently permanent or stable to permit it to be perceived, reproduced, or otherwise communicated for a period of more than transitory duration".[172]

It is generally agreed that all forms of computer programs, whether in the form of object code or source code, whether application programs or operating system programs, and whether stored on disks or tapes or embedded in memory devices, are automatically subject to copyright protection when the programs are fixed in a tangible form. Microcode (instructions that control microprocessors) is also subject to copyright protection.[173]

A computer program can be protected in source code form or in compiled form (eg, object code).[174] To protect source code, the first and last 25 pages (or 50 pages that reasonably represent the first and last portions of the code) must be submitted.[175] If the source code contains trade secrets, there are some options for protecting those secrets:

- submitting the first 10 pages and the last 10 pages with no redactions;

---

167 17 USC Section 102(b).
168 17 USC Section 408(a).
169 17 USC Section 411(a); see also Section 411(c) (permitting the filing of a suit for infringement for works first fixed at the same time as transmission).
170 17 USC Section 408(b).
171 17 USC Section 101.
172 *Id.*
173 *NEC Corp v Intel Corp*, 645 F Supp 590 (ND Cal 1986).
174 *Apple Computer, Inc v Franklin Computer Corp*, 714 F2d 1240, 1249 (3d Cir 1983).
175 US Copyright Office, Circular 61, "Copyright Registration of Computer Programs" (September 2017).

- submitting the first 25 pages and the last 25 pages with redactions, so long as the redactions do not amount to more than 50% of the deposited code;
- submitting the first 25 pages and the last 25 pages of the object code, along with 10 or more consecutive pages of unredacted source code;
- for programs with less than 50 pages of source code, the entire source code with redactions, so long as the redactions do not amount to more than 50% of the deposited code; or
- 20 to 50 pages that reasonably represent the first and last portions of the code, if the source code does not have a precise beginning, middle or end.[176]

Derivative computer programs can be protected by including portions of the source code that include new or revised source code.[177] Programs fixed on CD-ROM media may be submitted for registration by including the source code and the entire CD-ROM package (including instructional material).[178]

Computer programs are copyrightable subject matter, but certain aspects of a computer program may not be. For example, in a 1995 case that was affirmed by the Supreme Court,[179] a menu structure on a user interface was held not to be copyrightable subject matter, as it constituted a "method of operation" rather than an original expression.

(b) Compilations of data

Compilations, including compilations of data known as 'databases', are also protectable. The Copyright Act (17 USC 101) defines a compilation as a:

> *work formed by the collection and assembling of preexisting materials or of data that are selected, coordinated, or arranged in such a way that the resulting work as a whole constitutes an original work of authorship.*

Earlier copyright law protected all compilations in which the compiler invested time, effort, and resources in the process of creating the database. The Supreme Court changed this in 1991 in the *Feist v Rural* case.[180] In *Feist*, the court considered a telephone directory, finding it ineligible for protection under the copyright laws, and held that a modicum of creativity in the compilation's selection, coordination or arrangement was required for a compilation to qualify as an original work of authorship. Most expression meets the (very low) bar of originality for copyrightable subject matter. As the court stated, "the requisite level of creativity is extremely low; even a slight amount will suffice".[181]

---

176 *Id.*
177 *Id.*
178 *Id.*
179 *Lotus Dev Corp v Borland Int'l Inc*, 49 F3d 807 (1st Cir 1995), aff'd mem by an equally divided court, 116 S Ct 804 (1996).
180 *Feist Publications v Rural Tel Serv Co*, 499 US 340 (1991).
181 499 US 340 at 345 (internal citations omitted).

Despite this low bar, not all data collections, such as databases, are copyrightable. The following factors may be relevant to determine whether a database warrants copyright protection:

- the total number of options available;
- external factors that limit the viability of certain options and render others non-creative; and
- prior uses that render certain selections 'garden variety'.[182]

For example, in *Matthew Bender*, the Second Circuit Court of Appeals found that a database of legal opinions was not entitled to copyright protection for its formatting of case titles and party names, the deciding courts, and other minor ordering and formatting of other case information.[183]

The scope of protection for databases is generally narrow; a court is unlikely to find infringement unless the defendant has virtually identically copied the copyrighted work.[184]

5.2 Limitations on protection

(a) First sale

One limitation on copyright protection is commonly referred to as the 'first sale doctrine':

> *Notwithstanding the provisions of section 106(3), the owner of a particular copy or phonorecord lawfully made under this title, or any person authorized by such owner, is entitled, without the authority of the copyright owner, to sell or otherwise dispose of the possession of that copy or phonorecord.*[185]

The Section 109(a) first sale doctrine comes up in the context of licensing; a copyright licensor may wish to limit or control a licensee's use, but once the original purchased items change hands from the licensor to the licensee, the licensor can no longer control what the licensee does with those items.[186]

If a copyright owner wishes to authorise a licensee to use copyrighted computer software only for the licensee's use, but then sells copies of the software to the licensee, the copyright owner cannot then prevent the licensee from reselling the purchased (original) copies of the software. A licensor who wishes to subject licensees to collateral restrictions or to restrict licensees' use or disposal of the licensed work generally must retain ownership of the physical copies of the work.[187]

---

182 *Matthew Bender & Co v W Pub Co*, 158 F3d 674, 682–683 (2d Cir 1998), *cert denied*, 526 US 1154 (1999).
183 158 F3d at 683.
184 *Apple Computer, Inc v Microsoft Corp*, 35 F3d 1435, 1446 (9th Cir 1994).
185 17 USC Section 109(a).
186 See 17 USC Section 109(a).
187 See, eg, 17 USC Section 109(d)

Capitol Records v ReDigi[188] presents an interesting discussion on the first sale doctrine as it relates to digital works. ReDigi uploaded digital song files to a computer storage cloud where others could download the songs, and once a song was downloaded, the file would disappear from the cloud and original computer. The court held that the first sale doctrine did not apply to the above circumstance because a new copy had to be created. The first sale doctrine only applies to a particular copy and the owner's sale of that copy. Because a new copy was created, the first sale doctrine did not apply even though a resale of the song occurred from one person to another.

(b) The computer program copying limitation

A licensee might wish to obtain ownership of copies of software, to take advantages of other provisions allowing the making of copies or adaptations of the program for certain purposes, such as archival purposes or maintenance/repair.[189] This computer program copying limitation is codified in Section 117(a):

> *Notwithstanding the provisions of section 106, it is not an infringement for the owner of a copy of a computer program to make or authorize the making of another copy or adaptation of that computer program provided:*
>
> - *that such a new copy or adaptation is created as an essential step in the utilization of the computer program in conjunction with a machine and that it is used in no other manner, or*
> - *that such new copy or adaptation is for archival purposes only and that all archival copies are destroyed in the event that continued possession of the computer program should cease to be rightful.*[190]

At least one court has read this section broadly, permitting the owner of a lawfully owned copy of a computer program to authorise a third party to modify the program where the enhanced program was used only in-house by the owner of the copy.[191] Another court has held that the creation and distribution of a program that permitted the copying of copy-protected programs by the owners of the copies of such programs was not a contributory infringement because it facilitated an essential step in the utilisation of the copyrighted program and permitted the owner to make archival copies.[192]

However, the Federal Circuit limited the scope of Section 117 in *DSC Communications*.[193] The court, applying Ninth Circuit law, found that the terms of the licence determine whether or not the licensee is considered an owner of

---

188 *Capitol Records, LLC v ReDigi, Inc*, 934 F Supp 2d 640 (SDNY 2013).
189 17 USC Section 117.
190 35 USC Section 117(a).
191 *Foresight Res Corp v Pfortmiller*, 719 F Supp 1006 (D Kan 1989).
192 *Vault Corp v Quaid Software Ltd*, 847 F2d 255 (5th Cir 1988).
193 *DSC Commc'ns v Pulse Commc'ns, Inc*, 170 F3d 1354 (Fed Cir 1999).

the software. It found that the licensee in question was not an owner because its use exceeded the scope of its licence, which included language that limited use of the software only "in the equipment for which it was intended". The use in question was not "in the equipment for which it was intended", the court found. Therefore, the licensee was not considered an owner for purposes of Section 117.

Another aspect of Section 117 concerns repair/maintenance. The Digital Millennium Copyright Act of 1998[194] amended Section 117 to add additional protection against infringement based on machine maintenance or repair.[195]

5.3 Acts of infringement

Copyright infringement occurs when the exclusive rights of the copyright owner are violated. This includes reproduction, preparation of derivative works, distribution of copies, and public performance, display, or transmission.[196]

Establishing infringement requires demonstrating two things: that the plaintiff is the owner, exclusive licensee, or beneficial owner of a copyright, and that the defendant infringes the work by copying those elements that are original.[197] A certificate of copyright creates a presumption of validity[198] but it can be rebutted.[199]

Direct infringement of a copyright occurs when anyone violates one of the exclusive rights mentioned above. There are also two types of indirect infringement: contributory infringement, where one is "intentionally inducing or encouraging direct infringement",[200] and vicarious infringement, where one is profiting from direct infringement and does not exercise the right to stop or limit it.[201] An example of a direct infringer is a user who copies a movie from a DVD to his or her computer; an example of an indirect infringer is a company operating a file-sharing service or website that receives such a movie and allows users to download that movie by using the service.

Knowledge of the copyright registration or the intent of the alleged infringer is not relevant to a determination of liability.[202] A copyright owner may show infringement based on identical copying (verbatim infringement) or by demonstrating that the infringer had access to the work and that the works are substantially similar (derivative infringement).[203]

---

194 Digital Millennium Copyright Act, Pub L 105–304, 112 Stat 2860 (codified as amended at 17 USC Sections 101, 104, 104A, 108, 112, 114, 117, 512, 701, 1201–1205, 1301–1332, and 28 USC Section 4001).

195 17 USC Section 117(c)–(d).

196 17 USC Section 106.

197 *Feist Publications, Inc v Rural Tel Serv Co*, 499 US 340, 361 (1991).

198 17 USC Section 410(c).

199 *Gates Rubber Co v Bando Chem Indus*, Ltd, 9 F3d 823, 832–833 (10th Cir 1993).

200 *Metro-Goldwyn-Mayer Studios Inc v Grokster, Ltd*, 545 US 913, 930 (2005) (citing *Gershwin Pub Corp v Columbia Artists Mgmt, Inc*, 443 F2d 1159, 1162 (2d Cir 1971)).

201 545 US 913 (citing *Shapiro, Bernstein & Co v HL Green Co*, 316 F2d 304, 307 (2d Cir 1963)).

202 *Fitzgerald Pub Co v Baylor Pub Co*, 807 F2d 1110, 1113 (2d Cir 1986).

203 *Atari Games Corp v Nintendo of Am Inc*, 975 F2d 832, 844 (Fed Cir 1992).

An infringement suit must be brought within three years of the claim accruing.[204] The statute of limitations starts to run when the plaintiff learns, or should as a reasonable person have learned, that the defendant was violating the copyright.[205]

5.4 Software infringement

In a 1992 case, the Second Circuit Court of Appeals considered a situation involving a programmer who wrote a program that copied approximately 30% of another program's code.[206] The infringing party rewrote the program to avoid infringement.[207] The court used a three-part infringement analysis for computer software:[208]

- Abstraction – separating idea from expression, by dissecting the original program's structure and isolating each level of abstraction within;
- Filtration – examining structural components at each level of abstraction to determine whether their inclusion was protectable or unprotectable, and filtering out elements dictated by efficiency, external factors, and elements taken from the public domain; and
- Comparison – determining whether any protectable expression remains, and if so, whether any of that was copied.

5.5 Criminal aspects of infringement

The Copyright Act imposes criminal liability for certain types of infringement and other acts. For example, the act covers any "person who willfully infringes a copyright" if the infringement was for the purpose of "commercial advantage" or "private financial gain, involves the reproduction or distribution of more than $1,000 worth of copies", or by leaking a work "being prepared for commercial distribution".[209] In addition to forfeiture and destruction of the unauthorised copies,[210] the law provides for up to 10 years' imprisonment.

Other acts are punishable as well – placing a false copyright notice, fraudulently removing or altering a copyright notice from a copyrighted work, and falsely representing that a copyright registration has been filed are all punishable by a fine of up to $2,500.[211]

5.6 DMCA

The Digital Millennium Copyright Act (DMCA) contains certain provisions

204 17 USC Section 507(b).
205 *Gaiman v McFarlane*, 360 F3d 644, 653 (7th Cir 2004).
206 *Computer Assocs Int'l, Inc v Altai, Inc*, 982 F2d 693 (2d Cir 1992).
207 982 F2d at 700.
208 982 F2d at 706.
209 17 USC Section 506(a)(1)(A)–(C).
210 17 USC Section 506(b).
211 17 USC Section 506(c)–(e).

relating to the intersection of high technology with copyright. Section 1201(a) of title 17 proscribes anyone from circumventing technological measures that effectively control access to copyrighted works.[212] (This includes technology like Digital Rights Management (DRM) systems.) The DMCA prohibits the manufacture, importation, sale, or provision of products that do the same.[213]

Civil remedies include awarding of actual damages and statutory damages – between $200 and $2,500 "per act of circumvention, device, product, component, offer, or performance of service".[214] Criminal penalties may include a fine of up to $500,000 and/or five years in prison for first-time offenders (and $1 million and/or 10 years in prison for repeat offenders).[215]

The DMCA does provide certain exceptions, such as for law enforcement,[216] reverse engineering,[217] and encryption research.[218,219]

5.7 Permitted acts – fair use

The most notable exception to copyright protection is called 'fair use'. It is codified in Section 107 of the Copyright Act. It is both a *before-the-fact* limitation on the copyright owner's bundle of exclusive rights and an *after-the-fact* defence to a charge of infringement of such rights.

Notwithstanding the provisions of Sections 106 and 106A, the fair use of a copyrighted work, including such use by reproduction in copies or phonorecords or by any other means specified by that section, for purposes such as criticism, comment, news reporting, teaching (including multiple copies for classroom use), scholarship, or research, is not an infringement of copyright. In determining whether the use made of a work in any particular case is a fair use the factors to be considered shall include:

- the purpose and character of the use, including whether such use is of a commercial nature or is for non-profit educational purposes;
- the nature of the copyrighted work;
- the amount and substantiality of the portion used in relation to the copyrighted work as a whole; and
- the effect of the use upon the potential market for or value of the copyrighted work.

The fact that a work is unpublished shall not itself bar a finding of fair use if such finding is made upon consideration of all the above factors.[220]

---

212 17 USC Section 1201(a)(1)(A).
213 17 USC Section 1201(a)(2).
214 17 USC Section 1203(c).
215 17 USC Section 1204(a).
216 17 USC Section 1201(e).
217 17 USC Section 1201(f).
218 17 USC Section 1201(g).
219 See also 17 USC Section 1201(h)–(k).
220 17 USC Section 107.

Two specific types of fair use are worth mentioning briefly. The first is truthful comparative advertising embodying the copyrighted work of another – generally understood as criticism or comment.[221] The second is videotape time-shifting (whereby television programs are videotaped at one time for viewing at another) for non-commercial home use.[222]

Given the relative flexibility of these four factors, the fair use analysis can be unpredictable. For example, reverse engineering of software is frequently done by competitors, academics, students, and others, to discover the workings of a computer program, including the ideas contained therein that are in the public domain and not subject to copyright protection. However, reverse engineering a program necessarily involves reproduction and adaptation of the underlying computer program, which would appear to meet the requirements for copyright infringement. Applying the four factors, it does not appear that reverse engineering is a fair use. Reverse engineering is undertaken for a commercial purpose (factor 1). A published piece of software (factor 2) is reproduced and adapted (factor 3). The effect on the market for the product is potentially quite large (factor 4). But numerous courts have held that reverse engineering constitutes fair use.[223]

5.8 Ownership and authorship

(a) *Joint works*

Ownership of a copyright vests initially in the author(s) of a work.[224] (A joint work is "a work prepared by two or more authors with the intention that their contributions be merged into inseparable or interdependent parts of a unitary whole".[225])

(b) *Work for hire*

When the work is made for hire (meaning that it was prepared by an employee in the scope of his or her employment or that it was specially ordered or commissioned for use as a contribution to a larger work[226]) the "employer or other person for whom the work was prepared" is considered the author and, unless otherwise contracted away, owns all of the rights that normally vest in the author.

---

221 *Triangle Publications, Inc v Knight-Ridder Newspapers, Inc*, 626 F2d 1171 (5th Cir 1980).

222 *Sony Corp of Am v Universal City Studios, Inc*, 464 US 417 (1984) (sometimes referred to as the "Betamax" case).

223 See, eg, *Sega Enterprises Ltd v Accolade, Inc*, 977 F2d 1510 (9th Cir 1992), *cert denied*, 531 US 871 (2000); *DSC Commc'ns v Pulse Commc'ns*, 170 F3d 1354 (Fed Cir 1999); *Sony Computer Entm't, Inc v Connectix Corp*, 203 F3d 596 (9th Cir 2000).

224 17 USC Section 106.

225 17 USC Section 101.

226 *Id.*

6. Trade secrets

In 2016, Congress passed the Defend Trade Secrets Act (DTSA)[227] as a way of unifying a patchwork of state laws relating to trade secrets. It does not supersede state laws on trade secrets, but it establishes a new cause of action. To bring a DTSA claim in a federal court, the standard federal pleading rules apply: a plaintiff must allege a plausible nexus to interstate commerce,[228] must establish personal jurisdiction over each defendant, and must have standing to sue.[229]

Trade secrets include confidential, protected information that has economic value. They are defined as:

> *all forms and types of financial, business, scientific, technical, economic, or engineering information ... if*
>
> *(A) the owner thereof has taken reasonable measures to keep such information secret; and*
>
> *(B) the information derives independent economic value ... from not being generally known to, and not being readily ascertainable through proper means by, another person who can obtain economic value from the disclosure or use of the information.*[230]

This includes, for example, scientific, technical, and engineering information (such as code, programs, methods, prototypes, and designs), as well as financial and business information.[231] A trade secret cannot be misappropriated if it was obtained through "reverse engineering, independent derivation, or any other lawful means of acquisition".[232]

If a competitor independently develops a technology similar to a trade secret, there is no misappropriation. If that competitor also obtains a patent for the technology, it could sue for patent infringement for using the same technology.

Under the DTSA, "an owner of a trade secret that is misappropriated may bring a civil action ... if the trade secret is related to a product or service used in, or intended for use in, interstate or foreign commerce".[233] It is possible to bring an action for acts of trade secret misappropriation that occur outside the United States if: "the offender is ... a citizen or permanent resident alien of the United States, or an organization organized under the laws of the United States or a State or political subdivision thereof"; or "an act in furtherance of the offense was committed in the United States".[234] (Such suits are, as with all federal cases, contingent on being able to exercise personal jurisdiction over the defendant.)

---

227 Defend Trade Secrets Act, Pub Law 114–153, 130 Stat 376–386 (codified as amended at 5 USC Sections 1051 and 1116, 18 USC Sections 1833, 1836, 1838, 1839 and 1961).

228 *Hydrogen Master Rights, Ltd v Weston*, 228 F Supp 3d 320 (D Del 2017) (dismissing claim without prejudice for failing to allege interstate commerce nexus).

229 *Gold Medal Prod Co v Bell Flavors & Fragrances, Inc*, No 1:16-CV-00365, 2017 WL 1365798 (SD Ohio 14 April 2017). Note that some states' laws enable past owners of trade secrets to sue.

230 18 USC Section 1839.

231 18 USC Section 1839.

232 18 USC Section 1839(6)(B).

233 18 USC Section 1836(b)(1).

234 18 USC Section 1837.

7. Other forms of protection

7.1 Design rights

US law provides for design patents which protect ornamental characteristics applied to or embodied in an article of manufacture.[235] This includes, for example, shape or configuration of the article and ornamentation on the surface of the article. In contrast, utility patents protect how something is used or how something operates.[236] An article of manufacture that is unique in both its utility and its ornamental appearance could be protected with utility patents and design patents.

Design patents, when issued, last for 15 years from the date of grant, and do not require maintenance fees to maintain.[237]

Computer-generated icons, which include full-screen displays and individual icons, can be protected as surface ornamentation, so long as they are embodied in an article of manufacture (eg, a computer screen).[238]

While design patents are not as litigated as often as utility patents, they still can capture quite a bit of value. The well-known *Apple v Samsung* case involved both utility patents and design patents. The jury found three of the four design patents to be infringed, finding that Samsung owed $1.049 billion dollars for infringement of the design *and* utility patents.[239]

7.2 Trademarks

A trademark is a word, phrase, symbol or design (or combination thereof), that identifies a source of a good to distinguish it from goods from another source. Brand names are an example of a trademark.[240] To merit protection, a trademark must be distinctive and carry meaning indicating the source of the good or service.[241]

Trademarks can be registered at the federal or state level. A trademark is infringed when a party uses an identical mark or one that is confusingly similar to the registered mark on products or in relation to services that are covered by the registration.

Another class of marks is known as 'service marks'. Instead of protecting goods, these marks protect services by identifying the source of a service. Service marks are protected in the same manner as trademarks.[242]

Marks make appearances not just as marks for products and services. The

235 35 USC Section 171.
236 35 USC Section 101.
237 35 USC Section 173.
238 MPEP Section 1504.01(a)(I).
239 *Apple Inc v Samsung Elecs Co, Ltd, et al*, 5:11-cv-01846-LHK, at 15 (ND Cal 24 August 2012).
240 See, eg, 15 USC Section 1052.
241 *Qualitex Co v Jacobson Prod Co*, 514 US 159, 162 (1995) (citing 15 USC Section 1127).
242 15 USC Section 1053.

dynamic nature of software and electronics means that infringement can pop up in previously unexpected places, such as inside of the universe of a video game[243] or by framing content on a website (ie, viewing one website inside of another website).[244]

7.3 Database rights

Unlike some other jurisdictions (eg, the European Union), there is no separate federal protection for databases. A database may be protected under copyright law as a compilation of data, but the protection may be narrow in scope. The underlying data, moreover, is not protected – only the compilation itself.[245]

While database information may not be protected under traditional IP laws, protecting information in a database is possible with contract law. A 1996 case from the Seventh Circuit Court of Appeals[246] considered a shrinkwrap licence included with software that prohibited distribution of telephone directory information.[247] The court found that federal copyright law does not pre-empt the contract made by the licence because the licence was not equivalent to any of the exclusive rights within the general scope of copyright.[248]

8. Enforcement

8.1 Venue

The venue statute for patent infringement actions provides that: "[a]ny civil action for patent infringement may be brought in the judicial district where the defendant resides, or where the defendant has committed acts of infringement and has a regular and established place of business".[249] Until the beginning of 2017, one jurisdiction – the Eastern District of Texas – received nearly 40% of all patent cases.[250] In the 2017 *TC Heartland* case, the Supreme Court held that, under the venue statute, a domestic corporation resides only in its state of incorporation.[251]

The Supreme Court's decision only applied to domestic corporations – not foreign ones. In 2018, the Federal Circuit considered *TC Heartland* in determining venue for a foreign corporation.[252] It determined that venue is proper as to a foreign defendant in any district.[253]

243 *ESS Entm't 2000, Inc v Rock Star Videos, Inc*, 547 F3d 1095 (9th Cir 2008).
244 *Futuredontics, Inc v Applied Anagramics, Inc*, No CV97-6991 CM MANX, 1999 WL 34840608 (CD Cal 8 March 1999).
245 17 USC Section 103(b).
246 *ProCD, Inc v Zeidenberg*, 86 F3d 1447 (7th Cir 1996).
247 *ProCD*, 86 F3d at 1450.
248 *ProCD*, 86 F3d at 1455.
249 28 USC Section 1400.
250 See https://lexmachina.com/q4-litigation-update/.
251 *TC Heartland LLC v Kraft Foods Grp Brands LLC*, 137 S Ct 1514, 1517 (2017).
252 *In re HTC Corp.*, 889 F3d 1349 (Fed Cir 2018).
253 889 F3d at 1357.

8.2 Discovery

Before filing suit, a plaintiff should undertake a pre-litigation investigation. A patentee planning to take action against a suspected infringer should be especially attentive to Rule 11(b), which states that a person, by presenting to the court "a pleading, written motion, or other paper is certifying that to the best of the person's knowledge, information, and belief, formed after an inquiry reasonable under the circumstances":

- the purpose of the paper or pleading is not improper;
- the legal assertions are not frivolous; and
- the factual contentions have or are likely to have evidentiary support.[254]

After filing suit, the plaintiff is entitled to a process known as 'discovery'. Discovery in the United States, unlike other jurisdictions, can be very broad. The standard for discovery is that parties may obtain non-privileged material that is "relevant to any party's claim or defense and proportional to the needs of the case".[255] Notably, information does not need to be admissible as evidence in order to be discoverable.[256]

Discovery includes things like documents[257] as well as requests for admissions of certain facts,[258] depositions,[259] and interrogatories.[260] Documents include things like source code – a court may enter a protective order to protect the disclosure of highly sensitive information like this.[261] Protective orders can limit the disclosure of certain information to parties. For example, a protective order may restrict the dissemination of source code to anyone other than the parties' outside counsel, the parties' experts, and other necessary personnel (eg, court personnel or court reporters).

8.3 Relief/remedies/damages

Damages may be assessed at a rate no less than a "reasonable royalty for the use made of the invention by the infringer", together with interest and costs.[262] Expert testimony may be used to determine the amount of damages and/or royalty.[263]

In certain exceptional cases, the court may award reasonable attorney fees to the prevailing party. Cases can be found exceptional when they "stand ... out from others with respect to the substantive strength of a party's litigating

254 Fed R Civ P 11(b).
255 Fed R Civ P 26(b)(1).
256 *Id.*
257 Fed R Civ P 34.
258 Fed R Civ P 36.
259 Fed R Civ P 30.
260 Fed R Civ P 33.
261 The Northern District of California has model protective orders on its website (see www.cand.uscourts.gov/model-protective-orders).
262 35 USC Section 284.
263 *Id.*

position (considering both the governing law and the facts of the case) or the unreasonable manner in which the case was litigated".[264]

In addition to monetary damages, injunctions are available in patent cases in certain instances. A preliminary injunction may be granted during the pendency of the case, but the bar is high. The court uses a four-factor test that considers:

- the probability that the movant will succeed on the merits;
- the threat of irreparable harm to the movant should a preliminary injunction be denied;
- the balance between this harm and the harm that granting the injunction will cause to the other parties litigant; and
- the public interest.[265]

After judgment, a permanent injunction is also possible. A patent owner must show that:

- it has suffered an irreparable injury;
- remedies available at law, such as monetary damages, are inadequate to compensate for that injury;
- considering the balance of hardships between the plaintiff and defendant, a remedy in equity is warranted; and
- the public interest would not be disserved by a permanent injunction.[266]

Satisfying the first factor for a permanent injunction in a patent infringement suit requires a demonstration of a causal nexus between allegedly irreparable harm and the alleged infringement.[267]

9. Hot topics

One hot topic is renewed Congressional interest in patentability and 35 USC Section 101 is on the mind of many stakeholders in the patent system – including patent owners, defendants and trade groups. In June of 2019, the Senate held a series of hearings on amendments to this much-maligned section of the law. While no consensus was reached, the fact that Congress is still discussing how to fix the situation caused by *Alice* gives hope that there may soon be some clarity on this important area of law.

Another topic of interest is the sunsetting of Covered Business Methods (CBM) proceedings. When the Leahy-Smith America Invents Act (AIA) was signed into law, it created the CBM proceeding (along with other types of

264 *Octane Fitness, LLC v ICON Health & Fitness, Inc*, 572 US 545 (2014)

265 *Nutrition 21 v United States*, 930 F2d 867, 869 (Fed Cir 1991).

266 *Apple, Inc v Samsung Elecs Co*, 909 F Supp 2d 1147, 1150 (ND Cal 2012), *aff'd in part, vacated in part*, 735 F3d 1352 (Fed Cir 2013) (citing *eBay v MercExchange, LLC*, 547 US 388, 393 (2006)).

267 *Apple, Inc v Samsung Elecs Co*, 735 F3d at 1359–1360.

proceedings), and allowed petitioners to challenge a patent under any of Sections 101 (ineligibility), 102 (anticipation), 103 (obviousness), and 112 (written description, enablement, and indefiniteness). Such CBM proceedings, however, require that the petitioner has already been sued for or charged with infringement of the challenged patent, and that the patent "claims a method or corresponding apparatus for performing data processing or other operations used in the practice, administration, or management of a financial product or service".[268] These proceedings have been of great interest to financial institutions, software developers, and others accused of patent infringement because they provide a venue for a broad array of challenges that are less expensive than traditional litigation. The law was written with a sunset clause that automatically repeals the ability to file a CBM petition in September of 2020.[269] However, there have been murmurs that Congress will step in to extend the proceedings' lifespan.

268 Leahy-Smith America Invents Act, Pub L No 112–129, 125 Stat 331.
269 *Id*, at 330.

About the authors

Pierre-Olivier Ally
Chief IP counsel, Bitfury Group
pierre-olivier.ally@bitfury.com

Pierre-Olivier Ally serves as the Bitfury Group's chief intellectual property counsel and Bitfury Surround's legal counsel. Pierre-Olivier has handled complex national, pan-European and international patent litigation matters and also advised clients on software-related agreements as well as on the structure and negotiation of IP-related aspects of M&A and private equity transactions, joint ventures, IPOs, licensing, and public and private collaborative research transactions.

Prior to joining the Bitfury Group, Mr Ally was of counsel at Dentons and an associate at leading international law firms in Paris, Dechert LLP and Linklaters LLP. He has a Bachelor of Science (BSc) from McGill University and his Bachelor of Laws (LLB) and Juris Doctor (JD) from the University of Montreal. He is admitted as an attorney in the State of New York and the Province of Quebec in Canada.

Nick Bassil
Partner and patent attorney,
Kilburn & Strode LLP
nbassil@kilburnstrode.com

Nick Bassil is a European patent attorney and a chartered patent attorney in the United Kingdom with BSc and MSc degrees in biochemistry from the University of Birmingham. Mr Bassil's practice concentrates on pharmaceutical chemistry, regenerative medicine, immunotherapy, genome mapping and synthetic genome assembly. He has been significantly involved in patent applications relating to cancer immunotherapy/CAR T cells, and human embryonic and parthenogenetic stem cells. His practice ranges from drafting patent applications and devising filing strategies, including due diligence on IP rights, and obtaining Supplementary Protection Certificates for pharmaceuticals.

Mr Bassil has received the following industry recognition for his work within the sector: IAM Patent 1000 2017 quotes that he "is sought after for his SPC knowledge and understanding of stem cells and regenerative medicine". Mr Bassil is individually ranked in band 1, patent attorneys, Chambers and Partners 2016. His profile in that ranking reads: "Nick Bassil is experienced at all stages of patent matters, from drafting through filing to oppositions."

Michael Burdon
Partner, Simmons & Simmons
michael.burdon@simmons-simmons.com

Michael Burdon has more than 30 years' experience and is a widely acknowledged expert in the field of intellectual property, especially in relation to complex, high-value, multi-country patent disputes. He has represented clients in numerous reported cases, in a diverse range of technologies from pharmaceuticals, biotechnology and medical devices through to electronics, software and communications. He has significant expertise and experience of SEPs and FRAND issues.

He also has extensive experience of assisting clients with their IP strategy and structure and establishing best practice for the management of their IP portfolio and IP risk.

Mr Burdon is the elected chairman of the Intellectual Property Lawyers Association which represents about 65 law firms with substantial IP practices and is secretary of the Intellectual Property Court Users' Committee.

Kenneth K Cho
Partner, Wiggin and Dana LLP
kcho@wiggin.com

Kenneth K Cho is a partner at Wiggin and Dana LLP specialising in patent licensing, litigation and counselling matters. Previously, Mr Cho was a senior US attorney with Kim & Chang before returning to the United States to join Wiggin and Dana LLP in 2019. Mr Cho was a chief patent counsel for a private equity fund dedicated to investing in intellectual property, and also the IP counselling and acquisition manager of Philips IP&S supporting the consumer electronics product division of Philips Electronics. Before joining Philips, Mr Cho worked for AT&T Corp, where he was responsible for offensive patent assertion matters.

Mr Cho is a former member of the American Intellectual Property Law Association Board of Directors, former President of the International Association of Korean Lawyers (overseas chapter), and former member of the Board of Directors for the Korean American Lawyers Association of Greater New York. He is a trained mediator and is admitted to practise in New York, New Jersey, and Washington DC, and before the US Patent and Trademark Office.

Natalie Coan
Associate, CMS Cameron McKenna Nabarro Olswang LLP
Natalie.Coan@cms-cmno.com

Natalie Coan is an associate in the IP team at CMS. Ms Coan advises on both contentious and non-contentious intellectual property matters and has experience in multi-jurisdictional patent litigation, advising on e-cigarette technology and medicines regulatory matters.

Nicholas Fox
Partner, Finnegan Europe LLP
Nicholas.Fox@finnegan.com

Nicholas Fox is a partner at Finnegan Europe LLP specialising in intellectual property law with a focus on electronics, telecoms and software patent litigation. In addition to being qualified as a solicitor, he is also qualified as a European and Chartered British patent attorney, with full rights of audience to appear in intellectual property proceedings at all levels in the English High Court. He is also qualified and admitted as an attorney in New York.

With degrees in mathematics, computer science and law, Mr Fox has technical experience in the areas of electronics and software, having acted on behalf of a wide variety of multinational firms dealing with subject matter ranging from image processing and voice recognition to telecoms and e-commerce.

He has experience handling multinational patent litigation. At his previous firm, he was part of the litigation team acting on behalf of Samsung in the *Apple v Samsung* mobile phone patent and design litigation in 2012. He also assisted AMSL in its multinational patent litigation defending against allegations of patent infringement by Nikon in the field of photo-lithography. Mr Fox also deals with non-contentious intellectual property matters, most recently assisting radiotherapy company Vision RT with its acquisition by Williams Demant.

Mr Fox is the author of the practitioners' work, *A Guide to the EPC 2000*, has lectured on litigation and intellectual property matters for the European Patent Academy and has been a guest lecturer at the University of London on patent matters. He has also served as a board member of IPReg, the regulatory body for patent and trademark attorneys in the UK.

Erik Gelke
Associate, Leinen & Derichs Anwaltssozietät
erik.gelke@leinen-derichs.de

Erik Gelke is a German attorney in business law and advises national and international technology companies. He pursues his focus on copyright law in practice and as a lecturer at the University of Applied Sciences Dusseldorf. Before joining Leinen & Derichs, he worked for the law firm Reimann Osterrieth Köhler Haft (now part of Hoyng Rokh Monegier) in all fields of IP law.

Mr Gelke studied in Cologne and Freiburg and wrote a thesis at the Institute for Copyright and Media Law in Munich, which was taken into account by the German Federal Constitutional Court in its first decision about sampling.

Caitlin Heard
Partner, CMS Cameron McKenna Nabarro Olswang LLP
caitlin.heard@cms-cmno.com

Caitlin Heard is a partner in the patent litigation team at CMS. She is an experienced IP litigator, whose principal area of expertise is multi-jurisdictional patent litigation. Ms Heard represents clients on complex, often business-critical, litigation. Ms Heard has acted on a number of high profile pharmaceutical and technology disputes, across a range of technical areas including telecommunications, quantum computing, construction, medical devices, information security and biotech. She has particular expertise advising clients on their IP strategy in respect of Standard Essential Patents and FRAND licensing disputes, and has acted on several standard essential patent disputes across a number of jurisdictions.

Nobuhito Ishihara
Patent attorney, Kyowa Patent and Law Office
ishihara-nobuhito@kyowapatent.co.jp

Nobuhito Ishihara obtained his master's degree (2002) and a doctoral degree in engineering (2008) at Waseda University. During this time, he worked as a researcher at the Japan Patent Office and as research fellow at the Japan Society for the Promotion of Science. He then taught optics, information processing, electromagnetics, and electronics as an associate professor, while continuing his own research in optics and image processing. As a PhD, he worked for seven years at a software company developing image processing technologies (medical instruments, onboard cameras, etc) and program optimisation.

After passing the bar examination in 2014, Mr Ishihara joined Kyowa Patent and Law Office where he is involved in domestic and international patent acquisition and examination for leading general electronics manufacturers, software companies, and venture businesses. His technical specialisations are image processing, optics, information processing (optimisation, AI, software, computer architecture), semiconductors, and other electric, electronic, and informational technologies.

Christopher Johns
Associate, Finnegan LLP
Christopher.johns@finnegan.com

Christopher Johns is an associate in the Washington DC office of Finnegan LLP. Mr Johns maintains a diverse patent practice, working on ITC and district court litigation, PTAB trials, patent prosecution, global portfolio development and strategy. He has more than a decade of experience in the patent field. He speaks and writes regularly on topics such as patent eligibility and prosecution practice.

Before joining Finnegan, Mr Johns served as a US patent examiner for nearly five years, where he examined applications related to business methods and software. Mr Johns also had experience as a network technician at a research university.

Gregory B Kang
Partner, Nath, Goldberg & Meyer
gkang@nathlaw.com

Gregory B Kang is a partner at the firm of Nath, Goldberg & Meyer and is in charge of managing the electrical practice group. Mr Kang practised in the US for over 10 years before coming to Korea in 2007 where he served as in-house counsel for Samsung Electronics and was with Kim & Chang until rejoining Nath, Goldberg & Meyer in 2018.

In more than a decade of experience as an attorney in the United States, Mr Kang worked on International Trade Commission patent infringement actions, assisted on multiple other IP-related litigations, was responsible for overseeing IP issues for mergers and acquisitions, and has written and negotiated IP issues for multiple types of contracts.

In the area of patent prosecution, Mr Kang was an examiner at the US Patent and Trademark Office. In private practice, he prepared and prosecuted patent applications in the electrical and mechanical arts.

Nobuyuki Kashiwa
Attorney-at-law, Kyowa Patent and Law Office
kashiwa-nobuyuki@kyowapatent.co.jp

Nobuyuki Kashiwa graduated from the University of Tokyo (department of pharmacy) in March 2002, and from its master's course in March 2004. He was registered as an attorney-at-law in December 2008 and a patent attorney in April 2010. His areas of specialisation are litigation and contracts relating to IP rights.

Takashi Katagiri
Patent attorney, Kyowa Patent and Law Office
katagiri-takashi@kyowapatent.co.jp

Takashi Katagiri obtained his master's degree in science in global information and telecommunication studies at Waseda University. He worked at a major system integration company in Japan for eight years, engaging in projects on IT infrastructure (such as Authentication system, Groupware, and Wide Area Network) of big corporations. His main duties were suggesting solutions to customers' problems and planning and managing the projects. During this time, he got a high-level IT national qualification and vendor certifications (eg, Microsoft, cisco).

Mr Katagiri joined Kyowa Patent and Law Office in 2014 and passed the bar examination the following year. He is involved in domestic and international patent acquisition for leading general electronics manufacturers and software companies.

Martin Köhler
Partner, Hoyng Rokh Monegier
Martin.Koehler@hoyngrokh.com

Martin Köhler has been an attorney since 1998. He specialises in patent infringement litigation and is regularly active in trademark and design rights enforcement as well as in unfair competition litigation. Outside court, his other work includes providing advice on licensing, research and development agreements as well as on employee invention matters.

Mr Köhler regularly represents clients in the areas of electronics, telecoms, mechanical and chemical engineering, automotive technology, pharmaceuticals and medical devices.

In addition to his legal studies at the Universities of Saarbrücken/Nancy and Trinity College Dublin, he has completed a foundation degree (DEUG) and postgraduate studies (DESS) in French law.

Mr Köhler publishes regularly on IP issues and is a speaker at conferences on patent law issues. He is also a visiting lecturer at the Hagen Law School.

Hyejung Lee
Patent attorney, Kim & Chang
hyejung.lee@kimchang.com

Hyejung Lee is a Korean patent attorney in the firm's integrated technology practice group. Ms Lee is also admitted to practise in New York, and has expertise in a wide range of technologies and industries including construction, architecture and automotive.

Ms Lee worked at Samsung Electronics as an in-house counsel, where she focused on patent filing and patent infringement analysis. Ms Lee also worked at Hyundai Motor Company, where she focused on patent disputes and licensing.

Ms Lee received her undergraduate degree in architecture from Seoul National University (2003) and her law degree from the University of California, Davis School of Law (JD, 2011).

Yoon Sang Lee
Patent attorney, Kim & Chang
yslee@kimchang.com

Yoon-Sang Lee joined Kim & Chang in 2006 and is a patent attorney in the firm's electrical practice group. Mr Lee received his BS in electrical engineering (1996) from the University of Toronto and his LLB and LLM from Korea National Open University and Georgetown University Law Center, respectively. He was admitted to the Korean Patent Bar in 2004 and was also admitted to practise in Illinois in 2016. Before joining the firm, he worked at AIP Patent & Law Offices from 2004 to 2006 and served in the Republic of Korea Air Force from 1997 to 2000. Mr Lee's practice includes a wide range of IP matters, including patent prosecution and counselling involving patents in communications services and technology. He is a member of the Korean Patent Attorneys Association.

Peter Leonard
Principal, Data Synergies; Professor of Practice (IT Systems and Management and Business Law), UNSW Business School; and Consultant, Gilbert + Tobin
pleonard@datasynergies.com.au

Peter Leonard is a data and technology business consultant and lawyer advising data analytics and AI&ML businesses. Mr Leonard chairs the IoT Alliance Australia's Data Access, Use and Privacy work stream, the Law Society of New South Wales's Privacy and Data Committee and the Australian Computer Society's Artificial Intelligence and Ethics Technical Committee. He is a former global chair of the International Bar Association's Technology Committee.

Mr Leonard was a founding partner and is now a consultant to Gilbert + Tobin, now a large Australian law firm, and principal of Data Synergies. He holds economics and master of laws degrees from the University of Sydney.

Kendrick Lo
Founder, Law Office of Kendrick Lo
mail@k-lo.ca

Kendrick Lo specialises in advising companies on IP matters relating to software with a particular interest in artificial intelligence technologies. Previously, Mr Lo practised at a leading Canadian IP firm for nearly two decades.

Mr Lo received his undergraduate degree in engineering science, and joint law and MBA degrees from the University of Toronto. More recently, he completed a master of laws at Osgoode Hall Law School (York University) and a graduate degree in computational science and engineering at Harvard University.

Mr Lo is a Certified Specialist in Intellectual Property (Patents) with the Law Society of Ontario, Fellow of the Intellectual Property Institute of Canada, Canadian and US patent agent, and former Canadian patent agent examiner. He has taught computer science at the University of Toronto, and patent law at Osgoode Hall Law School.

Joanna Ma
Associate, Bereskin & Parr LLP
jma@bereskinparr.com

Joanna Ma advises high-tech companies on developing IP strategies that are aligned with their business goals. She is an IP adviser at several innovation centres and takes pride in applying a practical approach to intellectual property.

Ms Ma received her undergraduate degree in electrical engineering from the University of Waterloo and a law degree from the University of British Columbia. She is also a registered Canadian and US patent agent.

Elena Martini
Founding partner, Martini Manna Avvocati
elena.martini@martinimanna.com

Elena Martini is a founding partner of Martini Manna Avvocati, an Italian boutique law firm specialising in intellectual property, IT data protection and commercial law.

Ms Martini specialises in contentious and non-contentious intellectual property in a wide range of industry sectors including entertainment, design furniture, pharmaceuticals, food, mechanical engineering, new media, and information technology.

She acted as leading counsel in several high-profile cases, which have formed current Italian case law on issues like copyright protection of design works, the limits of the

Italian Bolar clause and know-how infringement.

Ms Martini has published numerous articles in international legal journals and regularly contributes to the online legal section of the Italian financial newspaper *Il Sole 24 Ore*.

Timothy May
Partner, Finnegan LLP
timothy.may@finnegan.com

Timothy May is a partner in the Washington, DC office of Finnegan LLP. Mr May has more than 20 years of experience in the field of IP law, and focuses his practice on advising high-tech companies on US patent matters. He devotes a significant portion of his practice to patent litigation and post-grant proceedings before the US Patent and Trademark Office (USPTO).

Mr May was a partner in the firm's European office from 2006 to 2011. He has substantial experience advising European high-tech companies on US patent matters. He currently manages the US patent portfolios for a number of European clients in various electrical and mechanical technologies. He is also a frequent lecturer on US patent law, particularly software, and has spoken at numerous conferences throughout Europe.

Before joining the firm, Mr May served as a patent examiner at the USPTO, where he examined patent applications relating to digital communications and image analysis, and where he regularly dealt with software patentability issues.

Heather Meeker
Partner, O'Melveny & Myers
hmeeker@omm.com

Heather Meeker advises clients on technology transactions and IP matters. Ms Meeker is an internationally-recognised specialist in open source software licensing. She received the prestigious IP Vanguard Award for private practice from the Intellectual Property Section of the California state bar for 2016. Best Lawyers named her an IT lawyer of the year for 2018.

The *Daily Journal* named Ms Meeker in its "Top 100 Women Lawyers in California" in 2013, and has named her twice in its "25 Top IP Portfolio Managers (Licensing) In California". Her latest book, *Open Source for Business*, is a definitive handbook for lawyers, engineers and businesspersons on open source licensing in business. Ms Meeker is an adviser to the American Law Institute's ongoing project on the restatement of copyright law.

Louisa Mottaz
Associate, CMS Cameron McKenna Nabarro Olswang LLP
louisa.mottaz@cms-cmno.com

Louisa Mottaz is an associate in the competition team at CMS. She advises clients on a variety of EU and UK competition matters including investigations, merger control, behavioural advice and compliance training. Ms Mottaz also has a particular interest and experience in the consumer goods sector advising on e-commerce and cross-border distribution. Her most recent experience includes obtaining phase 1 merger clearance before the CMA, assisting clients in relation to a European Commission investigation and advising clients on complex multi-jurisdictional merger control issues.

Prashant Phillips
Partner, Lakshmi Kumaran & Sridharan
prashant.phillips@lakshmisri.com

Prashant Phillips works in the design, patent prosecution, patent litigation and related practices at Lakshmi Kumaran & Sridharan. He is involved in drafting and prosecuting design and patent applications. He has expertise in inventions related to software, telecommunications, electronics and electrical engineering, and wind energy. Mr Phillips appears regularly at hearings before the Indian Patent Office in both patent and design matters. He also advises clients in relation to various IT and data privacy laws.

Mr Phillips regularly assists clients with patent-related strategies and provides advisory as to how to handle portfolios relating to different technologies. This has helped IP holders to identify valuable IP within their existing portfolio so as to minimise IP-related costs in maintaining IP that would be required for achieving the desired business objectives. His practice areas include industrial designs, patents with particular focus on telecommunication, software, electronics, non-renewable energy, electrical and mechanical engineering, and data privacy.

Handong Ran
Partner, Maucher Jenkins
Handong.ran@maucherjenkins.com

Handong Ran is a patent partner at Maucher Jenkins. He holds degrees in electronic engineering and IP law, and is specialised in patent prosecution and advisory work in Europe and China. Mr Ran has looked after UK, European and Chinese patent portfolios for both multinational corporations and start-up companies in the fields of telecommunications, home appliance and computing technologies and advised clients on numerous freedom-to-operate projects.

After practising in the London office for six years, Mr Ran relocated to China in 2012 to set up and manage the Beijing, Shenzhen and Nanjing offices for Maucher Jenkins. While continuing to handle European, UK and Chinese patent matters in China, he has also advised several international clients on technology transfer projects.

Marianne Schaffner
Partner, Europe Head of Patent Litigation,
Reed Smith LLP
MSchaffner@ReedSmith.com

Marianne Schaffner is the European head of Patent Litigation and the national head of the Intellectual Property Group of Reed Smith. She is a recognised trial lawyer in national pan-European and international patent', trademark and trade secrets disputes in the pharmaceutical, chemistry, agri-business and electronics and telecommunications sectors. Ms Schaffner also advises clients on intellectual property agreements, mergers and acquisitions, leveraged buyouts and capital market transactions. She has handled a wide range of litigation and strategic IP matters for a diverse array of companies, in particular in the life sciences and telecoms sectors. The vast majority of the cases she has handled have been pan-European, including before the European Patent Office (EPO) and often with a US counterpart. On these occasions, she has led the litigation and patent prosecution teams in Europe and in the US to ensure a full coordination of the litigation and patent prosecution teams and a true consistency of the court and EPO papers filed. Ms Schaffner is a frequent conference speaker and is a lecturer at the University of Paris, as well as the Paris Bar School.

Chang Hwan Shin
Attorney, Kim & Chang
chshin@kimchang.com

Chang Hwan Shin is an attorney at Kim & Chang in Seoul. Mr Shin is a member of the game contents rating board, and he formerly served as a member of Korea Copyright Commission and a policy adviser to the Office of the President.

Keisuke Soeda
Patent attorney, Kyowa Patent and Law Office
soeda-keisuke@kyowapatent.co.jp

After graduating from Waseda University (faculty of letters, arts and sciences) in 2007, Keisuke Soeda started working at Kyowa Patent and Law Office. He became a registered patent attorney in 2014 specialising in design and trademark cases.

Shoji Yoshida
Patent attorney, Kyowa Patent and Law Office
yoshida-shoji@kyowapatent.co.jp

After receiving his master's degree in engineering from the University of Tokyo in 1998, Shoji Yoshida worked for almost nine years as an engineer in the research and development department and in the management department of a leading telecommunication systems company. Mr Yoshida was then engaged in the development of optical communication apparatus for a major domestic carrier, as well as that of firmware and ASIC devices. He was a member of Japan delegation at ITU-T (International Telecommunication Union Telecommunication Standardization Sector) for four years.

After passing the Japan patent attorney examination in 2006, Mr Yoshida joined Kyowa. At Kyowa, his duties mainly cover acquiring and maintaining of IP rights in Japan and abroad.

He acted as a member of the patent board of the Japan Patent Attorney Association in 2016 and 2017.

Chris Watson
Partner, CMS Cameron McKenna Nabarro Olswang LLP
chris.watson@cms-cmno.com

Chris Watson is one of the world's leading communications law experts. He has been recognised in *The International Who's Who of Telecoms & Media Lawyers 2018* as a 'Most Highly Regarded Individual' in which he is described as "exceptionally bright", and boasts an "amazingly international practice". He is also included as one of only 15 Global Elite Thought Leaders in the *Who's Who Legal Data – Telecoms & Media 2019*. Mr Watson focuses his international practice, particularly in the areas where competition law and regulation overlap, and on EU law arising in cross-border and complex electronic network industry transactions. He has particular experience in relation to the interaction of IP and competition law, having advised over a period of time in relation to technical standards on 3G, 4G, DVD manufacture, margin squeezes and mobile pricing and a number of other related sector competition issues.

Chris Williams
Partner, Gilbert + Tobin
CWilliams@gtlaw.com.au

Chris Williams is an IP litigation partner at Gilbert + Tobin with specialist expertise in patents, who has considerable experience representing clients in the electronics and software industries. Mr Williams has litigated patents across a range of industries including electronics, e-commerce and life sciences. He has an excellent track record of winning at both trial and appellate level and is able to draw upon his extensive litigation experience across IP issues, traversing copyright, trademarks, trade secrets and consumer protection matters. Mr Williams is admitted to practise in New Zealand and Australia and continues to represent clients in both jurisdictions.

Min Xu
Partner, Huahe IP Limited
min.xu9@hotmail.com

Min Xu is a partner at Huahe IP Limited, a Chinese patent attorney and a Chinese attorney-at-law.

Mr Xu obtained his European patent attorney qualification when he was working in a British patent attorney firm in London between 2006 and 2010, where he advised many multinational tech companies on patent filing strategy such as Canon and Intel etc. He moved back to China in 2010 and joined a leading Chinese law firm, where he handled a number of complex litigation cases as a Chinese attorney-at-law. Those cases include landmark patent and antitrust litigations for Apple, Sony and InterDigital etc. Mr Xu joined Huahe IP Limited in 2015 as a partner and continues working on patent prosecution and litigation matters for his clients.

Mr Xu received his master's degree in engineering from Imperial College London in the United Kingdom and his LLM degree from UC Berkeley in the United States.

Adelaide Yu
Partner, Rouse Legal Hong Kong
ayu@rouse.com

Adelaide Yu is a partner at Rouse Legal Hong Kong, part of the leading global IP consultancy, Rouse. Rouse provides the full range of IP services, from portfolio management, disputes and enforcement through to strategy, consulting and commercialisation.

Ms Yu's practice covers all areas of intellectual property, including contentious and non-contentious matters. She is particularly skilled and experienced in the enforcement and protection of intellectual property, including trademarks, copyright, patents and registered designs in Hong Kong. She has extensive experience in handling China-Hong Kong related infringement cases. For China, she advises on IP enforcement strategies and has considerable experience in handling IP litigation, administrative and criminal raids and customs seizures.

Ms Yu has been recognised internationally as a leading IP lawyer and was recommended by World Trade Mark Review (2010–2018), Legal 500 (2010–2018) and Managing Intellectual Property (2014–2018).

About Globe Law and Business

Globe Law and Business was established in 2005, and from the very beginning, we set out to create law books which are sufficiently high level to be of real use to the experienced professional, yet still accessible and easy to navigate. Most of our authors are drawn from Magic Circle and other top commercial firms, both in the UK and internationally.

Our titles are carefully produced, with the utmost attention paid to editorial, design and production processes. We hope this results in high-quality books which are easy to read, and a pleasure to own. All our new books are also available as ebooks, which are compatible with most desktop, laptop and tablet devices.

We have recently expanded our portfolio to include a new range of journals, Special Reports and Good Practice Guides, available both digitally and in hard copy format, and produced to the same high standards as our books.

We'd very much like to hear from you with your thoughts and ideas for improving what we offer. Please do feel free to email me on sian@globelawandbusiness.com with your views.

Sian O'Neill
Managing director
Globe Law and Business

www.globelawandbusiness.com